The Real RFK JR.

The Real RFK JR.

TRIALS OF A TRUTH WARRIOR

DICK RUSSELL

NEW YORK TIMES BESTSELLING AUTHOR

Skyhorse Publishing

Skyhorse Publishing books may be purchased in bulk at special discounts for sales promotion, corporate gifts, fund-raising, or educational purposes. Special editions can also be created to specifications. For details, contact the Special Sales Department, Skyhorse Publishing, 307 West 36th Street, 11th Floor, New York, NY 10018 or info@skyhorsepublishing.com.

Skyhorse® and Skyhorse Publishing® are registered trademarks of Skyhorse Publishing, Inc.®, a Delaware corporation.

Visit our website at www.skyhorsepublishing.com.

10 9 8 7 6 5 4 3 2 1

Library of Congress Cataloging-in-Publication Data is available on file.

Print ISBN: 978-1-5107-7609-8
eBook ISBN: 978-1-5107-7610-4

Cover design by Brian Peterson

Printed in the United States of America

DEDICATION

This book is dedicated to reclaiming the futures that honor the many sacrifices of our past and the dreams that were deferred.

CONTENTS

Note to Readers   ix
Prologue   x

Chapter 1   A Man for Our Time   1
Chapter 2   A Role Model Named Rachel Carson   19
Chapter 3   A Father's Legacies   25
Chapter 4   When Darkness Falls   32
Chapter 5   Initiatory Schooling in Latin America   38
Chapter 6   Discovering Populism: A Southern Odyssey   47
Chapter 7   Whitewater Testing Grounds   58
Chapter 8   A Personal Breakthrough   61
Chapter 9   Becoming a Riverkeeper   67
Chapter 10   A Fish Story   77
Chapter 11   Ecuador Debacle: A Lesson in Petro-Politics   83
Chapter 12   Family Matters   93
Chapter 13   Power Play on Native Ground   100
Chapter 14   Midlife Transition   113
Chapter 15   Luck of the Irish   119
Chapter 16   Watershed Moments   128
Chapter 17   Riverkeeper Meets the Styx   144
Chapter 18   Up against the Navy in Puerto Rico   150
Chapter 19   Justice for Michael Skakel   164
Chapter 20   Legal Eagle: All in a Day's Work   182
Chapter 21   Themes of the Future: Origins   191
Chapter 22   Mercury in Vaccines: Opening Pandora's Box   200
Chapter 23   Investigating the 2004 Election   219
Chapter 24   Taking on Industrial Agriculture   225
Chapter 25   Shifting Tides   237
Chapter 26   Tragedy Strikes   248

Chapter 27   A New Lease on Life   250
Chapter 28   The Assassinations: Questioning the
               Official Narrative   254
Chapter 29   Encountering Sirhan   262
Chapter 30   Once More into the Fray   268
Chapter 31   Courtroom Drama: From Dupont to Monsanto   281
Chapter 32   Changing the Energy Climate   295
Chapter 33   At War over Vaccines   303
Chapter 34   "Ich bin ein Berliner"   314
Chapter 35   Behind the Censorship of RFK Jr.   325
Chapter 36   Undressing the Emperor   339
Chapter 37   Gathering of the Troops   355
Chapter 38   The Instilling of Values   364

Epilogue: Toward Healing the Divide   367
Acknowledgments   371

NOTE TO READERS

The quotations from RFK JR. that are woven through the manuscript come from a wide variety of sources including my interviews with him, articles and op-eds he wrote, and of course his books.

PROLOGUE

It would be difficult for you to have lived in America and not have a negative picture of Robert F. Kennedy Jr. You may recall him as the country's leading champion of the environment for most of his adult life, but this has been obfuscated in recent years by the mass media's labeling of him as an "anti-vaxxer" and "peddler of misinformation."

No longer does his name appear on op-eds, nor are his views ever aired on the nightly news programs. As he said in a nearly two-hour oration in April 2023 announcing his intention to run for president: "This is what happens when you censor somebody for eighteen years: I've got a lot to talk about."

Indeed he does and has for a long time. In this book, you're going to read the story of the real RFK Jr. Not the flimsy caricature, but the story of a robust, passionate leader who has overcome personal tragedy to become a tireless warrior for justice—justice for children, for every ethnicity, for the evaporating middle class, and for both Republicans and Democrats. You will come away with a more complete portrait of someone who represents the future and, most importantly, our children's future.

Following his announcement, he sat at the desk in his book-lined study and filmed a video describing why he was, at sixty-nine, finally running for high office. He began with a clip of his father, Robert F. Kennedy, addressing a crowd during his 1968 presidential campaign that was cut short by an assassin's bullet:

> What we need in the United States is not division. What we need in the United States is not hatred. What we need in the United States is not violence and lawlessness, but love and wisdom and compassion toward one another.

Bobby then said: "Every nation has a darker side and the easiest thing for a politician to do is to appeal to our hatred, our anger and our bigotry and greed and xenophobia, and all the alchemies of demagoguery. My father and my uncle had a vision for America, a vision of racial harmony, of prosperity for all Americans, of peace in the world, and honest government. Their lives were tragically cut short, and America took a different path. Yet the possibility they foresaw is still alive, the America that almost was and yet may be.

"I've been fighting corporate corruption my entire life. But I understand that today the problem is much larger than a few crooked individuals. The problem is a system that no longer serves the people, and a people who are so divided and so fearful that they are easily ruled. It's time to unlearn the reflexes of fear and blame and find ways to unify ourselves and turn our country around.

"I won't pretend to you that it will be easy, but I know what it will take. My father said it—love, wisdom, and compassion toward one another. And that's where we need to start. We will scale down the war machine and bring our resources home. We will rebuild our water systems, repair our roads, modernize our railroads, and clean up our environment. We will also clean up government and earn back the people's trust. We will end the secrecy, the censorship, and the surveillance. We will again be a fearless land of liberty."

The cost of freedom is always high, but Americans have always paid it.
                                        —President John F. Kennedy

"We will face honestly the darker parts of our history and the genocide and the racism, not to shame or blame or punish, but to repair as best we can in the spirit of compassion and kindness toward all. I'm inviting all of you to join me, to create an America that we can believe in and be proud of again. I'm Robert F. Kennedy Jr., and I'm running for president of the United States."

CHAPTER ONE

A MAN FOR OUR TIME

Few are willing to brave the disapproval of their fellows, the censure of their colleagues, the wrath of their society. Moral courage is a rarer commodity than bravery in battle or great intelligence. Yet it is the one essential, vital quality for those who seek to change a world that yields most painfully to change.

—Robert F. Kennedy, June 1966

On April 19, 2023, the son and namesake of the man who made that statement stood before a standing-room-only crowd of some two thousand people inside the grand ballroom of Boston's Park Plaza Hotel. Speaking without notes for almost two hours, Robert F. Kennedy Jr. declared his candidacy for the Democratic nomination in the next presidential election. "My mission over the next eighteen months of this campaign and throughout my presidency," he said, "will be to end the corrupt merger of state and corporate power that is threatening to impose a new kind of corporate feudalism in our country."

As his audience cheered and applauded, he continued: "We have a polarization in this country today that is so toxic and so dangerous. People are preparing for a sort of dystopian future." He would seek to end the division "by encouraging people to talk about the values that we have in common rather than the issues that keep us apart . . . I'm going to do that by telling the truth to the American people."

Robert F. Kennedy Jr. has that moral courage that his father, Senator Robert Kennedy, wrote about. Many controversial figures who go against the grain disappear when faced with widespread attacks and retreat into their comfortable lives. Kennedy has endured hit pieces from leading newspapers and magazines. His books have been censored more than any books in recent memory. His words have been taken out of context and used to attack him. He's been vilified, deplatformed, ostracized, and named as part of a so-called "disinformation dozen."

But Bobby Kennedy, as he (like his father) is widely known, isn't going anywhere. He has spent the past forty years taking on the country's worst polluters and standing against the corporate capture of regulatory agencies and media monopolies. Now he has also emerged as a champion of democracy when it has never been under such threat. He's brought hundreds of lawsuits. He's stood up, spoken out, and never backed down.

Gavin de Becker, bestselling author of *The Gift of Fear* and widely considered the nation's leading private security expert, put it like this: "Democrats need a candidate. If they want to get serious, it's not going to be Biden or the vice president. And they ought to be knocking on the door of people like Bobby who have shown their courage, who know how broken things are and could really fix them."

Nobody has spoken out so vociferously about the takeover of America that is underway. As Bobby wrote, "the Big Tech, Big Data, Big Pharma, Big Carbon and Chemical-Industrial Food plutocrats and their allies in the Military Industrial Complex and Intelligence Apparatus now control our government. These plutocrats have twisted the language of democracy, equity and free markets to transform our exemplary democracy into a corrupt system of corporate crony capitalism. The tragic outcome for America has been a cushy socialism for the rich and a savage and bloody free market for the poor. America has devolved into a corporate kleptocracy addicted to a war economy abroad and a security and surveillance state at home. The upper echelons of the Democratic Party are now pro-censorship, pro-war neocons who wear woke bobbleheads to disguise and soften their belligerent totalitarian agendas for our country and the world."

Two weeks before his filing to run, Kennedy and Children's Health Defense (CHD)—the nonprofit organization he founded in 2018—mounted a class action lawsuit against President Joe Biden, Dr. Anthony Fauci, and other top administration officials and federal agencies, accusing them of waging "a systematic, concerted campaign" to compel the country's three largest social media companies to censor constitutionally protected speech. The 120-page complaint was submitted on behalf of the more than 80 percent of Americans who access news online, principally from Facebook, YouTube and Twitter.

According to the plaintiffs, these companies had been induced "to stifle viewpoints that the government disfavors, to suppress facts that the government does not want the public to hear, and to silence specific speakers—in every case critics of federal policy—whom the government has targeted by name." As Kennedy wrote, this collaboration "is an assault on the most fundamental foundation stone of American democracy," the First Amendment in the Bill of Rights.

Kennedy and CHD filed another recent suit to apply antitrust laws against the Trusted News Initiative (TNI), an approach launched by the BBC in 2020 to a dozen news and social media networks in the US, asking them to pretend not to be competitors but come together and censor information from alternative sites that represent a threat to their business model. "We need to choke them off, we need to crush them," an internal TNI memo said.

Two landmark legal victories preceded those efforts to protect free speech. In January 2023, CHD prevailed when the New York Supreme Court ruled that a mandate requiring health workers to receive a COVID vaccine was "null, void, and of no effect." Two days after that, a federal court in California granted CHD's request for a preliminary injunction blocking a California law that would have allowed punishment of doctors by the state's medical boards for allegedly spreading "COVID-19 misinformation."

As Bobby says: "We have the forms of democracy. We have elections, but do individuals really have any bearing over most of their political officials or the agencies of government? I would say no, these are controlled by money. It's not really a democracy anymore. It's more of an oligarchy or a plutocracy that only responds to the needs of the rich and of corporations paying the lobbying election costs of the politicians, who then become their indentured servants on Capitol Hill."

The result is not only the rapid erosion of our time-honored civil liberties. Greed and corruption dominate our politics, our leading businesses, and our sources of information. Fear, anxiety, and mistrust run rampant in the population. Some question whether the Republic can even survive. At this precipice, leaders with passion, purpose, and perseverance are rare. Demagogues trumpet false promises while enriching themselves with contributions from the top 1 percent. This is not about left or right. It is about right or wrong. We are destroying the very fabric of our country.

In Bobby's words, "We have kids who are going to dilapidated schools. We have our infrastructure falling apart. We have a middle class in this country [that] has been hollowed out and destroyed and we need to start paying attention to these problems here at home and solving them."

During a period when income inequality soared to unprecedented levels, Bobby's research found these figures: "During COVID, in the US alone, there was a 3.8 trillion-dollar shift in wealth upward, largely from the middle class and the poor in our country, to at least, in part, this new oligarchy of billionaires. There were nearly five hundred new billionaires created during the pandemic. A lot of the money ended up concentrating ironically or coincidentally in the social media companies that were actually benefiting from the lockdowns, and that were simultaneously censoring criticism of lockdowns. Those companies have strong entanglements with the Pentagon and also the intelligence agencies."

Ten days before announcing his presidential candidacy, Kennedy journeyed to the same neighborhood in Watts where his father had addressed a teeming crowd of predominantly African Americans shortly before the California Democratic primary in 1968. Watts had given 95 percent of its vote to the senator, helping put the nomination in his grasp shortly before he was assassinated.

"My dad was very proud of that," Bobby explained to a new generation. He'd come here to join a group of African American students from the University of Southern California and several dozen others, planting a "healing garden" for the Watts community's Legacy Project. Wearing a short-sleeved white shirt and dark jeans, Kennedy was introduced with these words: "Today we have somebody who represents a legacy and a history who is also a survivor."

He had recently traveled to East Palestine, Ohio, where the disastrous wreck of a Norfolk Southern train carrying hazardous chemicals meant "the people are now drinking poisoned water and their cats are dying and their cows are dying and their children are now exposed to dioxin, the most toxic molecule we know that's not radioactive. And it's spread over the landscape." He said it didn't have to happen—if the train had carried two engineers instead of one and had used electronically controlled brakes and sensors on every wheel, the catastrophic incident could have been avoided. Bobby wants to put people over profits.

He described to the crowd how he became an environmental lawyer, targeting the toxic chemicals dumped by large corporations into the Hudson River that were wiping out a subsistence fishery. "I discovered early on that environmental injury was theft and disproportionately harmed people who were black and poor," he said. His first client had been the NAACP, a successful legal action on behalf of a small town that "did not have the political power to defend themselves from this onslaught." But he also understood that it's our environment, belonging to all of us, regardless of race, color, or creed. He went on to form the Waterkeeper Alliance, which has grown to become the largest water protection group in the world.

Bobby talked of his pivotal involvement in the 1990s helping to forge an agreement to protect the New York watershed. In the course of that, he'd sued the city over the hundred-plus sewage treatment plants discharging into the drinking water supply for Harlem, the South Bronx, and the Lower East Side. But this was hardly the only impact on the nation's most vulnerable populations.

Kennedy addressed the lockdowns and school closures during the COVID-19 pandemic: "The playgrounds that they couldn't close, they removed the basketball hoops. If you lived in a poor neighborhood, you got locked indoors. For a lot of those kids, the only hot meal that they had was at school." He added: "It was a war on the poor all over the world. There were ten thousand kids in Africa that were dying of starvation every month because of lockdowns . . . [and] lack of access to medication for dysentery and malaria." The middle class didn't do much better.

The name "Children's Health Defense" is not coincidental. Two decades ago, Bobby said: "My job is to be able to look myself in the mirror and say that I spent my short time on this planet trying to make it a better place for my children." When he began raising questions about the wisdom of subjecting kids to vaccines that were not adequately safety tested, he asked himself: "Am I going to ignore what's being done or does this knowledge compel me to go into the battlefield on an issue that I really don't want to have anything to do with?" Based on the science that he'd studied, he felt obligated to follow what he believed was right, no matter what.

* * *

I've known Bobby Kennedy for more than two decades as a colleague and a friend. We met in 1998, amid our involvement in a successful campaign to preserve the last pristine birthing lagoon of the California gray whales from industrial saltworks along Mexico's Baja peninsula. Our mutual concern and love for the natural environment brought us together for fishing and falconry trips. Together we then investigated voter fraud in the 2004 presidential election for a magazine article and cowrote a book called *Climate in Crisis* on the greed and corruption of the oil and coal oligarchs.

The Real RFK Jr. is the product of dozens of interviews with Bobby, his family and longtime friends, and his army of like-minded associates in the battles he's fought. I was also given unprecedented access to his private archive including personal journals. I approached my publisher about writing this book at a time when Bobby was being excoriated as an "anti-vaxxer," becoming *persona non grata* to mainstream media and deplatformed by social media giants, with the contagion spreading virus-like to longtime coworkers and even family members.

This portrait is an attempt to set the record straight about a man who does not fit conventional definitions. His outside-the-box thinking has seen him cross over into what others consider "enemy territory" on numerous occasions. He's never towed the party line. Moral courage requires both nuance and the ability to change when situations change or when new information becomes available.

"There is no stronger advocate for free market capitalism than myself," Bobby has said, "because I believe it's the most efficient and democratic way to distribute the goods of the land. While free markets tend to democratize a society, unfettered capitalism invariably leads to corporate control of government." Bobby has dedicated much of his energy toward combating agency capture at the Environmental Protection Agency (EPA), Centers for Disease Control and Prevention (CDC), US Food and Drug Administration (FDA), and other government agencies.

In Bobby's view, "The best solution to the climate crisis is democratizing the energy grid, ending the obscene subsidies to Big Carbon, and rationalizing markets to turn every American into an energy entrepreneur and every home into a power plant. But Bill Gates and the Deep State are advocating top-down capital incentive geoengineering solutions that expand authoritarian controls." Again, these issues are important to all Americans. They transcend politics, and they must.

It is no accident that highly unlikely allies have discovered a surprising commonality. One of these is Catherine Austin Fitts, former managing director of the investment firm Dillon, Reed Inc. who served under President George H. W. Bush as assistant secretary of Housing and Urban Development. She left all that behind to become a crusader against government fraud.

Fitts told me: "Every once in a while, I've met someone who truly believed in and understood the power of the law, had incredible integrity and was willing to do the unbelievably hard work. Bobby Kennedy Jr. defines that."

* * *

While he's not the only person fighting the greed that imperils our future, he is a passionate, sincere, and tireless advocate for these distinctly American values. Wade Davis, anthropology professor and the award-winning author of twenty-three books on natural history, said of Bobby after rafting the Colorado River in his company: "I had never met a person whose love of country was so sincere and yet so free of chauvinistic cant. I found him to be an extraordinary man, charismatic and brash, decent and true, with a terrific sense of humor and immense reservoirs of physical energy and strength."

All but forgotten in the torrent of negative publicity surrounding Bobby in recent years is that no one else has proved so successful as an environmental advocate and champion of the powerless. "We're not protecting the environment just for the sake of the fish and the birds," he said in his Boston announcement speech. "We're protecting it for our own sake because those things enrich us. They enrich us economically, spiritually, culturally. . . . If we want to meet our obligation as a generation, as a civilization, as a nation, which is to create communities for our children and provide them with the same opportunities for dignity and enrichment and prosperity and good health as the communities that our parents gave us, we have to start by protecting our environmental infrastructure, the air we breathe, the water we drink, the wildlife."

He has won or favorably settled more than five hundred environmental lawsuits, almost all involving scientific controversies. The Waterkeeper Alliance now has 350 chapters on waterways and more than a million volunteers in forty-six countries. In its fifth year of existence, the nonprofit Children's Health Defense has more than fifty active lawsuits and twenty chapters globally. In the process of building these organizations, Kennedy has written four bestselling books. His latest, *The Real Anthony Fauci*, sold more than a million copies despite being completely ignored by reviewers.

This biographical analysis opens with a look back at his childhood, where Bobby grew up debating history and current events at the Kennedy dinner table. His fascination with animals both wild and domestic intersected with the family's support of Rachel Carson, much maligned by Monsanto and the chemical company's media allies for her campaign to ban DDT. The scientist became a kind of emblem for Bobby's initial understanding of injustice as well as environmental degradation. He would eventually be part of successful legal efforts against Monsanto's malfeasance that resulted in unprecedented jury awards to citizens injured by its Roundup weed killer and its active ingredient glyphosate.

Following the traumatic assassinations of his uncle President Kennedy, and five years later his father, Bobby's rebellious teenage years found him leaving privilege behind and learning lessons riding freight trains with hoboes and among indigenous tribes in Latin America. In his twenties, he developed organizational skills in planning complicated river trips in remote reaches of the Amazon. Enrolled at Harvard, he immersed himself in studying the life of civil rights legend Judge Frank Johnson, moving to Alabama for a year of populist education that led to publication of his first book.

In law school, facing a crisis with addiction, Bobby came to an awakening that "anything I do that is durable or persistent or meaningful has to come from a spiritual place." That rebirth allowed him to set out on a journey of helping others. His lawyer friend Chris Bartle told me: "In our small circle, I'd say there are twenty-five people who got sober because of Bobby. And there are literally hundreds of people that Bobby has helped," while remaining to this day an active participant in Alcoholics Anonymous. His desire to help others suffering from addictions led naturally to a broader sense of purpose.

Following his personal recovery in the early 1980s, Bobby joined Hudson Riverkeeper and the Natural Resources Defense Council (NRDC) as a primary litigator and served as founding professor for Pace University's Environmental Law Clinic. There he obtained a special court order allowing his students to try cases in state and federal courtrooms—and won landmark statutory decisions forcing municipalities and industries to cease their polluting discharges. Once a national disgrace, the Hudson River and its watershed were transformed into flourishing fisheries and swimmable waters, inspiring the creation of "riverkeepers" all across the country and eventually the globe.

New York environmental police captain Ron Gatto, who went up against his bosses to expose illegal dumping, recalled becoming part of the extended Kennedy clan: "Going to his house for Capture the Flag games, Bobby not only invited famous people like Alec Baldwin, but also the plumber, the carpenter, and the mechanic," Gatto recalls.

Bobby's campaign to curtail the insidious practices of the factory farming giant Smithfield Foods found him shadowed by a hired goon in rural America. His Riverkeeper partner in North Carolina, an ex-Marine turned commercial fisherman named Rick Dove, told me: "What impressed me most about Bobby was his fearlessness. I think he would've made a great Marine."

Alongside Gatto, Dove, and others, Bobby also became an avid student of science. "What makes Bobby so unique is his desire to understand all the technical aspects," said NRDC Executive Director John Adams, "not just the legal questions involved in environmentalism. But the science and nature, as well."

Bobby spearheaded development of the NRDC pioneering international program for environment, energy, and human rights, traveling to Canada and Latin America to assist local people in safeguarding their homelands and opposing large-scale energy and extractive projects in remote wilderness areas. Bobby elaborated: "My father believed that America should be an exemplary nation, but we would never live up to our promise as a moral authority around the world unless we first went back and corrected or made meaningful amends not only for slavery but the genocide of the American Indian."

No environmental group had ever explored the possibility of working with a major oil company in creating a regulatory structure beyond the governmental. In Ecuador, Bobby proposed to Conoco's lawyers: "In addition to doing the independent environmental oversight and management plan, would you also be willing to give a portion of your profits to the Indian community for rainforest preservation and development projects?" The company's attorneys gave a resounding yes, but a consortium of environmental organizations still sought to ban Conoco's drilling rights altogether and saw no need to involve the locals. A stalemate resulted. As one observer described the situation, "Bobby was moving beyond the conventional wisdom of the environmental movement. He was risking his reputation in seeking a new, enlarged, and some believed a dangerous agenda."

In French-speaking Canada, Bobby joined with tribal leaders in organizing river expeditions to oppose the Hydro-Quebec utility's plan to flood more than 1,300 square miles with several dozen massive dams. Bobby faced the company's helicopters monitoring his every move. Eventually, he used his political clout to influence New York State to cancel billions in contracts for purchasing the electricity. "For the first time in the Americas, Indians have been able to assert control over their land," Bobby said at the time.

Much of his energy in the 1990s went into the pioneering agreement to preserve the New York watershed. This also represented a historic transfer of wealth from city water consumers to fund economic vitality and environmental protection in the Catskill Mountains. Here Bobby often antagonized Establishment politicians but forged new alliances to develop creative strategies. Richard Coombe, an upstate farmer and Republican state senator, recalled of working with Bobby: "We could have remained mortal enemies, and instead it became a partnership."

Bobby later wrote: "Oftentimes I had the [news]papers against me and I felt isolated, and that this was an uphill and unwinnable battle. But I resolved it by knowing that I was fighting for something that was critical. The reservoirs were sacred." The watershed agreement marked his entry into high-level politics, and, after it was finalized in 1997, New York's Republican Governor George Pataki called Ethel Kennedy to say he could hardly wait for Bobby to run for his job.

Instead, in 2001 came Bobby's first (and only) incarceration for an act of civil disobedience. He spent thirty days in a Puerto Rican jail after joining a protest against the US Navy's using the impoverished island of Vieques for a bombing range. "I hadn't been able to get them their day in court," Bobby wrote. "With my hands tied and the Navy poised to open fire, I felt all I could offer my clients was to join them." Vieques became an international cause célèbre, ultimately forcing the Navy to leave the battered island.

Also in the early 2000s, Bobby embarked on a personal crusade aimed at exonerating his cousin Michael Skakel, imprisoned for a murder he didn't commit. "I support him not out of misguided family loyalty, but because I am certain he is innocent," Bobby wrote in an *Atlantic Magazine* article, "A Miscarriage of Justice." Eventually he wrote a book about the case called *Framed*, and, in 2018, Skakel's conviction was overturned, and he was set free.

Bobby also supported the release of Paul Hill, falsely convicted of murdering a British soldier during the Northern Ireland troubles. Hill had married Bobby's sister Courtney, and his confession was found to have been coerced by three Court of Appeals judges in Belfast. You will learn in these pages about the Kennedy family's near-fatal encounter with terrorists while attending the trial.

Reflecting on the assassinations of both his uncle and his father, Bobby set down in 2011: "It is my native inclination to give wide berth to conspiracy buffs of all ilk. But as an attorney and litigator with a passionate love for science, empiricism, and logic, I believe that facts are the foundation stones of all understanding and that we have an obligation to pursue them fearlessly, even when they collide with conventional wisdom and even when their implications meet with disapproval or disfavor."

This was certainly what he encountered as he came to realize that neither assassination could have been the work of a lone gunman. And after he took the remarkable step of visiting his father's convicted killer, Sirhan Sirhan, in prison in 2018, he incurred the wrath of the majority of his siblings. Bobby supported Sirhan's release on both compassionate and legal grounds. It was approved by the parole board before being overturned by California Governor Gavin Newsom.

* * *

During the two terms George W. Bush served as president, Bobby addressed hundreds of audiences citing the administration's abysmal environmental record and the millions invested over the previous two decades to corrupt science and "distort the truth about tobacco, pesticides, ozone depletion, dioxin, acid rain, and global warming." He also published two exposés after investigating computerized vote stealing that enabled Bush to win the 2004 election. And he lambasted the increasingly corporate-controlled media for failure to report what was really going on.

Mercury, one of the most potent neurotoxins known, had long been at the top of Bobby's agenda. He had observed firsthand the birth defects caused by mercury pollution among Native Canadian children in Quebec villages. He'd witnessed the Bush Administration's scrapping of regulations to limit mercury emissions from coal-burning power plants. And in 2005, he learned that mercury used as a preservative for childhood vaccines could be responsible for an epidemic rise in autism.

"Privately, I was skeptical," Bobby wrote in his first article about the fraught subject. "I doubted that autism could be blamed on a single source, and I certainly understood the government's need to reassure parents that vaccinations are safe." But ultimately, he couldn't ignore the dozens of mothers with injured children who came to his speeches, urging that he look beyond the known sources of mercury poisoning. Eventually, one of those women refused to leave his doorstep until he examined the evidence these mothers had been gathering. What he discovered in study after study was a whitewash by the Centers for Disease Control and Prevention about the potential dangers of the mercury-based preservative thimerosal.

That marked the beginning of a deep dive into a realm that no other public figure had been willing to undertake. In the early 1980s, American preschoolers had received only three vaccinations for diphtheria, pertussis, and tetanus (DPT); measles, mumps, and rubella (MMR); and polio. However, after manufacturers received blanket immunity following passage of the National Childhood Vaccine Injury Act in 1986, a gold rush to create new vaccines by pharmaceutical companies quickly followed. Today, in addition to the DPT, MMR, and polio vaccines, children receive vaccines for haemophilus influenza type B, hepatitis B, rotavirus, pneumonia, varicella, hepatitis A, meningococcal B, human papilloma virus, annual influenza, and COVID. As the number of vaccines increased, so did the rates of neurodevelopmental disorders and chronic diseases in our children.

"Autism was really the tip of the iceberg," said Lyn Redwood, one of the mothers who approached Bobby and later went on to work for his organization. "Today one in every five children suffers with obesity and mental health disorders, one out of six have developmental disability, one in ten have asthma, one in eleven with ADHD, and one in thirteen has food allergies. In addition to vaccines, we have to look at all environmental toxins that contaminate our food, water, and air as potential culprits destroying our children's health."

Bobby learned that the Centers for Disease Control, the nation's purported guardian of public health, had replaced an environmental health advisory panel with industry representatives—a move denounced by ten leading scientists who wrote that "regulatory paralysis appears to be the goal here, rather than the application of honest, balanced science." It was déjà vu of what Bobby had observed with the Environmental Protection Agency's cozy relationship to the chemical giants it was supposed to be regulating. At the same time, in 2000 the CDC released a study in conjunction with Johns Hopkins revealing an 80 per cent drop of all infectious diseases in the twentieth century—attributable not to medical interventions with antibiotics and vaccines, but to better nutrition and sanitation.

Freed to advertise on TV by another law passed in the 1980s, "Big Pharma" (as it came to be known) went beyond doling out huge sums to the news networks. The industry dispatched more lobbyists on Capitol

Hill than the number of congresspeople, senators, and Supreme Court justices combined, giving twice as much money to federal officials as fossil fuel corporations. "We use more pharmaceutical drugs than any other country, pay the highest prices, and have the worst health outcomes in the developed world," Bobby came to realize. Today, chronic disease has increased from affecting 12.8 percent in 2006 to 54 percent among our nation's children. Autism cases have soared from one in ten thousand in RFK Jr.'s generation to an astounding one out of every thirty-six children in 2023. Food allergies have escalated from one in 1,200 to one in twelve. The richest country on Earth now has the sickest children.

As Bobby told his audience in Boston in April: "When I'm president of the United States, I am going to end the chronic disease epidemic. If I have not significantly dropped the level of chronic disease in our children by the end of our first term, I do not want you to re-elect me."

Yet until now, the more Bobby called attention to this situation, the more isolated he found himself. Once a regular contributor to the editorial pages of the *New York Times* and other outlets, the last time Bobby published an op-ed was for the Huffington Post in 2009. After he joined forces with prominent businessman J. B. Handley (the father of a child diagnosed with autism) to prevent a vaccine mandate from passing the Oregon State Legislature in 2015, he began to be labeled as the leader of an "anti-vax" movement.

Bobby summed up what he was experiencing: "I've spent thirty-seven years trying to get mercury out of fish. Nobody calls me anti-fish. I've spent thirty-seven years trying to get pesticides out of food. Nobody calls me anti-food. I spent thirty years trying to improve fuel efficiency and get carbon fuel out of automobiles. Nobody calls me anti-energy. My partnership was the first and biggest investor in Tesla, and I was part of another partnership that built the two biggest solar plants in our country. I'm anti-stupidity. I'm against bad science, and for good robust science, and for honest regulators. And we don't have any of those things. People call me anti-vaccine because it's a way of marginalizing me, making people think I'm a crackpot and keeping me silent."

J. B. Handley described his relationship with Bobby like this: "I was wary because he's a Democrat and I'm not. But I really believe that Bobby has completely absorbed the code that his dad lived by and views one of his roles as dismantling corrupt structures. He has an extraordinary intellect; his capacity to understand and assimilate, process, and change his mind is rare. The final quality that makes Bobby extraordinary is just his humanity. Below all that, he's a street fighter, the guy you want beside you in the foxhole. You cannot intimidate or shake him. He could be on his last gasp and still throwing punches."

* * *

Bobby has never abandoned the environmental cause. He's just broadened it. The same kinds of people who pollute the environment for money are poisoning people for money. He said in announcing his candidacy: "You'll hear this trope from the big polluters and their indentured servants on Capitol Hill that we have to choose between economic prosperity and environmental protection. And that's a false choice. We ought to be measuring our economy based upon how it produces jobs and the dignity of those jobs, and how it preserves the value of the assets of our community."

As a governor-appointed member of an advisory panel to determine whether hydraulic fracturing (fracking) of natural gas could continue in New York, he worked closely with an executive for Southwestern Energy named Mark Boling. "I think what Bobby and I learned from each other," Boling later said, "is that the categories we tend to put people in—'that's an industry person' or 'that's an environmentalist'—are unfortunate." In this instance, even though it cost him his job, Boling sided with Bobby in pointing out that fracking was a bad idea. And state health authorities ended up banning the practice, a trend that has since spread to other states.

In the Appalachians of West Virginia, Bobby's legal skill forced the DuPont corporation to pay out $196 million in punitive damages, among the highest compensatory verdicts in the state's history. He had worked for thirty years to bring General Electric and the Monsanto chemical company to justice for dumping PCBs into the Hudson River that put some 2,800 fishing families out of business. And in what might be called his "legacy project," Bobby continued to fight Monsanto. In 2018, he joined with a legal team involved in mass tort suits over cancer-causing glyphosate in Monsanto's Roundup weed killer, which the company had claimed was "as safe as salt" for gardeners to spray. The National Trial Lawyers

Association designated them "Trial Team of the Year" after a jury awarded groundskeeper DeWayne "Lee" Johnson an unprecedented $289.2 million in damages. Monsanto's new owner, Bayer, later announced it would stop selling glyphosate-based weed killers in the US residential market.

Bobby's willingness to engage across party lines reached a breaking point following Donald Trump's election in 2016. The incoming president invited him to participate in a new commission on "vaccine safety and scientific integrity." Bobby accepted, but this never transpired (as we shall see, after Pfizer reportedly gave Trump a million-dollar contribution for his inaugural), nonetheless costing Bobby his longstanding ties to the NRDC and Pace University, which deemed him too controversial for further association.

Then came the pandemic in 2020, when he began to ask questions that made him anathema to many fellow liberals amid the increasing climate of fear. Bobby wrote: "Are powerful state and corporate entities using the current crisis to remove basic rights, and intensify pressures to promote vaccines and surveillance? Does anyone else feel the suffocating darkness of tyranny descending on our nation?"

Over time, Bobby found another "strange bedfellow." He has appeared for extensive interviews with Tucker Carlson, formerly of Fox News. "Tucker and I don't agree on 90 percent of the things he talks about," Bobby said. "But he loves the First Amendment, and he's willing to stand up for it." Similarly, Bobby has found common ground with Steve Bannon on issues relating to Big Pharma, government corruption, and the widespread censorship plaguing this country.

Bobby scrutinized the abandonment of therapeutics to combat COVID-19 and came to have doubts about the safety of rushed-to-market, government-subsidized novel vaccines as well as the efficacy of lockdowns and masks. He pointed out that the relationship between Bill Gates and Dr. Anthony Fauci was more about collusion and corruption than true collaboration. He then found himself "de-platformed" by Instagram and Twitter and not allowed even to publish a letter-to-the-editor response to a series of media attacks on his integrity.

When he decided to examine Dr. Fauci's career in light of the coronavirus crisis, he was shocked to discover cruel experiments by Dr. Fauci's National Institutes of Health on HIV-positive Black and Hispanic children in the New York foster care system. Bobby was also alarmed at how closely "America's doctor" aligned with the biodefense industry after 9/11. For Bobby personally, realizing how deeply, but quietly, the CIA took part in pandemic response mirrored what he'd painfully come to realize concerning the Agency's likely role in his uncle's and father's assassinations.

Skyhorse publisher Tony Lyons recalls: "As I learned more and more about the book, I began to feel it was the most important that we'd ever published. But Bobby wouldn't cut corners. He was completely invested in making sure it was impossible to impeach what was in the book" (altogether, 2,194 fact-checked footnotes). But when *The Real Anthony Fauci* appeared late in 2021, the book was ignored by the mass media (not a single review appeared) and boycotted by bookstores; Amazon sales accounted for the overwhelming majority of hardcover and e-book sales that quickly soared into the hundreds of thousands. Bobby donated all his royalties to Children's Health Defense, which in the course of the pandemic reached a peak of 4.7 million visitors per month.

* * *

Over the two decades that I've known him, Bobby has had a number of opportunities to seek high office. Already in the early 1990s, he'd been touted as the best candidate to fill a vacant US Senate seat in New York. Later he considered not only the Senate, but perhaps pursuing the governorship or becoming head of the EPA under Obama. Former Minnesota Governor Jesse Ventura asked Bobby late in 2006 to run with him for president on an independent ticket. Each time, he decided against it while raising six children. Now all those children are grown, democracy is further under siege, and like his father, Bobby believes he can offer the American people a positive choice despite what many pundits consider insurmountable odds.

While considering whether to do this, he addressed nearly a thousand attendees at CHD's first national conference last October in Knoxville, Tennessee. "I watch the Earth become biologically impoverished," he said. "The world my kids are growing up in, 80 percent of the insects and 85 percent of the songbirds that I knew are gone. This year, all the snow crabs in the Bering Strait disappeared. . . . In the environmental movement, every victory is temporary and every defeat is permanent. If a species or a landscape is destroyed, you're never getting it back. And with us, the same is true. If you lose a child, things are never going to be the same."

A silence settled over the auditorium as Bobby continued: "God speaks to us through Creation. And as we diminish the abundance, the diversity, and richness, our relationship with our Creator also gets diminished, along with our capacity for imagination and self-fulfillment.

"We are all foot soldiers in an apocalyptic battle against the forces of ignorance and greed and totalitarianism. And it's really important to maintain your spiritual center if you're going to be an effective warrior. We are not going to win this by beating them with force of arms—but from our integrity, our compassion, our kindness, and our humanity."

As he spoke those words, I remembered a story related by Kent Correll, a friend from Bobby's law school days. It was a rescue story, about a child playing in the yard of a friend's house. Correll told me: "Bobby looked up and saw a hawk that he could tell had spotted the child and was in a hunting mode. There were about sixty or seventy yards between him and the child. He took off running, and the hawk went in at the exact same moment. It struck a talon into the kid's cheek, but Bobby swept him up, and the kid was fine with only a scratch. That kind of observation, sensitivity, and knowledge of his environment and animal behavior enabled him to react and get there before the child was harmed. You hear about this happening with oncoming cars—but how many people have grabbed somebody out of the way of a diving hawk?"

Time after time, we see Bobby speaking truth to power, fighting corruption, exposing greed, without any fear of the personal consequences. He has refused to bow down to what William Faulkner described as "the end of man." Like Faulkner, he believes that we will prevail because of our "compassion, and sacrifice, and endurance," and he has dedicated his life to being a leader, a healer, and a warrior for justice. You may not always agree with Robert F. Kennedy Jr., but never doubt his honesty, his sincerity, or his dedication.

CHAPTER TWO

A ROLE MODEL NAMED RACHEL CARSON

The roots of Robert Kennedy Jr.'s environmental activism can be traced both to his forebears and his upbringing. It is not often remembered that his uncle, President John F. Kennedy, made ecology a keystone of his 1960 campaign for the presidency—a full decade before the inauguration of Earth Day, the founding of the Environmental Protection Agency, and the passage of numerous landmark environmental laws. As Bobby described in his 2018 memoir *American Values*, JFK had bucked the conservative hierarchy of his own Catholic church to warn about the population explosion as well as the grave threats posed by air and water pollution. As president, he often quoted Theodore Roosevelt that "the nation behaves well if it treats natural resources as assets which it must turn over to the next generation increased and not impaired in value." JFK created a vast National Seashore and appointed ardent conservationist Stewart Udall as Secretary of the Interior.

Born January 17, 1954, Bobby was the third of Robert and Ethel Kennedy's eleven children. His first memory is of sitting below a statue of Saint Francis holding his father's dying black Labrador, Charcoal, in the garden of their Georgetown home. Bobby later wrote: "My heroes as a boy were the Catholic saints, like Saint Francis. The Franciscans saw nature as the vector by which God communicates most clearly with human beings."

His bedroom walls contained more than forty framed pictures portraying events from Saint Francis's life.

When Bobby was three, the family moved to Hickory Hill, in the farm country of northern Virginia, a six-acre antebellum estate that had served as General McClellan's Civil War headquarters. Bobby recalls: "By 1963 we had ten horses, eleven dogs, a donkey, two goats, pigs, my 4H cow, chickens, ducks, geese, forty rabbits, and a coop of Hungarian homing pigeons, along with the hawks, owls, raccoons, snakes, lizards, salamanders, and fish in my menagerie." His mother, Ethel, had him catch live bullfrogs on St. Patrick's Day so she would have something green to put on the table. "I was happy to help," Bobby remembered, "but only after she promised the frogs would be kept moist and returned afterward to the pond."

When school ended in June, Bobby "hated to leave Virginia this time of year when the forests were exploding with life and there were so many animals to see." He was allowed to linger for a few days before driving to the Hyannis Port compound with the family's groom, Bill Shamwell, a two-day trip hauling the horse trailer. On the sea at Cape Cod, Bobby was essentially raised communally with not only his siblings, but the progeny of twenty-nine grandchildren of Joseph and Rose Kennedy. During JFK's years in office, the compound became a summer White House where all nine families had homes. Here, Bobby snorkeled for scallops to feed the family and "swam with my little spear gun into dark caverns in the wrinkled rocks below the mile-long Hyannis Port jetty" stalking local fish.

When they went sailing, the kids received a direct lesson in taking care of the less fortunate. "Each day we brought a sandwich for 'Putt,' a disabled World War II vet made deaf and mute by mustard gas, who spent most days in a tiny motorized dinghy fishing for scup and pulling lobster pots on Nantucket Sound."

After his parents returned from a trip to Africa "to carry Uncle Jack's message of support for independence movements," Bobby spent hours studying photos of villagers and wildlife—and becoming "particularly agitated at age seven to learn that animals were going extinct. I was angry at the men who had exterminated passenger pigeons and the dodo. Air and water pollution seemed to me a brand of theft, and killing off the last members of a species felt like a mortal sin."

After announcing to his father that he wanted to write a book about pollution, wildlife, and the environment, Bobby began setting down his thoughts. He then wrote a letter to his uncle on embossed stationery he'd received for Christmas. "Dear Jack, I would like to see you soon," was all it said. (A framed copy still hangs on Bobby's living room wall.) Early in his presidency, the president invited his nephew to a private audience in the Oval Office. Bobby brought along a gift—a seven-inch spotted salamander he'd captured in the family swimming pool. The president then arranged for Bobby to interview Interior Secretary Udall, bringing along his bulky reel-to-reel tape recorder to ask questions about pollution and extinction.

For his ongoing wilderness education, Bobby had another mentor—Justice William O. Douglas, a staunch environmental advocate on the Supreme Court. In 1963, Douglas took his father, Bobby, and other Kennedy kids and their Shriver cousins on a ten-day horseback trip through the most remote regions of Washington State's Olympic Peninsula to Whiskey Bend. Bobby remembers: "Justice Douglas taught me how to fly-fish. He had a long, graceful cast that I've spent a lifetime emulating."

When Bobby was in fifth grade, his father arranged for him to work in the Washington Zoo for a few dollars per week, his duties including feeding the raptors. Bobby later recalled: "A kind of iconic bird for me was the Eastern Peregrine falcon, which was the eastern version of the most spectacular predatory bird in the world. They flew 240 miles per hour, and I could watch them come down Pennsylvania Avenue at these extraordinary speeds and pick pigeons out of the air right in front of the White House, fifty or sixty feet above. For me, seeing a sight like that was more exciting than visiting the president at the White House. But that bird went extinct from DDT spraying in 1963, the same year that my uncle was killed."

The DDT compound had been invented by a Swiss chemist in 1939, and the Monsanto corporation manufactured it under the trade name Satobane. DDT killed mosquitoes and particularly in the Pacific during World War II was considered a miracle chemical because it eliminated yellow fever and malaria that had bedeviled soldiers from the beginning of time.

"As a kid, I remember a DuPont chemist on the cover of *Time Magazine* as 'Person of the Year.' Americans believed that, having won the war against the Nazis, now we were going to win the war against the bugs and the pests. These agricultural pesticides would amplify the food supply in America, and we were going to feed the world."

Then along came Rachel Carson, an extraordinary woman from Reading, Pennsylvania. She was a marine biologist who never saw the ocean until she was twenty-two years old. She turned out to be a writer whose prose poured from her like poetry, and she wrote two bestsellers about the ocean in the 1950s. This was before Jacques Cousteau, and she made the ocean's mysteries come alive like few ever had.

Her third book in 1962 was called *Silent Spring*, about the chemical industry and particularly DDT. She made the connection that the industry internalizes its profits and externalizes its costs—and there is a heavy price to pay. Among the costs were the songbirds that Americans love. Even people's house pets and farm animals were being impacted in ways that human beings would find unimaginable if they understood what was happening.

Carson pointed out the salient fact that, if you eliminate bugs, the birds have nothing to eat. And the bugs are so good at mutating their breeding cycles, they will quickly figure out ways around the DDT molecule. Then they will come back, and with no birds to eat them, and the diversity of the soil biome will be destroyed.

As Bobby remembers, "Rachel Carson wrote about not only the disappearance of wildlife, but the relationship between the pesticides and cancer as they waged chemical warfare. She laid all this out in beautiful language and in words that people could understand. It was serialized first in the *New Yorker*, and then *Silent Spring* became an instantaneous bestseller, one of the biggest in history.

"Monsanto immediately launched a vicious attack against her, a highly personal attack because they couldn't win on the science. But they sent out a small army of white-coated scientists to preach to the garden clubs and the Audubon societies and the women's forums. They went across the country and appeared on TV to persuade the country that DDT was actually a really good thing."

In a mirror image of what would happen to Bobby some sixty years later with the pharmaceutical companies and their allies, Monsanto was able to mobilize the major medical institutions and the media behind them. All the popular magazines owned by Henry Luce—*Time, Life*, even *Sports Illustrated*—went after Rachel Carson with a vengeance. The American Medical Association and the Audubon Society were co-opted by the industry and deployed against her. The press releases and the articles all used the same language. They referred to her as a spinster, the contemporary euphemism for lesbian, seeking to discredit her in the most damaging way possible in that era. She never rose to defend herself, while fighting cancer, which ultimately killed her.

Bobby, not yet ten, felt proud that President Kennedy as well as his parents read *Silent Spring*. "They invited Rachel Carson to dinner at my house at Hickory Hill, and it was one of the great thrills of my life to meet her. And my uncle went to bat for her. He went to war with his own Department of Agriculture (USDA), because just as the Centers for Disease Control and Prevention (CDC) today is captive of the vaccine and pharma industries, the USDA is and has been for fifty years a captive of agribusiness and the chemical industry.

"My uncle asked his chief scientist, Jerome Wiesner, to assemble a group of the best in the country, toxicologists, environmentalists, to read *Silent Spring*. Rachel Carson had been meticulously careful when she wrote it. She had three references for every fact in the book, because she knew she was going to be attacked. So when this panel of scientists went through the book, fact by fact, sentence by sentence, assertion by assertion, they came back with a report that vindicated Rachel Carson in the last months of her life. She died in April 1964, six months after my uncle Jack."

The smear campaign orchestrated against Carson's pathbreaking book "laid the strategic groundwork for later battles by the tobacco and carbon industries." Rachel Carson's fact-checked work was labeled a collection of lies, while the chemical industry and its allies called for more studies. Yet the independent commission formed by President Kennedy issued a highly critical report that raised conflicts of interest within his own USDA. Bobby remembers the seminar hosting Carson at their home in 1962 as "one of the treasured moments of my youth."

Other moments weren't so much treasured as seared into his consciousness. "I wasn't allowed to swim in the Hudson, the Potomac, or the Charles Rivers," Bobby recounted in his 2004 book, *Crimes Against Nature*. "I remember dusting our home daily for soot from the black smoke that billowed from stacks around Washington, DC. I was aware that thousands of Americans died each year during smog events. I never saw herons, ospreys, and bald eagles. Pesticides had mostly extinguished their mid-Atlantic populations and were doing the same to our songbirds."

In late September 1963, President Kennedy embarked on a whistle-stop conservation tour across eleven states to warn Americans about looming environmental threats and to gather public support for land conservation. "He was frustrated," Bobby recalled, "that at each stop reporters refused to engage him on those subjects, asking him instead about topics ranging from the threat of Soviet expansion to the upcoming elections."

Less than two months later, he was assassinated. In 1972, thanks to the actions of Bobby's uncle a decade earlier, America banned the use of DDT. And Bobby would not soon forget his first impressions of Monsanto.

Bobby was at his Quaker-modeled grade school, Sidwell Friends, when his mother rushed to pick him up early on November 22, 1963. He slid into the backseat of the Chevy Station Wagon alongside his older siblings Joe and Kathleen. "A bad man shot Uncle Jack," Bobby would remember his mother saying, "and he is in heaven." He had arrived home to see his father standing alone under a white walnut tree with the family's three dogs. All weeping, the three children ran over to hug him. "He had the most wonderful life," the boy would remember his father saying, "he never had a sad day."

Bobby would also remember his father walking in a downpour at Arlington National Cemetery to choose his older brother's gravesite. He would remember the small morning mass around his uncle's coffin the next day, and the ocean of a million mourners who lined the funeral route from the Capitol back to the White House. And he would remember Black Jack, the riderless ebony horse that never stopped bucking as a soldier struggled to hold onto his bridle.

After the burial, his father wrote Bobby and each of his siblings a letter, urging them to "remember all the things that Jack started—be kind to others that are less fortunate than we—and love our country."

CHAPTER THREE

A FATHER'S LEGACIES

Following the assassination of his brother, Robert Kennedy "tried to keep us from the vortex of his anguish by taking us on hikes and engaging us in contests, sports, and games, but we felt his desolation," Bobby would recount in his book *American Values*. "My father read us the scriptures but he also read us the Greek myths, urging us to find in them important teachings about courage, fate and the virtues encoded in the poetic language, metaphors, symbolism and allegory."

The following summer of 1964, ten-year-old Bobby went along with his cousin Bobby Shriver and his father, Sarge (the first director of the Peace Corps under JFK), on a tour of East Africa. They drove a Land Rover through the vast Serengeti Plain, "marveling at the vast herds of wildebeest, impalas, and elephants." Camping in tents in the Ngorongoro crater, "lions and hyenas serenaded our tent with their barks and coughs, and a vexed rhino charged our jeep. It was beyond my wildest dreams."

Soon after they returned, his father emerged from seclusion to run for the US Senate in New York. Bobby accompanied him on some of the campaign trips, especially captivated by the throngs of Black and Hispanic kids chasing after the motorcade in Harlem and Brooklyn shouting, "Kennedy, Kennedy, you my main man!" Sometimes RFK lifted them into the convertible to ride along for a few blocks. He would go on to win the senatorial election.

Another quotation savored by Robert Kennedy came from Emerson: "In the presence of nature, a wild delight runs through the man, in spite of real sorrows." RFK did his best to follow such wisdom, taking his family bobsledding and skiing in the Catskills and Adirondacks and then on a whitewater rafting expedition down the Colorado River through the Grand Canyon—an expedition that, at the time, only a few hundred people had ever taken.

"This was before the Glen Canyon Dam transformed the river into a giant plumbing conduit," Bobby later wrote. "There were still wide beaches on which to camp, and the Colorado's warm, muddy waters were home to large schools of eight native fish, four of which have since gone extinct, with the other four on the way. . . . The guides were alarmed by my father's inclination to ride the rapids on an inflatable sleeping mattress, or in a life jacket, and by our practice of leaping in to join him. My father saw wilderness as the furnace of America's character. He prodded our guides to tell us about the region, its geology and natural history."

During the often arduous hikes into the canyon, the guides pointed out ancient Hopi petroglyphs. Bobby recalls their stopping at a Navajo reservation near Red Rock and visiting a medical clinic. "I saw the crushing poverty—and how grateful people were for my father's attention. Even the children my own age seemed empty-eyed and hopeless. After we left the reservation my dad was quiet for a while, then he talked about how ashamed all of us should be—every American—that the Navajo lived in such desperation."

Bobby also remembered how his father "would ask us each morning to survey the horizon for a high peak, which we would then climb before breaking camp." On the last day, it took six hours to journey up a trail in blazing 115-degree heat. Bobby would write in *American Values*: "We all had the sense that in our adventures, we were being prepared for some larger role—I don't mean a hereditary post in politics, but that we were being schooled in gallantry and courage, prepared to take risks and make sacrifices to accomplish something for humanity. . . . I understood that . . . [this was] boot camp for the ultimate virtue—moral courage."

Back on Cape Cod that same summer, while playing a game with his siblings, Bobby jumped from one roof to a nearby garage roof and severed a tendon in his leg swinging through a closed window on the way. As he recovered in a hospital, his father bought him a red-tailed hawk from a pet store. Bobby named the bird Morgan LeFay after the sorceress in the King Arthur legends and began to train it under the tutelage of Alva Nye, an expert in falconry who lived near Hickory Hill.

As his own father had pursued, Robert Kennedy engaged the children in dinner debates about current events where they all had to choose up sides. He made sure they knew about the children he'd seen in Appalachia and Mississippi with rickets and swollen bellies, beseeching his kids to "imagine what it would be like to go to bed hungry, or to have no education and no hope of a good job . . . He took us repeatedly to Bedford-Stuyvesant in Brooklyn, the toughest ghetto in America, which became a second home to him." He told them at the dining room table: "When you grow up, I want you to help these people."

Years later, Bobby would write of his own journey into Appalachian coal country: "I've met the children who sleep fully clothed out of fear that the poorly constructed slurry ponds that perch precariously in the hills over their homes will rupture and flood their valleys and bury their homes under a tidal wave of toxic sludge."

Pollution had been high on Senator Kennedy's agenda. "My father was furious that state officials weren't doing anything about it. Soon after his election, he made a trip up the Hudson to inspect the pollution. Dead for twenty-mile stretches north of New York City and south of Albany, the river caught fire regularly and turned whatever color they were currently painting trucks at the GM plant in Tarrytown." Years later, in his office at the Waterkeeper Alliance, Bobby would keep two photos of his father examining dead fish along the riverbanks near Newburgh, New York.

In early 1967, construction began on the Dolley Madison Boulevard that would connect Washington to northern Virginia's woodland and farm country. Twelve-year-old Bobby and his younger brother David became outraged witnessing huge bulldozers clear-cut forests, level hillsides, bury streams, and uproot trails to the pond where they captured frogs, snakes, and tadpoles. They decided to act. Bobby would write in his book *The Riverkeepers*, which made it onto the *New York Times Best-Seller List*, that "in protest," they "dislodged some stacked highway culvert pipes and smashed them down an embankment, a crime for which we were soon caught." A local construction engineer at one point fired his shotgun at the boys, hitting them with rock salt. Their father meted out a punishment for their transgression, ordering the two to sweep the boardwalks at Cape Cod. It would not be his last act of civil disobedience for a cause.

When Robert Kennedy made the fateful decision on March 16, 1968, to challenge Lyndon Johnson and Eugene McCarthy for the Democratic presidential nomination, he knew he was up against the party faithful— including his brother Teddy, who strongly opposed what he was doing. With Ethel the strongest advocate behind his running, Kennedy nonetheless faced boos from a predominantly Irish crowd as he walked in the city's St. Patrick's Day Parade the day after his announcement. Young Bobby didn't think his father had a prayer. Less than three months later, he would win the California primary that likely assured him the nomination.

That was June 5, 1968. Bobby was asleep in his dorm room at Georgetown Prep, the nation's only Jesuit boarding school, located in the Washington, DC, suburb of Bethesda. Father Dugan suddenly awakened him at six a.m., saying brusquely that a car was waiting outside to take him home. Upon arriving, his mother's secretary Jinx Hack told him that his father had been shot.

Accompanied by his siblings Joe and Kathleen, as he'd been when his uncle was assassinated almost five years earlier, Bobby boarded Air Force Two for the cross-country flight to Los Angeles. When he overheard someone order all the campaign offices to close, Bobby knew. His father was not going to make it. Thousands of silent people stood outside Good Samaritan Hospital when they arrived, some holding "Pray for Bobby" signs.

It was the middle of the night on the West Coast. Bobby would remember his father "lay on a gurney with his head bandaged and his face bruised, especially around the eyes. A priest had already administered last rites. My mother sat beside him, holding his hand. . . . I prayed and said goodbye to him, listening to the pumps that kept him breathing. Each of us children took turns sitting with him and praying opposite my mom. My dad died at 1:44 a.m., a few minutes after doctors removed his life support." In the morning, his eldest son Joe came into the ward where all the children were sleeping and said simply, "He's gone." They all wept.

JFK, serving his first term as president, had died at forty-six. Bobby's father was forty-two. Bobby and his siblings flew with his body back to New York, "taking turns to sit with him in the stern cabin, never letting him be alone. I stared at his coffin, thinking it seemed too small to contain him."

During the wake at St. Patrick's Cathedral, while a crowd in the hundreds of thousands filed past, the boy would remember standing vigil. As the funeral procession in Washington slowly moved by the Mall, he would remember "thousands of homeless men organized by the late Martin Luther King Jr.,'s poor people's campaign stood there quietly, heads bowed and holding their hats against their chests."

Bobby helped carry the casket from the hearse to the grave, as a brilliant moon rose in the night sky and "mourners with candles turned Arlington Cemetery into a city of lights." At the crest of a hill, his father was laid to rest alongside his brother, "between two magnolias, beneath a small black stone simply engraved with his name and the years he lived."

When he got home that night, Bobby went alone into his father's office and closed the door. "He kept a single bed there for nights he had to work late, and I could smell the faint aroma of his favorite cologne." The fourteen-year-old gazed at the framed photos on the wall - his aunt Kick, his uncles Joe and Jack, "thinking they all looked so young and they were all dead. Then I lay down on my dad's bed and wept until my tears soaked the pillow."

An hour, maybe more, passed before his father's closest friend, Dave Hackett, came in and joined him. "He just sat there silently for I don't know how long," the boy would remember. "Then he said to me, 'He was the best man I ever knew.' That's the only thing that I remember. I don't remember after that."

* * *

So many legacies . . . The importance of Catholicism to the family cannot be overemphasized. It provided a central organizing principle and doubtless enabled them to survive the trauma of the assassinations. Bobby's parents had supplemented his parochial school instruction in the Catholic creed by emphasizing the lives of the saints and particularly the joys of martyrdom, secular or religious. Each night the children gathered in their parents' room to say the rosary and listen to RFK read the Bible.

Bobby was partial to the Old Testament. "I loved hearing about Samson slaughtering forty Philistines with the jawbone of an ass and then having his eyes plucked out by his captors, before crushing them beneath the rubble of their own palace! There was David and Goliath, and Noah with all those animals! God picked favorites and wasn't always fair." Bobby's first theological lesson of consequence was that Jesus's gospel marked a historic break with the Old Testament's murderous tribalism. Christ brought religion into the age of ethics, "telling us to love our enemies and 'turn the other cheek.'" A central preoccupation of the New Testament was Christ's attacks on religious fundamentalist and hierarchical religion. "He said that God is love. He endlessly rebuked the Pharisees, the Saducees, and the Scribes for their hypocrisy, rigidity, and venality and, most of all, for making religion the tool of wealth and power."

All these things remained a central tenet of Bobby's philosophy of life. So did the ways his father made them manifest in his own shortened existence. In 2011, Bobby would write of how for the previous twenty years he had sat on the board of the Bedford-Stuyvesant Restoration Association, replacing his late uncle Steve Smith, who had taken Robert Kennedy's place on the board. "During that time I've watched the boarded up storefronts along Fulton Street give way to open stores, restaurants, and booming street commerce." His father's "efforts made Bed-Stuy the anchor that has led to the renewal of much of downtown Brooklyn."

In 1986, Bobby would work with his father's friend Cesar Chavez in the "Wrath of Grapes" final campaign against pesticides poisoning 156,000 farm workers annually. He would be a pall bearer at Chavez's funeral and would march with his mother in South Florida in 2012 protesting the unacknowledged slavery in America's tomato fields.

When Robert Kennedy died, Native Americans from hundreds of tribes across America "wrote mournful letters to Ethel speaking despair at losing the first white leader who they regarded as their own." Bobby's youthful exposures to the plight of Native Americans had a profound impact on him. "I have spent a good deal of my career representing the interests of indigenous Americans in court against governments and mining, timber, hydro and oil industries endeavoring to steal their resources and destroy their land and culture."

And there was a book bequeathed by Robert Kennedy to Bobby, who reflected in 2018, "one day he came into my bedroom and handed me a hardcover copy of Camus's *The Plague*. 'I want you to read this,' he said with particular urgency. It was the story of a doctor trapped in a quarantined North African city while a raging epidemic devastates its citizenry; the physician's small acts of service, while ineffective against the larger tragedy, gave meaning to his own life, and, somehow, to the larger universe." In the epic battle for justice, for the future of mankind, Camus wrote that "while unable to be saints but refusing to bow down to pestilences, [we must] strive [our] utmost to be healers." Bobby continued, "I spent a long time thinking about that book over the years, and why my father gave it to me. I believe it was the key to a door that he himself was then unlocking."

CHAPTER FOUR

WHEN DARKNESS FALLS

In the aftermath of Robert Kennedy's assassination, many of the senator's friends chipped in to ensure that the older children disperse with something of interest to occupy their summer, so their mother could focus on the little ones while reorganizing her shattered life. The family's longtime friend Lem Billings volunteered to take Bobby on a second trip to Africa, starting in Egypt and then traversing the wildlife reserves of Kenya.

Billings had been JFK's roommate at Choate back in the early 1930s and spent just about every weekend in the White House after Kennedy became president. "My earliest memories are of hunting for salamanders and crawdads with him in Pimmit Run near Hickory Hill," Bobby later recalled. "Lem was the most fun person I ever met in my life." Now Billings, who envisioned Bobby as the most likely of the kids to rise to prominence in political life, would become a surrogate father to him.

That summer of 1968, despite his notoriously poor eyesight, Billings was the first to spot poachers on the Serengeti Plain. Bobby later chronicled how "we accompanied park police as they arrested ivory and rhino horn traffickers and raided poacher camps in the Serengeti on Kenya's northern frontier, confiscating wire snares, wooden bows, and poison-tipped arrows. Most of the poachers were impoverished, hungry men clad in tattered burlap that hardly covered their bodies. . . . The uniformed park guards treated them roughly, and despite my low opinion of poaching, I felt sorry for them."

The trip lasted more than a month. A photo of Bobby shaking hands with a smiling Kenyan President Jomo Kenyatta appeared in the *Boston Globe*. Bobby never forgot that "everywhere we went we found a universal sense of loss from the deaths of President Kennedy and my father. . . . Passing Lem and me on the street, knowing only that we were American, strangers would smile and greet me saying, 'Ah, Kennedy milk.' The Food for Peace program had helped nourish a generation of the world's poorest people and inspired warm feelings for the United States in far-flung corners of the globe. . . . As part of their daily routine, schoolchildren in Kenya sang a patriotic song that referred to John Kennedy as their father. We found his photo inside many huts and homes."

In September 1968, Bobby enrolled in the ninth grade at Millbrook, a private all-boys boarding school located on a rural eight hundred acres in upstate New York's Hudson Valley. Environmental stewardship was among Millbrook's five core values, counting conservation biologist Thomas Lovejoy among its noted alumni. It is likely Bobby's mother thought Millbrook would be right up his alley, as it had natural history and ornithology programs and contained the only zoo in the United States actually located inside a high school. The Trevor Zoo housed almost two hundred exotic and indigenous animals representing eighty different species, ten of them endangered.

While other students chose activities like art or theater for their three afternoon hours of "community service" elective activity, Bobby quickly found "a tight-knit cadre of boys who shared my devotion to raptors." Falconry, the art of taking quarry with trained birds of prey, dates back three thousand years (Genghis Khan kept a thousand falconers with war eagles in his front ranks).

There was something rapturous about being with the raptors. "We talked about hawks every spare moment—at meals, between classes, and after chapel. We marked in our memories the raptor nests we found during our daily winter hunting excursions when no leaves obstructed our view of the upper canopy. In the spring we climbed those nests to band baby red tails, crows, and owls. We learned to use ropes and climbing spurs to scale the tallest oak and ash trees. We even began a raptor breeding project, one of the first in history, and were successful, persuading a golden eagle and red-tailed hawk to lay eggs in captivity." Asked about the role this played at that critical life juncture, Bobby told me: "I would say the falconry gave a gravity to my life that was not destructive, and everything else happening was." Bobby's love of falconry would last the rest of his life—raising and training dozens of birds.

* * *

In the campus zoo, Bobby kept a lion cub named Mtoto Mbaya, which means "Bad Boy" in Swahili. Toto was a gift to him from Jack Paar, the era's leading talk-show host who was close to the Kennedy family. On a summer trip to Africa, Paar had saved Toto and two other cubs from being put to sleep after their participation in the movie *Born Free*. Only ten weeks old and unweaned when it arrived, Toto grew steadily through the school year to 130 pounds, more than Bobby weighed. Paar came regularly to check in on Bobby and Toto. But as Bobby walked Toto through the zoo one day, the lion spotted a deer and bolted, dragging Bobby holding the leash. "After Toto mauled the deer," Bobby recalls, "they shipped him off to the Lion Country Safari Park in Florida." Perhaps it was an omen.

The previous summer, Bobby and Lem had stayed in Nairobi for several days with Tom Mboya, a charismatic Kenyan labor leader who played a key role in the liberation war against British colonial rule. Bobby had first met Mboya when his parents hosted him in Hyannis in 1960. "It was a high point of my young life to meet a real African leader. He wore a dashiki and spoke kindly to us. He was from the Luo tribe," Bobby recalled. He and the Shrivers had then stayed in Mboya's home on their visit to Kenya in 1964. It seemed fitting he and Lem do the same in 1968, because "Mboya and my father had found kindred spirits in each other and developed a strong friendship."

Now, early in July 1969, an assassin opened fire and escaped in the ensuing confusion as Mboya fell victim to a single bullet outside a Nairobi pharmacy. "Only thirty-eight, the handsome, articulate Mboya embodied many of the qualities so urgently needed by the fledgling nations of black Africa," *Time Magazine* reported. "Grieving Kenyans soon gathered in such numbers at the hospital that baton-wielding police were called out to keep the crowd at bay."

For Bobby, it must have seemed like a déjà vu. "My father had been assassinated and then I spent the summer with Mboya, who was assassinated," he told me in 2022. Yet there was a mysterious coda. It happened almost precisely forty-five years later, when Bobby was to give a speech on environmental issues at the 2004 Democratic Convention. Waiting in the Green Room beforehand, he was introduced to a state senator from Illinois named Barack Obama.

"After I spoke, he went onstage and gave a speech that blew the roof off the convention," Bobby recalled. "That night he happened to be going to Martha's Vineyard, where I was staying with my friend Larry David. So we arranged to meet him for dinner. I sat next to him. I was asking him about his life, and he told me his father was Kenyan. I asked what tribe he was from, and Barack said he was Luo. Then I asked had he ever heard of Tom Mboya. And he said, 'Tom Mboya is the reason I'm in this country.'"

Obama's father had come to America in 1961—the same year Barack was born—via a program organized by Mboya that enlisted American colleges including Harvard to provide scholarships to gifted Kenyan students. When the Republican State Department wouldn't help with their flights, Martin Luther King suggested to Mboya that he call JFK, who headed up a subcommittee on Africa in the Senate. Harry Belafonte ended up bringing Mboya to meet the family at Hyannis, and then-presidential candidate Kennedy was so impressed by the man that he gave him a private donation of $100,000.

When word leaked out, Republicans tried disparaging the "Kennedy Airlift" of all these Africans. But JFK proceeded to win the election - and almost half a century later the African American baby who followed in its wake became president of the United States.

* * *

Aside from the falconry, life at Millbrook was a struggle for Bobby. Understandably, he was still in a lot of pain. "They clearly did not want me to come back," Bobby says. "My mom got one of those letters—we don't think Bobby is happy here and we would be happier if he were somewhere else."

In his personal battle between light and darkness, returning to the family compound that summer he joined with some of his teenage relations

and led a local gang of kids dubbed the Hyannis Port Terrors in skir-
mishes between Irish and Yankee teens. An undercover narcotics officer
proceeded to penetrate the HPTs, resulting in several squad cars of police
arriving at the compound on July 10, 1970, to bust Bobby and his cousin
Bobby Shriver for marijuana possession. It was the height of President
Nixon's "war on drugs." The raid, which also rounded up some forty other
teenagers from all around the area, was front-page news in the *New York
Times*. "It seemed like a slow motion nightmare," Bobby remembered.
"I would have rather cut my arm off than tarnish my family." The boys
appeared in the Barnstable County Court in August, along with Bobby's
mother and uncle Ted, and the judge agreed to dismiss the charges if they
stayed out of trouble for a year.

Attorney John W. Douglas advised Bobby to be "particularly careful
to comply with the laws" in anticipation of his case being dismissed the
following year. The probation officer wanted Bobby to write every month
"telling him of your whereabouts and in general what you are doing. . . .
No matter what some misleading and unfair press reports said or what
anybody else may have said or implied, you have not been found by the
court to have violated any law."

Bobby had transferred to the Pomfret School, a private boarding
school in rural Connecticut in September 1970. Pomfret seemed a hope-
ful choice for Bobby's 11ᵗʰ-grade year. Notable alumni included Edward
Stettinius, FDR's last Secretary of State and instrumental in the founding
of the United Nations, and William Casey, a future director of the CIA
under Ronald Reagan.

Appropriately for Bobby, Pomfret's mascot was the mythical griffin,
part lion and part eagle. There, perhaps he could soar if not roar. Bobby
joined the football squad. Most of the team consisted of African American
teens from broken families in the New Haven projects, recruited under a
program called ABC (A Better Chance). Bobby immediately gravitated
toward them, recalling that "those were my friends there. You got to
choose your dorm and I was the only white person." He sat with his black
teammates in the back of the bus going to the games.

However, pot smoking had reached the Pomfret campus of 250 students. One night in the late spring of 1971, Bobby and his dormmates

barricaded the doors to the hallway and lit up some joints. When one of the administrators came banging, the smell permeated the room. The students relented and let him in. He demanded to know who was smoking. It was clearly an awkward situation. If the truth were told, Bobby and all the black students recruited for Pomfret's experiment would be expelled. So Bobby quickly decided to take the rap. He was the only one smoking, he said. The next day, he alone was expelled. These were tough times for Bobby, but he never lost his courage or his willingness to sacrifice for others. Nor had he lost his early interest in reading and ideas and philosophy.

Shortly before his dismissal, Bobby had written a paper analyzing "C. S. Lewis's Science Fiction Trilogy" (May 14, 1971). In later years, Bobby would be a student of the British theologian's famous *Screwtape Letters*, satiric exchanges where the Devil seeks to tempt a young Christian gentleman. In a curious synchronicity, Lewis had died from kidney failure the same day that President Kennedy was assassinated.

At seventeen, Bobby's careful analysis of C. S. Lewis's *Out of the Silent Planet, Perelandra*, and *That Hideous Strength* is a telling forecast of his own future philosophical bent. Seeing numerous Biblical allusions "if a reader looks hard enough," Bobby writes that Lewis "was able to balance both mysticism and realism. . . . Dr. Lewis fears the direction that scientific technology seems to be taking mankind [where] using science as his instrument and the Devil's temptation as his drive, [man] attempts to make himself invincible and immortal. . . . Lewis predicts that we as human beings are on our way to becoming dehumanized supermen, forgetting God and putting all our faith in science. . . . If one looks around oneself, Lewis becomes reality."

That same year, Bobby saw a TV ad that became a kind of internal guidepost for him amid the difficulties of his teenage years. The ad depicted a single tear rolling down the face of a stoic Native American. The "Crying Indian," as the iconic figure came to be called, was watching someone heave a bag of garbage from the window of a passing car. The tagline was: "People start pollution. People can stop it."

CHAPTER FIVE

INITIATORY SCHOOLING
IN LATIN AMERICA

The silver lining in all of Bobby's turmoil was his growing passion for a land as remote and impoverished as he often felt in the depths of his young being. "For all its troubles, I fell in love with the continent," he would write of Latin America in *American Values*, "its people and culture, its magical landscapes and wildlife."

His first trip to South America came in the summer of 1969, when Lem Billings had arranged for him to work on a ranch managed by his nephew in the Rio Meta region of Colombia. The family had history here. During the Kennedy administration years, Colombia had returned to a limited democracy with the assistance of JFK's Alliance for Progress. Since then, the country had descended into dictatorship and violence.

Bobby recounted: "For roughly two dollars per week I worked from dawn till nightfall, clearing fields with machetes and fire, branding skeletal Brahmin steers in rough-plank corrals, and riding fence lines to check for breaks in the wire, or for squatters. I broke horses, castrated hogs, drove cattle, and helped treat them for disease and the ubiquitous tropical worms and parasites. I learned to remove chiggers from my skin by suffocating them first with nail polish, and how to change and patch a tire without a jack by driving the hobbled car onto a log crossbar stacked on piles of stones. We slept in a tin-roof bunkhouse, sweltering in hammocks beneath a stifling web of mosquito netting. We ate yucca, yams, arepas and potatoes, fried plantains, and, only occasionally, beef or capybara—a giant aquatic rodent that we would hunt with a .22 rifle and dim flashlights in the evening from a dugout canoe on the lake, or along the riverbanks. Once in a while we had canned sardines, which I still consider a great delicacy. In the fields we drank water sweetened with slivers of brown sugar we shaved from brick-sized blocks with our machetes."

Bobby was again surrounded by nature full-time, and it nurtured his troubled adolescent soul. "I had never seen such an abundance of life. I loved the eerie songs of the howler monkeys that broke the daylight silence as we sweated and swung machetes in the forest-fringed fields. At twilight, thick flocks of spoonbills, red ibises, and white egrets abandoned the lake for their evening roosts in the jungle, passing over the farm amid rush-hour crowds of chattering toucans, macaws, and a dozen other parrot species. At night I sometimes left a kerosene lantern near the machine shack so I could collect the giant insects attracted by the flickering light." His days in Colombia were filled with hard work, natural beauty, and important life lessons that you can't learn in school. Bobby hung out with some very tough local characters and got into occasional brawls.

This was very different from white-water rafting with his father or taking pictures of animals in the African wilderness. This was toiling alongside the *campesinos*—the impoverished people whom the Alliance for Progress was intended to assist. Bobby recalled: "I learned in civics class that the United States is great because our people are resourceful and hardworking, ambitious and productive. Yet it was difficult for me to imagine anyone working harder or more resourcefully, or enduring greater difficulties, than these Colombian men and women. Every Colombian, it seemed, was an ingenious mechanic, able to keep rusting heaps of machinery running long after their ghosts had fled." He began to understand how suffering, insecurity, fear, grief, hunger, and helplessness could drive decent people to break the law. These people "lived day to day, bullied and robbed by ranch managers and absentee owners, paid meager salaries insufficient to clothe or properly feed their children. Illness or injuries, which were daily occurrences, could cost them their jobs, their homes, or their lives. Parents watched their children die, often in agonizing pain, without access to a doctor or medicine. Rendered voiceless by the political system." Here Bobby's political education had begun, an initiation at one of the world's rawest edges.

* * *

After Bobby came home to Hyannis Port, an incident occurred on Main Street. He'd gone into an ice cream parlor to get cones for himself and a girlfriend, when a passing summer policeman ordered the girl to stop leaning against the hood of his car. Bobby tried to explain, but the cop began writing up a ticket for "sauntering and loitering." Bobby some years later told the *Washington Post*: "He said something smart to my girlfriend and I probably said something smart-ass and he slapped the ice cream out of my hand, handcuffed me immediately, put me on the ground and then called a cruiser. When I got to the police station, he said I was drunk. I wasn't." There the policeman falsely accused Bobby of spitting ice cream into his face, the reverse of what actually happened.

Bobby spent the night in jail rather than call his mother. Bobby Shriver bailed him out the next morning. Going to court, the judge suggested he enter a plea of nolo contendre to the charges rather than not guilty, because Bobby hoped to end the situation without any press coverage. He paid a fifty-dollar fine and thought that was the end of it—only to find the driveway outside lined with reporters shouting questions at him when he emerged from the courtroom.

Tension between Bobby and his mother had been escalating ever since the assassination three summers before. In defiance of family rules, Bobby sometimes spent nights with his girlfriend at the compound. Now Ethel accused him of dragging the family's name through the mud and threw him out of the house. "In Twain's words, I 'lit out for the territories,'" he recalled. He drove with a friend to California carrying the notion of going on to Costa Rica or Panama. But after their used car broke down, Bobby spent the summer riding freight trains across the American West and the Great Plains.

It was more harrowing than anything he'd ever experienced: stuck and starving on a siding for days never knowing when the train was going to move, sometimes in heat that made him think he would die of thirst; being chased around the yards by the railroad "bulls" and awakened by their sticks and billy clubs; becoming covered in coal dust that wouldn't come off even with the most intense scrubbing. Along the way, Bobby spent two weeks replenishing his funds at a lumber camp in Colorado, until he had so many blisters on his hands he couldn't work anymore.

And yet: "I recall the feeling of exhilaration riding through the Colorado canyons under the clear blue sky, wearing my patched and faded blue jeans and my feet dangling out the boxcar door, sharing a gallon jug of Gallo wine with a carload full of hoboes. We built fires and heated cans of stew, baked beans, and spam. I was in paradise." After witnessing a shootout between a group of white men and a black man, he and his friend decided they'd seen enough. Bobby, seemingly needing even more adventure, even more gritty education, hitchhiked all the way home from Kansas City.

When he managed to reach Boston, flat broke, Ethel Kennedy was on Martha's Vineyard. Bobby called and explained his situation, asking if she could send him enough bus fare to get to Hyannis. She had no idea where he'd been. "Get a job!" his mother told him and hung up. "She invented tough love," he recalled. "Those rebukes stung at the time. But looking back, I recognize that she was doing her best to keep a world together for eleven children and my rebellious nature was not helping."

Bobby attended a small high school named Palfrey Street while he lived in a blue-collar Boston suburb for his senior year. He lived in the home of a family of bohemian liberals. Palfrey blessed Bobby with a teacher named Skip Lozell, whom Bobby remembered as "kind of a right-wing iconoclast." His encounter with Lozell opened his mind to the importance of bridging the divide and transcending ideological differences. Bobby remembered: "He was a stout extrovert rarely without a beer in his hand, and he made biology come vitally alive. He had encyclopedic knowledge and a brilliant grasp of the impacts of evolution."

Palfrey offered another opportunity, through a program allowing students to receive credit for working abroad. Bobby spent much of the school year living with Maryknoll missionaries in an Aymara Indian village in Peru. He traveled with a young Mexican American priest to rural farms and villages. He described assisting the priest "serving mass, performing baptisms and funerals, and running errands as he ministered to the sick." He witnessed the "back-breaking work," the subsistence living, the lack of doctors, and the funeral processions of young children. Fewer than 30 percent of villagers lived past age thirty. Late one night, Bobby heard frantic knocking. "A young weeping Aymara mother cradling her tiny baby, already dead," he remembered. She was "desperate to have it baptized" because she believed their "crops would fail if the little corpse was buried unchristened."

As he would see in many other countries around the world, pictures of his uncle and father often adorned the walls of the poor. They were reproduced on plates, carried in wallets, symbols of only one thing— hope. Seeing the impact that his family had on so many was a constant reminder that ideas mattered, that they couldn't be killed, that they changed the world.

Bobby journeyed to neighboring Bolivia, where the poverty was even more stark, and a handful of wealthy families controlled the natural resources in what was basically a police state. Bobby set down of his experiences: "The wretched poverty I saw nearly everywhere in Latin America contrasted starkly with the opulent estates of the superrich who ruled these nations and controlled their resources." While he spent much of his time with the poor, he was also able to meet with the most powerful people in the country, even a former president. As he traveled, he understood how American foreign policy had failed to live up to American ideals. "After Uncle Jack's death," Bobby wrote, "thirteen constitutional governments were overthrown in nine years." Military juntas ruled over eleven Latin American nations with the full military and financial support of the US government.

Yet amid all the injustice and corruption, magic could still permeate the air. Bobby remembered a moment in Cuzco, once an imperial city of the Incas, high in the Peruvian Andes: "I was this kid with long hair and chewing coca leaves, wandering around the central plaza. There was a sudden freak hailstorm, giant pieces between the size of a golf ball and a baseball, the biggest hail I'd ever seen, coming down with such ferocity that they were splintering some of the little wood slats that hold up the market stands. All the women in the square huddled together against the walls, shielded by the eaves. And in the middle of this ferocious storm . . . a tiny nun came out of the church, calmly shut the door, and walked down the steps. People were screaming at her not to come any farther. She was looking at the ground and praying her rosary, completely oblivious that there are rocks falling out of the sky. She walked all the way across the square, and then sunlight suddenly appeared and beamed down on her. As she went into a church, everybody in the square began shouting: 'Milagro! Milagro! Milagro!'" A miracle.

"That kind of stuff happens to you all the time down there. It was like a Gabriel García Márquez story."

* * *

In June 1972, having spent four turbulent years at three different secondary schools, Bobby graduated from Palfrey. Then, like his father and uncle before him, he enrolled at Harvard that fall. While he would major (and earn a degree) in American History and Literature, his science professors remained a major influence. One was the pioneering evolutionary biologist Robert Trivers, who "on the side" had spent time behind bars, driven a getaway car for Black Panther leader Huey Newton, and once formed an armed group in Jamaica that protected gay men from mob violence. Bobby took a course that Trivers taught with E.O. Wilson, one of the world's foremost naturalists whose specialty was the behavior of ants. Wilson's controversial philosophy of sociobiology intrigued Bobby, that inherited genetics are key to understand human behavior.

On campus, wrote Laurence Leamer in *Sons of Camelot*, Bobby "had a sheer physical presence that by itself called attention. He was six feet two, with a twenty-eight-inch waist and eyes of searing intensity. . . . Bobby rarely traveled alone but with an entourage." His closest friend at Harvard, and for years thereafter, was Peter Kaplan, a stringer for *Time Magazine* and a cartoonist for the *Harvard Crimson* who also ran a campus film society. Kaplan remembered: "He was out for freshman crew and he had really built himself up, getting real strength into that lean body." He'd also joined the rugby team.

This would stand Bobby in good stead toward the end of his freshman year, when he successfully pitched an article to the *Atlantic Monthly* that would take him to Chile, Latin America's longest-lived democracy. In 1971 a Marxist professor, Dr. Salvador Allende, had been elected the country's president and immediately begun seeking land reforms and nationalizing powerful American corporations like Anaconda Copper. The CIA, backed by the Nixon administration, plotted a coup with reactionary elements of the Chilean military. Initially the CIA financed a truckers' strike that paralyzed the nation for eighteen months, causing dire shortages of food and fuel. Coupled with other measures, the Chilean *escudo* currency plummeted in value. By the time Bobby got there in July 1973, the country's inflation rate had soared 1,000 percent annually, and the people were carrying their money around in paper bags.

"In a firsthand experience reminiscent of Germany's Weimar Republic before Hitler," Bobby later wrote, "I needed a suitcase to change a US hundred-dollar bill. In Santiago I saw endless lines form in front of any stores with stocks of bread, cigarettes, or even matches. Armed militias guarded gasoline pumps as long lines of cars awaited their rations of gasoline, often in vain. Daily headlines reported acts of sabotage, mass strikes, and political violence."

Bobby arrived in July at the height of the Andes ski season, his other reason for going. At the Portillo resort, he encountered another aspiring journalist, Blake Fleetwood, who like Bobby learned to ski as a youngster. Seven years older than Bobby, Fleetwood had been born in Chile to a South American mother before his family emigrated to America when he was four. Along with three others including a guide, they decided to make a day-long cross-country expedition ultimately bringing them into Argentina. They intended to ski-hike almost twenty miles and ascend some three thousand feet to see "Christ of the Andes." In his first published piece in the *New York Times's* travel section later that year, Fleetwood described it as "the famous but rarely visited statue forged from melted-down cannon barrels and placed in a remote mountain pass on the border between Argentina and Chile to commemorate the 1903 peace treaty between the two countries."

Starting off early in the morning from the Portillo lodge, the two climbed for six hours through glaciers and snowfields approaching the pass. "It is silent and desolate up here at thirteen thousand feet, with the far sound of the wind whistling through faraway valleys," Fleetwood wrote. As they trudged higher and the midday sun found them sweating and eventually gasping for breath in the thin air, Bobby assumed the lead, sometimes a half-mile ahead of the others. Then he spotted a uniformed patrol of a half-dozen Chilean soldiers following them up the narrow, snow-gorged valley. He knew that a mountain division conducted drills near Portillo, and this must be part of it. As it happened, there had been an attempted coup that morning by a group of colonels against the Allende regime. The border patrol was on high alert.

"They began shouting to us to stop," Bobby remembers. "But I pretended to not hear them, because I thought we could make it to the top of the final ridge and they wouldn't be able to catch us. As we edged toward the ridgeline, however, the officer had his men kneel and fire on us."

In his article, Fleetwood would describe hearing "an incomprehensible yell from off to our left. About six hundred yards away we see a Chilean airline trooper who suddenly, without further warning, raises his machine gun. Bobby, who is about fifteen feet in front of me, drops to the ground just as two or three shots ricochet off a rock not five feet away. At this distance the shots sound like someone quickly snapping his fingers. As Bobby falls, I think he is hit, and that we are all goners." Their ski patrolman companion, "directly behind Bobby, collapses in the snow as the shots keep crackling. I'm next in line. I throw myself down and try to bury my head in the snow."

Suddenly the shooting paused. Fleetwood peered up to see the standing commander had lowered his gun. Fleetwood called out to see if his companions were okay. They were. "Finally we put a white handkerchief on top of a ski pole, waving in the air," he recalled in our interview. The lead trooper then motioned the group to turn around and ski back the way they'd come. Fleetwood wrote in the *Times*: "After what seems like hours, but is actually more like sixty seconds, we are all poling our way across a ridge. . . . As our skis glide side by side, I shudder to think what was going on inside Bobby's head as he lay curled up, his face shoved in the snow, while those shots ripped overhead. . . . Strangely, he seems calmer than the rest of us right now."

Bobby's recollection is that "the soldiers detained us for violating curfew and escorted us back to the capital, Santiago, which was in chaos as tanks ringed the presidential palace." An apparent overthrow of Salvador Allende was underway. Fleetwood recounted in our interview: "Later they said they thought we were smugglers trying to flee the country on skis with Chilean gold. Either that or Socialist enemies of the people."

Although the first attempt to oust Allende was suppressed, a few months later General Augusto Pinochet led a successful coup that killed the president and launched a brutal seventeen-year dictatorship. "Upon returning to the United States," Bobby later wrote, "I briefed Uncle Teddy on the Chilean crisis and urged him to expand his involvement." Senator Kennedy, who had already held a hearing on the takeover, received credit from many Chileans for intervening to save their lives during the early days of the Pinochet regime.

Fleetwood's article, headlined "Skiing the Andes—and Drawing Fire From the Border Patrol," appeared in November 1973. He went on to a stellar career with the *Times,* while he and Bobby remained lifelong friends. As Bobby turned twenty, his first published article on foreign affairs, "Poor Chile," appeared in the February 1974 issue of the *Atlantic Monthly,* taking aim at the American role in destroying Latin America's flagship democracy.

Bobby recalled: "Teddy orchestrated the passage of the so-called Kennedy Amendments, first cutting off all military aid, and then, as Pinochet's violence escalated, all assistance, to Chile. I watched proudly from the gallery as Teddy grilled the Chilean junta leaders about human rights abuses. In late 1974, the generals returned to Washington for a second hearing. Afterward, my uncle's staff told me that the generals had pointedly warned them that both Teddy and I should stay out of Chile."

This marked Bobby's first time—but not his last—to be declared persona non grata by a region whose antidemocratic policies he was challenging.

CHAPTER SIX

DISCOVERING POPULISM: A SOUTHERN ODYSSEY

You cannot swim for new horizons until you have courage to lose sight of the shore.

—William Faulkner

In his history course at Harvard, Bobby came across some curious polling data. They indicated that the white mourners he had seen lining the tracks of his father's funeral train, his strong supporters in the 1968 primaries, had largely voted in 1972 not for Democratic candidate George McGovern—but for Robert Kennedy's polar antithesis and personal nemesis, Alabama Governor George Wallace.

"It struck me then that every nation, like every individual, has a dark side and a lighter side," Bobby said years later. "I became fascinated by populism, a democratic impulse that was often perverted by demagogues. The easiest thing for a politician to do is to appeal to our greed, bigotry, racism, xenophobia and crude self-interest. My father had endeavored to do something different—to appeal to the hero in each of us—to persuade us to transcend narrow self-interest and act on behalf of the community."

Bobby remembered well his father in 1962 having led the Justice Department through a series of events that marked the greatest confrontation between the states and federal power since the Civil War. That

September on the campus of the University of Mississippi, five hundred federal marshals sent by Robert Kennedy to protect James Meredith as he sought to enroll were attacked by a howling mob firing guns and hurling bricks and bottles. More than sixty were wounded after unsuccessfully begging RFK for permission to open fire to defend themselves. Instead, the Attorney General ordered army troops moved to the campus. Eight-year-old Bobby found a note from his father on his pillow the next morning. "Bobby, tonight we got James Meredith into the University of Mississippi. I hope these battles are over by the time you are grown."

Bobby reflected: "That vision, and something of my father's other qualities, had evoked a powerful populist response among all types of Americans." So it was that he decided to write his junior thesis at Harvard on Governor Huey Long, the Louisiana "Kingfish," and the lives of agrarian rebel Tom Watson and prairie fireball William Jennings Bryan. "I admired them because they represented what de Tocqueville had described as: 'Americans' insistence on limited government and culturally in their long-standing disdain for elites, that spirit has become one of this country's greatest gifts to the world.'"

Each of those men had begun their careers as rabble-rousing idealists, bringing together America's poor and working people, factory workers and farmers, black and white, behind an idealistic vision. But Long had fallen into an unquenchable thirst for dictatorial power that allowed him to tolerate a great deal of corruption. Watson devolved into a Klansman, supporting the lynching of blacks and Jews. Bryant became a religious fanatic, ending his career a laughingstock for his role as an attorney attacking evolution in the Scopes Monkey Trial.

Then, for his senior honor's thesis, Bobby wanted to find "a comparable southern figure with some kind of visibility or flamboyance." He received some advice from Harvard psychiatrist Robert Coles, who had been instrumental in his father's being drawn to issues of poverty and race, having traveled with RFK to Mississippi. Why not focus on George Wallace, another powerful populist who began his career as an idealist uniting blacks and whites before turning to the dark side and making himself as Alabama's governor the national symbol of racial hatred and discontent?

Bobby contemplated the idea. "Although my family is invariably associated with New England, I am, in some ways, a child of the South," he would later reflect. "I grew up in Virginia, so I had my own recollections of the injustice of Jim Crow. When I was a boy, I often rode in a station wagon pulling a horse trailer across Virginia and Maryland to hunts and horse shows with my family's groom, a black man named Bill Shamwell, who would die tragically following a fall from the roof of our house. Shamwell was a proud World War II veteran who stood six-foot-three and whom I recall for his tremendous physical strength and natural dignity. He enthralled me during those long drives with tales of dodging attacks by Japanese Zeros as he worked as a Seabee in Burma, on Midway and Iwo Jima. When we stopped for meals, he gave me money and I went in to a roadside restaurant to buy our food. We ate together in the car. I only understood years later why it had to be that way. In 1961, if you were a black in Virginia, Alabama, or any one of a dozen southern states, you had no more rights and no more protection of the law than did a black in South Africa under the system of apartheid."

Bobby had followed the story of his father visiting Montgomery in 1962 to confront Wallace on his refusal to desegregate the University of Alabama: The governor had been a master of symbolic imagery, standing in the schoolhouse door pledging "segregation now, segregation tomorrow, segregation forever." But when Wallace was shot and paralyzed while running as an independent candidate for president in 1972, Ethel Kennedy had traveled to see him and even invited him to stay at Hickory Hill. Wallace declined but didn't forget her gesture. Four years later he accepted Bobby's request to interview him.

Accompanied by his Harvard roommate Peter Kaplan, Bobby remembered the two "drove south in my roofless Jeep, camping in a tent by the roadside." Upon arrival in Montgomery, the onetime capital of the Confederacy, "I noticed on its dome that the stars and bars of the Confederacy still flew—illegally above the American flag. I stepped over the bronze star embossed in the marble atop the Capitol steps to mark the spot where Jefferson Davis had taken his inaugural oath as President of the Confederacy.

"I first met Wallace, seated in his wheelchair, in the foyer of the State Capitol Building in Montgomery. Over his shoulder I could see the

Dexter Avenue Baptist Church from which a young preacher, Martin Luther King, had directed the Montgomery bus boycott. 'Your daddy was a good boy,' the governor told me wistfully. 'They was both good boys,' he repeated to include Uncle Jack. 'They was good for this country.' With that peculiar phrase he acquitted my father of the years of vilification and acknowledged the manipulative nature of his own demagoguery. Then Wallace added: 'I could never figure out how to get to them.'"

Peter Kaplan remembered that Wallace "felt compelled to explain himself politically to Bobby, trying to reconcile himself to the Kennedys . . . [as if] he felt a tremendous bond with Bobby." Bobby later wrote: "Knowing the charged political terrain of the south, Wallace's warm endorsement of my father did not entirely surprise me. This was consistent with his revised views on civil rights, which were more appealing to Alabama's newly minted black voters. But he didn't even seem to have any particular convictions on the issues about which he'd been so outspoken. He used his demagoguery for self-promotion and easily switched to the other side when it availed him. I was thinking to myself as I shook his hand of the many people who suffered to feed his ambitions, and all for what? Here now he sits in a wheelchair."

Bobby and Kaplan had moved into a temporary house, and part of Bobby's itinerary involved a visit to the Alabama State Legislature. As Kaplan recalled it for the authors of *The Kennedys*, Bobby's entrance brought a burst of electricity to the staid proceedings. It was strikingly reminiscent of the enthusiastic reception RFK had received in 1966 at the University of Alabama. "Suddenly everyone was mobbing him, grabbing his hands and slapping him on the back" and exchanging reminiscences about his father. "I'd seen him in political situations before and he'd always acted intelligent but a little stiffly. This time he started flashing the Kennedy smile and pumping hands in a way I'd never seen him do before." Kaplan told another writer: "To me, that was the day of Bobby's coming of age. He had real rapport with them and he loved it. It was a completely political moment, and he saw what it was like to stand for something and be recognized for it. He was happy responding to the task of being a public person."

Bobby had planned to stay only a few weeks. But while Kaplan returned to Harvard, he'd found another figure around which to focus his thesis. "Wallace, I discovered, was not really running the state," he told me. Years earlier, when a grandstanding Wallace refused to integrate the state's public high schools, a federal judge took them over. The same thing happened with the state parks and pools, mental hospitals, and transportation facilities. When a federal court found that prison conditions violated constitutional guarantees against cruel and unusual punishment, Wallace refused to rectify them and once again the judge assumed control. In each of these cases, this man was Wallace's former University of Alabama law school classmate, Judge Frank M. Johnson, Jr.

To Bobby, Johnson's takeover represented "the predictable end game when the ginned-up hostility to government is taken to its extreme. The end of government! Frankly, there wasn't much left for Wallace to manage and he seemed a less interesting character to me. He was all about imagery and propaganda." Bobby decided to focus his research efforts on "the guy who was left holding the bag."

Bobby's father had known Judge Johnson, who had worked hand in glove with RFK's Justice Department. In 1954, Johnson had presided in Alabama over the landmark *Brown vs. Board of Education* decision by the Supreme Court to integrate the nation's public schools. He'd tried every major civil rights case in the state thereafter, including the Montgomery Bus Boycott, the Selma March, and the Freedom Riders. Considering himself a strict constructionist, Johnson took the Constitution and the Bill of Rights at their word, his courageous decisions making him anathema in Montgomery. Klansmen burned a cross on his lawn and firebombed his home. The city's wealthy denizens excluded him from the country clubs. Partially in response to the social isolation afflicting the family, Johnson's only son committed suicide.

"When I met him soon thereafter," Bobby recalls, "he was still quite isolated and happy for my company. Despite his reputation as a tough uncompromising jurist who so frightened limtigators that one had famously fainted in his courtroom, I found Johnson warm, kind, and vulnerable, and the best company."

Bobby went on to live in Alabama for a year, visiting Judge Johnson and his wife Ruth often. "He was a tall, wiry backwoodsman, with droll features, homespun manners, modest humor, and laconic style. He taught me to play golf and encouraged me to chew tobacco (Johnson kept a spittoon in his courtroom), a habit that made me a pariah when I finally returned home. We spent lots of time sharing our mutual passion for fishing, and driving together to the 'Redneck Riviera' near Pensacola and Gulf shores to tease the mangroves for speckled trout and snook."

Johnson also gave Bobby a lesson in how to effectively get things done. "I think he gave me a lot more faith in the system than I had," Bobby reflected to a radio interviewer five years later. "I was raised during the 1960s and participated in many of the protests. My generation grew up with the Vietnam War and very little sense of the system having any moorings in justice or equity. My work with Judge Johnson reinforced the view, which I think was also my father's, that in the end you really could get justice through the system and that's the only way to work."

* * *

In the process, Bobby ended up moving into the home of a fellow he met at a Montgomery honky-tonk, a rangy cowboy named John "Sweet Pea" Russell. It became an unforgettable friendship even though he was ten years Bobby's senior. Bobby recalled Sweet Pea having "a youthful outlook, a skilled storyteller with a great laugh—six feet three inches of beef jerky gristle, perpetual smile, and sunny disposition that a certain kind of woman found irresistible."

Sweet Pea owned the ancestral estate, a rundown antebellum mansion on almost a thousand acres in Hayneville, located halfway between Selma and Montgomery. Lowndes County had a predominantly black population, the descendants of plantation slaves. One of the first things Bobby noticed was a large portrait of Sweet Pea's grandfather on the mantle, uniformed in rebel gray. A Confederate General, he looked uncannily like Sweet Pea. In the attic was an old chest filled with stacks of Confederate currency. The General's descendant revealed to Bobby "that his high school history teacher had had his classmates tear Abraham Lincoln's picture out of the history textbooks, and confessed that his classmates had cheered at the P.A. system announcement when President Kennedy was shot." But according to Bobby's college friend Chris Bartle, "Sweet Pea treated everybody as if he was their brother . . . for Bobby it was a real breath of fresh air to meet these southern white guys who didn't have a prejudiced bone in their bodies."

Besides growing soybeans, Sweet Pea raised cattle, and Bobby often joined him on horseback to round up strays and rope any cows that looked to need medical attention. Using the skills he'd learned on the ranch in Colombia, Bobby was, according to Sweet Pea, "just one of the boys."

By that, Sweet Pea implied a wild streak that he himself cottoned to. Which included Bobby's being "very interested in trying how to ride a bull and a bucking horse." Sweet Pea had been taking part in southern "Buck Out" rodeos for some years. One of the nicest was a little more than two hours away in Bonifay, Florida. Sweet Pea loaded up the equipment needed, a different type of spur for bulls where "you just want to dig in a little bit, grab on with your heel, and hold on with one hand."

As Sweet Pea recounts, "Bobby did surprisingly well. Most people who first start are off the bull before it hardly gets outta the chute, with the first buck. But Bobby held on for two or three jumps and landed on his feet, which was a blessing. Because most of the time you hit on your butt or your back and you've gotta get up and away from that bull because he knows you're his antagonist. Well, Bobby made it to the fence and out of harm's way. I was real proud of him."

Bobby's southern education wasn't confined to outdoor activities. He spent a month observing the State Legislature, then Judge Johnson in court and having him over for dinner several times at Sweet Pea's, who found him "a great guy, we just sat down and listened to him talk." The judge himself came out of a wild youth filled with fast cars and bootlegged moonshine that landed him in a military academy, a background with which Bobby could relate. As an article in the *Miami Herald* put it, "some have even speculated that Kennedy and Johnson, an ironic juxtaposition of names, developed a father-son relationship during their formal year-long association." To Bobby, Johnson represented "a stance of defiance," with "a dedication to integrity and course that you seldom see in American politics today." Asked by an interviewer whether Johnson reminded him of his own father, Bobby replied: "I think in a lot of ways they have similar goals and ideals." Bobby even drew this comparison in 2011: "Like Abe Lincoln, Judge Johnson's courage and character and his idealistic view of our nation had plunged him reluctantly to the center of the historic [civil rights] conflict."

Bobby also came to know a number of unsung heroes of the civil rights movement. And he encountered attitudes he would never have known still existed. "Despite Wallace's conciliatory rhetoric and the outward appearance of racial harmony, I sometimes glimpsed the dark underbelly of racial fear and hatred waiting to be harnessed by the next demagogue. We were raised to loudly object if someone offered to tell an ethnic joke or used racial epithets and to leave the room if it persisted. In Alabama, most people generally respected my feelings or those boundaries."

At one time or another Bobby went to all 64 counties in the state. He visited the headquarters of the Alabama Ku Klux Klan "and met poor and bitter white men burning with tribal hatred," including one who told him that "Monkey Lucifer Coon" had it coming. At such times, it was a painful learning experience, but through it all, Bobby made a number of lifelong friends from Alabama. One was Harris "Sugar Boy" McGough (pronounced Magoo) to whom Sweet Pea introduced him three years earlier. He'd been a crew chief on a helicopter gunship in Vietnam, and his family owned an auto dealership in Montgomery. After spending a year at Sweet Pea's, Bobby moved into Sugar Boy's house.

Sugar Boy remembers: "In the morning when Bobby'd eat breakfast, he never looked at his food. He was reading the paper or something else all the time. If I needed to know something, all I had to do was ask him and he'd give me the history of it. He was mainly interviewing Judge Johnson, but on weekends he was always adventurous. Bobby wasn't scared of a thing! He wasn't trying to impress anybody or prove anything, but something new he just had to try it."

Nighttime coon hunting became a favorite pastime. A black friend of Bobby's transported his hunting hounds stacked like cordwood in the trunk of his smoking Buick LeSabre from which they would pour "when we arrived at the swamp to chase the raccoons with dime store flashlights." Bobby later wrote: "I loved the Spanish moss, running through the swamps and cotton fields at night with dim lights, following the baying hounds, swimming the creeks and coming back midday with the smell of the swamp. I loved the heat and humidity and climbing the moss-draped oaks in the middle of the night toward the shining eyes of a treed raccoon. I ate plenty of barbecued raccoon."

"Bobby didn't like to kill anything," Sugar Boy told me. "When he lived with me, he caught a hawk and trained it, and same with an owl. It's kind of a shock when you walk in the bathroom before going to bed and you look up and there's an owl lookin' at you."

Bobby and his visiting friends from back East heard the song of the South in its myriad of appearances. The Alabamans would have their eyes opened, too. Sweet Pea would be invited to visit the compound in Hyannis Port over the Fourth of July. Sugar Boy would accompany Bobby and his entourage on a couple of whitewater rafting adventures in South America and stayed at his new friend Lem Billings's apartment in New York.

After Bobby largely finished his three-hundred-page thesis on Frank Johnson, he received encouragement from Billngs to consider publishing it as a book, as JFK had done with *Why England Slept* in 1940. After he graduated from Harvard in the summer of 1977, the editor-in-chief of Putnam's signed Bobby to a publishing contract. The forthcoming book received a fortuitous boost when President Carter announced in August that Judge Johnson was his surprise choice to become the next head of the FBI. "I think he'll change the image of the bureau and its activities so that it'll be feared by criminals, not by the average citizen," Bobby told the *New York Times*. Although the judge's nomination had to be withdrawn after he suffered an aneurism, the advance publicity created a stir.

When *Judge Frank M. Johnson: A Biography* appeared, Bobby shone in his promotional appearances, including network TV's popular *Good Day* program, as a remarkably handsome, charismatic, and articulate twenty-four-year-old with a no-holds-barred future ahead. Asked by TV interviewer Bill Boggs about racism in America, he responded: "I really immersed myself in southern life for a year . . . I don't think it's a southern peculiarity . . . what a lot of northerners have thought for years. We're beginning to see that it's a nationwide problem. And something, basically, that my generation is going to have to deal with. I think it's going to be a lot tougher than the kind of job that my father or Frank Johnson was involved with. . . . I think that you have to ensure minority groups, or groups that have been deprived because of prior racial injustice, [be given] some outlet, some sense of self direction, some sense of self respect - and

I don't think you do this just by handing out jobs or welfare money. You've got to really strengthen traditional ties, family ties, community ties. What my family was involved with over in Bedford-Stuyvesant, with community projects and factories that actually have a base . . . not just have a nine-to-five job with no hope of mobility."

Bobby had enrolled for a brief stint at the London School of Economics, and the talk-show host presumed that like JFK he would go on to Harvard Law School. "No, I think I'm gonna go down South," Bobby replied. His father had obtained his degree from the University of Virginia Law School, which is what Bobby was contemplating. In the spring of 1979, he would rent a nondescript house in Charlottesville and commence his law school studies.

"Politics seemed to me a lot more lively down South," he said. So, when Uncle Ted challenged incumbent Jimmy Carter for the presidency in 1980, Bobby's existing connections found him placed in charge of the campaign in Alabama. He was welcomed into a number of appearances in black churches, where his speeches became legendary. But Ted Kennedy's challenge to Carter failed badly, and Ronald Reagan ended up defeating the Georgian easily in the 1980 presidential election.

Looking back, Bobby believes that he'd gone to live in the South during a time of historic realignment. Wallace's white followers, traditional Southern Democrats, were looking for a new home. Ronald Reagan announced his 1980 presidential campaign at the Neshoba County Fair in Philadelphia, Mississippi, on the site where three young civil rights leaders had been murdered in 1964. Poking the eye of the civil rights movement with this sharp stick, a Republican National Committee member later explained it would win over "George Wallace's inclined voters." Adopting Wallace's core view, Reagan proclaimed to a crowd of ten thousand white people, "I believe in states' rights." This, as Reagan's strategic guru Lee Atwater boasted, was a code word minted to mobilize bigots as part of the Republicans' southern strategy. Reagan called the Voting Rights Act of 1965, which he'd been opposed to at the time, "humiliating to the south."

Bobby pointed out the opportunism at play, setting down in 2011: "For fifteen years Reagan, whose acting career had fizzled, made his living

as a propagandist for corporate profit-taking as chief spokesperson for the
General Electric company. With his professional delivery, starry smile, and
bedside manner, he made the perfect vessel for delivering Wallace's message of hate, now wrapped in honey-style and homespun patriotism."

In our interview, Bobby's longtime friend Blake Fleetwood marveled
that Bobby possessed "a real affinity for middle class people, white southerners and black southerners, everybody who was not in the top echelons. It struck me that Bobby had inherited this. I don't know where his
father got that affinity, but Bobby could relate even *more* to people from
all kinds of different paths in life. That's what's missing from the whole
Democratic Party today. They've squandered it and given that affinity up
to the Republican Party."

CHAPTER SEVEN

WHITEWATER TESTING GROUNDS

As a boy, Bobby's heroes were not only the Catholic saints like Saint Francis. There were also the explorers: Alexander von Humboldt, Charles Darwin, Burton and Speke, Stanley and Livingstone. He read everything he could about them, including the Spanish conquistadores Cortez and Pizarro, rallying their armored men losing bowel and bladder control as they were about to attack Atahualpa's twenty thousand armed Inca warriors.

In his early twenties, Bobby traveled widely in Latin America as a guide for his brother Michael's whitewater company. They made first descents by raft and kayak of rivers in Peru, Colombia, and Venezuela. In 1974, he and his brother David, their cousin Chris Lawford, Lem Billings, and three friends built a raft of giant balsawood logs on the Apurímac River in Peru and floated the Amazon's headwaters with two Asháninka Indians. At the end of 1979, with a contingent including his sister Kerry and several of his new friends from Alabama, they made the first successful descent down the Atrato River headwaters in Chocó province along Colombia's Pacific Coast through the rainiest jungles on Earth. Bobby remembered: "There we encountered a tribe of black Africans who spoke no Spanish— probably descendants of slaves brought to the lower Chocó in the seventeenth century to pan for gold who had escaped into the jungle." Bobby's final voyage went down Venezuela's Caroní River in 1981.

Each of these trips required major team-spirited undertakings to bring together everything from tents to lights, clothing including heavy boots, not to mention food and transport. The elements were unpredictable. During one torrential downpour, several tents were swept away, and after the rafts were loaded in pitch darkness, some overturned in roller-coaster rapids sending all of the group's canned goods into the river. Blake Fleetwood's journal described how, in Peru, they beached the rafts at night "trying to find high ground for our camp to guard against the alligator-like caimans, poisonous snakes, and packs of river rats the size of cats." Chris Lawford recalled in his memoir being on the Apurimac for ten days and catching "a 100-pound catfish, which we dragged behind one of the rafts until it was decimated by piranhas when we hit some still water."

And it wasn't just the treacherous waterways. Fleetwood recalled renting a car with Bobby as the advance party. As they ascended into the high plains of Peru, Fleetwood was driving when a child dashed out onto the badly paved road in poor visibility—and the car struck the boy. "I was really shaken," Fleetwood told the authors of *The Kennedys*, "but Bobby behaved with perfect poise and calm, obviously working to live up to the Kennedy idea of grace under pressure in the middle of this crisis. He was just twenty . . . but he took charge right away. He jumped out and gave the kid first aid, splinted his broken leg and carried him into the car, then drove us back to the nearest town and took the kid to the hospital. When he found out that the hospital didn't have pain-killing drugs, he went to a local pharmacy and bought some himself. We didn't leave town until it was clear that this kid was going to be okay."

Later on that same trip, Billings tumbled into a ravine and badly cut his leg. Again, Bobby was able to handle crisis beautifully, using skills no one knew he had while sewing the wound up with twelve stitches. In crises big and small Bobby carried with him the lessons he and his siblings learned from their father's bootcamp for moral courage. The crew placed Lem in a proneposition luggage rack on the raft when they set out again in the morning.

In 1979 Fleetwood and Bobby made one more trip together, to an Ojibway Indian village on Canada's Wabigoon River. Fleetwood recalled a portentous moment "where we saw a whole tribe being poisoned by mercury dumped into the river from a paper factory one hundred miles upstream. We were shocked to see the cats dancing in circles driven mad by the mercury. And it made Bobby crazy that a corporation would be allowed to get away with killing so many indigenous families—men, women, and children—in the name of profits."

* * *

One day before what would have been Jack Kennedy's 64th birthday, Lem Billings died in his sleep on May 28, 1981, of an apparent heart attack according to the autopsy report. Billings had turned sixty-five a month-and-a-half earlier. His dying wish had been for the young Kennedy men to carry his casket to the burial site. But when they arrived at the cemetery, the casket was already in place to be lowered into the ground. Led by Bobby, they picked it up and carried it around the gravesite before bringing it back to the burial plot.

"Lem brought Bobby through some very, very difficult periods," says Blake Fleetwood. "He gave Bobby a sense of connection with his uncle and his father that Bobby did not have. He imbued [in] Bobby and to some extent [in] all of Bobby's brothers and sisters the sense that they were here, on this Earth, for a purpose." That purpose was to make life better—for his friends, his family, and for the world. "Lem really wanted to reinforce that," Blake told me. "And I think he did."

"In many ways Lem was a father to me," Bobby later wrote, "and he was the best friend I will ever have . . . I thank God that we had him because if we had not we would never begin to dream what we had missed."

CHAPTER EIGHT

A PERSONAL
BREAKTHROUGH

I was in the air, with outstretched arms, and floating fast. There was a
fearful dark river that I had to go over, and I was afraid. It rushed and
roared and was full of angry foam. Then I looked down and saw many
men and women who were trying to cross the dark and fearful river,
but they could not. Weeping, they looked up to me and cried: "Help us!"
But I could not stop gliding, for it was as though a great wind were
under me.

—Black Elk Speaks

When Ted Kennedy was running for president in 1980, Bobby had traveled to South Dakota to campaign for his uncle. There Bill Walsh had long been active in the state's Democratic Party and remained a staunch Kennedy supporter. Though Walsh was fifteen years older than Bobby, they took to each other right away. He arranged for Bobby to stay in Deadwood's historic Franklin Hotel that he comanaged, and where Bobby then established South Dakota's Ted Kennedy for President headquarters. Walsh also organized several local TV appearances for Bobby and accompanied him to the Pine Ridge and Rosebud Indian reservations that his father had visited during his presidential campaign. After they touched base again at the Democratic Convention, Bobby decided to return to Deadwood for the remainder of the summer.

A good-ol'-boy activist in a cowboy hat, Walsh had been the chaplain for the state prison and cochaired South Dakota's prison reform committee before resigning from the priesthood and becoming coordinator of social services for the Black Hills. "Bill was a big strong, sort of ham-fisted square-jawed Irishman," Bobby recalls. "A very funny unusual guy, who went from being a priest to running a bar and hotel. He continued to kind of minister to people from behind the bar."

At the time, Bobby had a drug problem. "I was what I'd call an addict for fourteen years," he later told an interviewer. "I became an addict for the same reasons most people do, which is they have a big empty hole inside, and that was the one thing that would fill it."

Walsh remembered in 2022, "We talked and Bobby was very upfront about his drug addiction but that he wanted to become a lawyer. It was very important for Bobby to get in physical shape so he could overcome his addiction. So we did a lot of running, and we worked out together."

Two summers later in 1982, Walsh put him up in one of his cabins in the Black Hills so he would study for retaking the bar exam. Bobby remembered: "I was there for three months. It was inaccessible by road. I bought a dirt bike, that was the only way you could get up to it. I would take my bike into town to get groceries and then cook back there. I had to bring water up from a well about a hundred feet from the cabin, so it was kind of primitive and a good place for me to be quiet."

He later wrote down: "Addiction baffled me. I was hardheaded and proud of my ability to bear pain and of my iron willpower in other areas of my life. But my will power was bafflingly ineffective against the compulsion of my drug addiction. The promises to quit that I made sincerely and earnestly would soon stop binding me, and the intervals between promise and drugs got shorter over time dropping from months to weeks to days and finally hours. I had to continuously lower my standards and my aspirations to keep pace with my declining conduct. Addiction is a disease of isolation and, although I was often surrounded by people, my world over time grew smaller and smaller."

His friend Chris Bartle remembers: "One thing Bobby said always stuck with me. It was 'I would rather be in a pit of poison snakes than to take drugs again.' There was nothing he wanted more in the world. But he just couldn't do it." Bobby recalls: "To me, the most demoralizing feature of this disease was my incapacity to keep contracts with myself." Determined to give Bill Walsh's form of therapy one more try, he was aboard a plane to South Dakota, on September 11, 1983, when he appeared to be suffering an almost fatal overdose. Arrested upon landing, he received help from Walsh to enter a rehabilitation facility in New Jersey. "I was completely broken," he says. "I knew I needed a profound spiritual realignment."

Bobby reflected back on his upbringing. Raised in a pious Irish Catholic family, he and his siblings had read *Lives of the Saints*. "These were men who'd had spiritual epiphanies—St. Francis at Assisi, St. Paul at Damascus, St. Thomas Aquinas. All of them had led dissolute lives as alcoholics or other forms of debauchery, St. Augustine particularly suffered sex and alcohol addiction—and then had white light moments when the compulsions completely disappeared and they were different people in the end. And I really wanted that to happen to me."

He started by attending church every morning, as he had in his youth. Then one day, he chanced upon a book written by the pioneering depth psychologist C. G. Jung. It was titled *Synchronicity*, and it contained the story of a troubled patient who dreamt of a golden scarab which didn't exist in that locality. Suddenly, that very flying insect tapped gently on the windowpane, then landed on Jung's palm, where he presented it to the woman.

"The chances of that happening were one in a trillion," Bobby realized. "Jung thought such things were interventions by God in our lives. They defied the laws of nature and of mathematics. He was a scientist and tried to reproduce these instances in a clinical setting but was very honest about the fact that he couldn't prove the existence of God using empirical and scientific skills. But Jung described having seen thousands of patients coming through his hospital, including a high percentage of alcoholics, who he could show based upon anecdotal evidence that if they believed in God, their recovery was more durable than those who did not.

"At that point, God had gotten a great distance from me. My dilemma was a very common one—how do you start believing in something that you can't see or smell or taste or hear or acquire with your senses? Jung solves that problem in the book, putting forward the German version of 'fake it till you make it'—you *act* as if you believe in God. And I started acting as if there were a God watching me all the time. I started breaking my day down into about forty different little choices and each one took on a spiritual dimension. Simple things: do I get up in the morning when my alarm goes off or do I sleep through or stay in bed with indolent thoughts? Do I hang up the towels, brush my teeth, make my bed as soon as I get up? Do I put water in the ice tray before I put it back in the freezer? When I go into the closet and put on a pair of pants and some of those wire hangers fall on the floor, do I shut the door like I used to and think I'm too busy to pick them up or that's somebody else's job? Do I put the shopping carts back where they're supposed to go?"

Bobby would later learn that Jung had had a powerful influence on the recovery movement, becoming a foundation stone upon which Alcoholics Anonymous was built.

I was about to leave one day after doing an interview with Bobby, when he suggested we sit outside on his porch. Then he began to talk with the benefit of thirty-eight years of hindsight into what he'd been through, and what he'd been able to make of it. "I have an addictive personality. Which means that the way I process the world, my mind is like a formulation pharmacy. It can turn anything into a drug. That could be exercise, ice cream, candy, my career, or whatever."

He continued: "One of the features of all addictions is this illusion that you can fix whatever pain is inside of you and fill whatever empty spaces are in your soul with things that are outside of you. There's been a long time in my life when I reacted to everything that way. In my teenage years, I was essentially just a bundle of appetites that constantly demanded to be fed, and I had no defenses against them—and no techniques for figuring out other options. What maturity and the twelve-step program brings is a series of techniques that give me a little interstitial space between my impulse and my action, that allows me to change my biological destiny in those little instances—by thinking through, by using prayer. The more you exercise those muscles, the stronger they get.

"The writer Susan Sontag once said that addicts have a unique opportunity for redemption because they've lived in hell. The way you're living is so wrong to yourself, to all of your senses and sensibilities, that you get desperate enough and powerless enough that you can surrender to a higher power. Nobody goes into twelve-step meetings when they're on a winning streak. They do it because what they're doing with their lives isn't working.

"I had an experience early in my sobriety. At one of the first meetings I ever went to, a guy talking was a big drug smuggler from the 1970s. He had an operation bringing pot up from Mexico with a big fleet of trucks and hundreds of people working for him. He was making a million bucks a week. Then he got busted and he lost everything. He'd gone into rehab and received treatment, and he was now working in a clinic for minimum wage, cleaning lab jars. He said that he came early, tried to clean the jars better than anybody else, and left late. And that he took joy and found peace in doing little things well - the kind of things that, in his former career, he was too much of a bigshot to do. But the humility and the simplicity of doing this was what gave him joy and peace.

"I got a lot of inspiration from that. At that time I was a lawyer but not practicing law. I'd started working for Riverkeeper, but I had a very humbling job, doing essentially janitorial work putting together a field station that would be Riverkeeper's headquarters up in the Hudson Valley. I painted that house myself, I swept the floors, all that kind of stuff. And I also took a vow I was going to keep my mouth shut for five years and become an expert at what I was doing.

"When the house was ready, I had to furnish it. I wanted to do a good job of finding cheap but nice furniture and getting it transported. So I borrowed a big pickup truck from one of my friends, and I found a warehouse sale in New Jersey, and I drove down with my list of all the necessary items—desks, lamps, rugs, beds, and so on. I packed them really well into the back of the truck, jampacked into the flatbed in a way that they wouldn't fall out.

"But one thing that I didn't do was tie them in. I was driving up the Garden State Parkway and I hit a big pothole. There was a racket from the back and I looked in the rear view mirror - and the bed of the truck was empty. The gate was open and there were foam rubber and splinters all over the highway. I got out. I'm not somebody who cries, but I was ready to. I felt as low as I could go. Here I couldn't even do this very humbling job right.

"I've packed hundreds of pickup trucks since then and I always tie stuff in. The way that we learn is through pain. When I was a teenager and into my twenties, my strategy was to anesthetize it. But when I got sober at twenty-eight years old, I had to learn all those lessons. And what I found was, when I was a teenager and in my early twenties, I was like a big truck stuck in a ditch with flame coming from the headers and the pipes, and sound and fury and roaring revving engines and smoke pouring out of the tail pipe. Going nowhere, spinning the wheels. All activity and no progress.

"One of the things I learned after that incident with the truck was just how to be still. There's a line from Isaiah that really helped: 'Be still and know that I am God.' In other words, God is in charge. I didn't have to fix what was wrong with something *outside* of me to expose the wrong *inside* of me. So I learned to be still and to listen—to other people, to the universe—and not react to everything. It says lot of times in the Bible 'It came to pass,' which means whatever you're feeling right now is going to pass. And many times it just means sitting on the train patiently and waiting for the scenery to change.

"Pain is the touchstone to spiritual growth. If you don't experience the pain, you don't have any spiritual growth. Hopefully what we get from enduring that pain is wisdom. The meaning of the word wisdom is 'a knowledge of God's will.' In other words, an intuitive sense of right and wrong in every situation . . . to understand what you're supposed to do and how you're supposed to react intuitively. I did not have that when I was young, and I have that now—through trying to discipline many of my impulses."

He would tell his AA group: "The question is, how do I stay in that posture of surrender even when everything's going well in my life? That's why I still go to meetings every day. When I first started, I asked someone how long do you have to go to these meetings? And he said, 'just keep coming until you like it.' I've been in it for thirty-eight years and I don't enjoy it or look forward to it. But I don't want to live with the consequences of what my life will look like if I stop. The twelve-step brought order out of the chaos of my life and turned all the mayhem of my addiction into capital that I could use to help other alcoholics or addicts."

CHAPTER NINE

BECOMING A RIVERKEEPER

Upon graduating from the University of Virginia Law School, Bobby started full-time employment as an Assistant DA in Robert Morgenthau's office after passing the bar exam in February 1983. There he "spent a year prosecuting misdemeanor cases, assault, prostitutes, and fare beaters." But after he entered the rehabilitation center, Bobby understood how much his life needed to change—and hopefully take on deeper meaning. In January 1984, the same month he turned thirty, Bobby commenced two years of court-ordered probation. He'd been ordered by a Rapid City judge to continue his treatment and go through periodic drug testing, while performing 1,500 hours of community service in New York State, where he resided. Bobby was in the process of moving from his apartment into a rented house near the Hudson River in Bedford, forty-five minutes outside the city.

"I met Bobby when he was really trying to refocus his life," remembers John Adams, part of a group from the US Attorney's office in Manhattan who had cofounded the nation's first environmental advocacy group, the Natural Resources Defense Council (NRDC), in 1970. "He had a determination to head in a direction that he cared about and, if possible, it was going to be the environment. We made a suggestion that he might want to start by working on Hudson River issues, because I was then on the board of the Hudson River Fishermen's Association."

Bobby took Adams up on accepting a volunteer internship to fulfill his court-ordered community service. "My decision to work on environmental issues was consistent with my other efforts to retrace my steps to that point where I had started off on the wrong path," he would recall. He began commuting to the bucolic hamlet of Garrison, where Adams lived. "He asked me to help him with a project to preserve the Osborn Estate," Bobby recalled. "He wanted to turn the old farmhouse into a research facility for scientists and environmental groups to use to study and protect the Hudson River. He asked me to go up there to prepare the farmhouse for occupation, which I did with the employ of about twenty prisoners from the Beacon Correctional Facility."

It was fitting that his assignment revolved around the Hudson River. Bobby would remember President Kennedy telling of the trip that Thomas Jefferson and James Madison made up the Hudson in 1800 on a botanical expedition, ending up at Tammany Hall in New York City and forming the Democratic Party. In the winter of 1965, eleven-year-old Bobby had accompanied his father to kayak the river's upper reaches in the Adirondack Mountains to paddle the Whitewater Derby. His dad "hoped the trip would help bring attention to the Hudson's spectacular headwaters, and help to derail a proposed dam on the Hudson River gorge." Senator Kennedy and his young relatives found the rapids "impossible to read, so we kept capsizing [and] ended up swimming several formidable rapids." While similar trips out West were marked by delicious river water to drink, Bobby was shocked when "these outfitters said the Hudson water was poison. Even at that age it occurred to me: 'There's a theft involved here,' and I took it personally. The outrage stayed with me."

His father had returned from the Hudson excursion to call the pollution he saw a disgrace to America that had to be cleaned up. The next year, a coalition of commercial and recreational fishermen mobilized. Their leader was a *Sports Illustrated* writer-turned-activist named Robert Boyle, who formed a Hudson River Fishermen's Association (he was the first journalist to call attention to PCB's contaminating local fish). The NRDC represented the Fishermen's Association in successful legal challenges, and the old farmhouse that the environmental group asked Bobby to turn into a research facility was designed for Boyle's organization.

Bobby had first become familiar with the Hudson Valley while attending Millbrook in his middle-school years. Now he devoured Boyle's book, *The Hudson River: A Natural and Unnatural History*, "which advocated an in-the-muck, waders-on environmentalism" and in which Boyle foresaw enlisting someone to be "out on the river the length of the year, nailing polluters on the spot." Boyle's models were the English riverkeepers, who protected private trout streams against poachers.

The Hudson hosted commercial fishermen who had been supporting their families off their catch since colonial times. In 1966, as Bobby recounted in the speech announcing his presidential candidacy, a leaky pipe from the Penn Central Railroad "sent diesel oil up the river and blackened the beaches, and their fish couldn't be sold." Several hundred men and women came together in an American Legion hall, only to realize that the agencies supposed to be protecting the river "had become the sock puppets for the industry they were supposed to be regulating."

Many of them combat veterans, the fishermen were talking about a violent response when Boyle told them about an ancient navigational statute called the 1888 Rivers Act that said it was illegal to pollute any waterway in the United States. "The law had never been enforced in eighty years, but it was still on the books," Bobby recalled. Boyle suggested: "We shouldn't be talking about breaking the law, we should be talking about enforcing it."

Out of this, Riverkeeper was born. Eighteen months later, they shut down the pipe and used the money they collected to go after other destructive corporations. "I wanted to be in the trenches working with people and engaged in hand-to-hand combat against the big polluters," Bobby said in his announcement. "And I wanted to particularly work with people who were most harmed by environmental injury but also were alienated or marginalized from the mainstream environmental community."

Fortuitously for Bobby, in 1983 Boyle had hired John Cronin to launch a twenty-five-foot patrol boat as the first full-time Hudson Riverkeeper. Cronin was a city kid from Yonkers, six years older than Bobby, who'd started out as a draft counselor for the Quakers and anti-war organizer during the Vietnam War era of the 1960s. He'd moved on to sleuth for Pete Seeger's Clearwater organization, collecting evidence that resulted in conviction of a Hudson River polluter for violating the 1972 Clean Water Act. Cronin also worked as a legislative aide on the New York State Assembly Committee on Environmental Conservation. Tiring of office routine, he became a commercial fisherman, learning to read tides and handle nets in all kinds of weather. "The river went from being something I acknowledged to something I understood," Cronin said.

Bobby met Cronin in mid-February 1984 and says he "liked him instantly." Bobby described him as "eloquent, thoughtful, and utterly committed," with a talent for tactics and strategy.

In the book he eventually coauthored with Cronin, Bobby described going on many outings with Bob Boyle that first spring, fly-fishing and beach-seining and scuba diving from canoes, along with their hiking together in the woods. "It was a way of getting my bearings, of adapting to a new home by systematically developing the same sort of familiarity with my surroundings and its indigenous plants and animals as I'd had with the Virginia and Cape Cod homes of my youth. . . . I learned to think of the Hudson Valley as my home, my place." And, although he and Boyle would never develop the intimacy of his relationship with Lem Billings, Bobby clearly enjoyed having a new older mentor. Boyle understood Bobby's struggles and inspired him to move forward. "You can seek a new life through the river, and through ecology," Boyle once told Bobby. Fittingly, like a master musician passing along his instrument, Boyle gave Bobby his time-honored fishing net.

While continuing to work under the auspices of the NRDC, before long Bobby would be partnering with those other two Irishmen (Boyle and Cronin), first as an investigator. Boyle would later remember Bobby saying to him: "Riverkeeper has saved my life. It's given me something to live for." Boyle concurred. "I think what gave meaning to his life was his growing realization that there was a lot to do for the Hudson River and the world. I saw almost an awakening. Look what I can do. That was my sense of what it could do for him. It turned him around."

Bobby remembers his first project came about like this: Joe Augustine, "a pawnshop operator with a giant Afro and energy as abundant as his notable girth, showed up at the office one day with horror stories about political corruption and environmental catastrophe in his community." He lived in the small town of Newburgh, sixty miles north of New York

City. It was poor, with a predominantly black and Hispanic population. And Bobby would soon realize, a town that "regularly leads the nation in virtually every criterion used to measure urban decay and is famed for producing local politicians with elastic ethics and a knack for plundering the public trust."

Bobby and Cronin were initially worried about Newburgh's plan to sell its boat ramp, the only public access to the riverfront, to a developer who planned to build a floating restaurant there. When they went to the developer asking if he'd still permit the public to use the ramp, the man's partner is said to have responded: "We are not going to have spades drinking and screwing and swimming around our new boat ramp, if that's what you mean by the public."

Bobby later wrote: "At that moment I knew I wanted to be an environmental lawyer. Any ambivalence I'd felt about having abandoned the battle for social justice when I took up arms for a clean environment was gone." Like his father before him, Bobby knew it was really all one thing, the pollution and the racism. And it infuriated him.

Joe Augustine also told Cronin about industrial sewage discharges at the foot of a minor stream that meandered along Newburgh's southern border, known as Quassaic Creek. "Our investigation," Bobby and Cronin wrote later in *The Riverkeepers,* "took us into pipes and culverts and underneath ancient factories. We mounted 24-hour surveillance of intermittent pipes. We sat for days on lawn chairs to sample factory and sewer plant outfalls, and set traps to catch liquids being dumped at odd hours. . . . We scuba dived in the Hudson to collect evidence of illegal dumping and donned wet suits to swim across a pond on a cold winter night to collect samples from an unpermitted pipe. We carried backpacks filled with sampling vials, flashlights, maps, note pads, a camera, and, occasionally, fishing rods to avoid suspicion."

It was as if the conquering of those Latin American rivers had morphed into an even deeper dive into primordial matter, one with representative consequences even in the smallest of tributaries. A profile of Bobby in *Outside Magazine* continued: "Donning waders, Kennedy walked every foot of the Quassaic's 'Dantean' seven-mile length. He seined for fish among 'brown trout' and 'river pickles'—human fecal matter—until the cuts on his hands began to fester. Seeking the source of the raw sewage, he discovered another pipe belching volatile chemicals, had Cronin lower him into the culvert, sampled the dregs, and discovered quantities of naphthalene, a toxic solvent known to cause mutations and blood disorders in animals. After tracing the effluent to a textile company that had illegally connected its discharge pipe into a city sewer line, he and Cronin scaled the company's roof to check its emissions sources and found twenty-four illegal pipes and stacks discharging liquid and chemical fumes. Eventually Kennedy and his coworkers identified two dozen separate polluters on the Quassaic. They brought sixteen lawsuits, all settled before trial, yielding pledges from each defendant to stop polluting and clean up the creek, and hundreds of thousands of dollars in penalties and fines."

Literally deep in the shit, in a microcosmic forerunner of what lay ahead, Bobby quickly made enemies. Newburgh's town supervisor eventually told *New York Magazine*: "Here was this Kennedy kid running all over town, sloshing around in the creek, talking to the press, using his fancy name. He was saying that the town had a sewer problem. Well, I *knew* that. I don't think there is anyone more concerned about the environment than me. Then he sued us without even coming in and talking first to the board—took the town to court over the sewage problem. Kennedy isn't from around here, and neither were the people working with him. To have to bargain with people of that type makes me sick."

The supervisor claimed that Bobby and Boyle's Fishermen's Association were committing "legalized extortion," alleging that the activists used "lawsuits and threats to shake the town down for money." This was after Bobby presented twenty strong cases to the US Attorney's office, which decided to prosecute four of them. Bobby's response? "He really hates my guts." As Cronin later summarized; "They were saying everything he did was politically motivated and that he planned to run for office. They painted him out to be a carpetbagger. But he never took the bait."

Indeed, Bobby stayed away from the press conference announcing the outcomes of the Battle of Quassaic Creek. This remained his policy during his first five years with Riverkeeper because he was afraid "press interest in my involvement might distract from the central issue." Cronin told *Outside* writer Barry Werth another reason for Bobby's avoidance of taking public credit. "Bobby didn't want it to appear as if he was using this issue as instant rehabilitation."

Asked what effect that first venture into tracking down polluters had on him, Bobby replied: "I saw the connection between poverty and pollution in addition to the connection between the erosion of democracy and the erosion of our environment and natural treasures." It was a motif his father had introduced, and one that would dominate the rest of his life.

Cronin recalled in 2022 that Bobby "became very interested in the law, the Clean Water Act specifically and its rules and permit system. He started finding things. One of the first was, 'wait a second, they [the company] have to do monthly reports, let's look at them.' So we started pulling some and he said, 'they're just reporting their violations every month and not doing anything about it.' That set out an agenda for us for virtually the next ten years, going after polluters who through subterfuge or lying or hidden activities were violating the Clean Water Act. We were each equally comfortable with investigative paperwork or pulling on a pair of hip boots and getting out in the water and the mud. So we formed an unexpected but natural partnership over the next year. And his original community service evolved into becoming our chief attorney."

* * *

In 1985, Hudson Riverkeeper promoted Bobby to senior attorney, the same year that Exxon agreed to cease its tanker traffic on the Hudson. This harkened back to the Riverkeeper beginnings in mid-1983, before Bobby came onboard. An NBC News crew had agreed to follow Cronin out to look for oil tankers said to be siphoning fresh water ninety miles upriver from their entry into New York Harbor. Over a three-year period, Cronin estimated Exxon had stolen 700 million gallons of water. The story ended up making news around the world. In 1986, Exxon paid $1.5 million to the state and $500,000 to the fishermen, with half that money going to fund Riverkeeper.

The Croton River feeds into the Hudson, not far from where Cronin grew up and now a favorite fishing spot for Bobby. On Christmas Eve 1985, a local fisherman alerted Cronin that workers planned to pump toxic liquid from Westchester County's Croton Point Landfill into the Hudson. "They chose Christmas," the whistleblower told Cronin, "because they figure you guys won't be around." So Cronin and Bobby spent Christmas Day "at a garbage dump, with a bone-chilling wind blowing across the river." Croton Landfill turned out to be "an environmental nightmare—hundreds of thousands of tons of domestic garbage and one billion pounds of industrial wastes, including a radioactive building, buried in two hundred acres of Hudson River tidal marsh." By the next spring, Riverkeeper "had collected enough evidence of unpermitted discharges from the storm sewer into the river to start our own citizens' suit."

Interest in Riverkeeper mushroomed. "We needed nothing less than a team of prosecutors to meet the public demand for our work," according to Bobby. Soon he would get just that. Bobby hadn't taken a single course in environmental law in law school "because there really wasn't a discipline called environmental law at the time." With a growing sense of purpose, Bobby attended Pace Law School and got an advanced degree, known as a Master of Laws or LLM in environmental law.

He did so well there that the Pace faculty asked Bobby to create an Environmental Litigation Clinic where second-and third-year law students would work on cases under his supervision. He said he'd do it on one condition: that this could happen in conjunction with Hudson Riverkeeper and a nascent Long Island Soundkeeper as the clients. Kennedy went to the New York State Court of Appeals to obtain a special order allowing his ten Clinic students to actually try cases in state and federal courtrooms. In a small building whose entryway held an aquarium full of Hudson River fish, the Clinic was soon suing dozens of municipal wastewater treatment plants to force them to comply with the Clean Water Act and taking on huge polluting corporations like General Electric.

When the Clinic opened in September 1987, the first case revolved around the Croton landfill. Bobby and his students filed a Clean Water Act citizens' suit against the county. A judge called the landfill a "time bomb." The county attorney stunned everyone by asking for a settlement. "The discharges were halted, tank trucks began to regularly pick up the leachate for disposal at a treatment facility, the landfill was shut down and a plan for its remediation agreed upon, and a program for the restoration of the remains of Croton Marsh were established," Bobby and Cronin wrote in *The Riverkeepers*. It was a massive victory.

As Riverkeeper's reputation spread, "Pace Clinic quickly evolved into a legal aid agency for many small community environmental organizations."

Bobby became a board member of Long Island Soundkeeper, bringing numerous lawsuits against cities and industries along the coastlines of Connecticut and New York. In 1986, he won a landmark case stopping the practice of lead shot being fired into the Sound. His federal lawsuits forced the shutdown of the Pelham Bay landfill and New York Athletic clubs for interfering with the public's ability to use the Sound. And Pace students "uncovered more than three thousand violations of the Clean Water Act by the cities of Norwalk, Bridgeport, and Milford," with indications of similar situations in a half dozen other cities.

While Bobby covered the legal waterfront, his Riverkeeper partner John Cronin did a lot of the on-site research—including sometimes anchoring and crawling through huge pipes seeking evidence of midnight dumping. A 1990 profile of Cronin in *People* magazine described how, in its first seven years, Riverkeeper had successfully brought forty enforcement cases and lawsuits against polluters. "The reason we have been able to save the river is that we've used public relations, politics and the courts," Bobby told the magazine. "John knows how to work them all."

Looking back, Cronin sees a misconception concerning Bobby's Riverkeeper efforts. "It's been told in such broad strokes that it does him a disservice," Cronin says. "The portrait that's often been painted is the courtroom attorney who saved the Hudson. No slam against litigators, but the work he did required so much more skill and determination than that. Litigation is a war of nerves: who's going to give in first, who's going to challenge who, who has the best information. I could recite to you a dozen times when that's exactly where he and I were with opposing forces."

Cronin remembers a specific instance with Mobil Oil's chief counsel. They were negotiating on a conference call when the lawyer informed them: "I'm about to get on a plane to Russia, and if I do I'll see you in court, there's going to be no settlement here and you're going to lose. You've got until tomorrow." When the call ended in a stalemate, Cronin recollects, "Bobby and I went over all our facts and then he said to me: 'What do you think? We're not gonna give in, right?' I said, 'No we're not.' And the Mobil attorney called us from the airport an hour before his flight took off and said, okay he'd schedule settlement negotiations for when he comes back. It never went to court."

Cronin further explained the complexity of the suits and the settlements. You had to be able "to muster the information to win against a seasoned corporate attorney." It's rarely some "Clarence Darrow closing argument in an open courtroom." Bobby needed a whole host of skills, and he developed them. "Those stereotypical portraits of Bobby don't really recognize what goes into sometimes very complex cases that don't just involve the Clean Water Act but property law contracts and corporate history."

Following the publication of Bobby and Cronin's book in 1997, the Riverkeeper movement spread West, seeing the launch of Baykeepers in San Francisco, Santa Monica, and Seattle. By then sturgeon had returned to the Hudson, along with shad, herring, eel, and blue crabs. So had the Atlantic striped bass, the vaunted migratory species that had enabled the Pilgrims to survive their first winters along Cape Cod and that, only a decade earlier, had seemed on the verge of collapse.

CHAPTER TEN

A FISH STORY

"Striped bass are a sacred fish here," Bobby said, speeding toward a Hudson River boatyard to pick up his seventeen-foot Boston Whaler before dark. "Since I was a little kid, this was the iconic fish, the one everybody wanted to catch. I'd be standing on the dock with a bucket full of scup, watching fishermen getting off their boats with big striped bass. I'd watch while they cleaned them on the pier. Those guys were my heroes. They'd caught the king."

He continued: "Stripers in many ways define the culture, the same way the salmon do on the West Coast. In the Hudson Valley, the Native Americans lived off of them. They were a major cultural and subsistence asset of the Algonquin and Delaware communities. Then the first Europeans to visit the valley were dumbstruck by the stripers. There's a great passage in the log of Robert Jewett, who was Henry Hudson's pilot, describing the thick schools that the prow of his ship cuts through like a comb." He gave a brief science lesson—"their life cycle is intertwined with the biological integrity of virtually every part of the river."

The Atlantic striped bass had been at the heart of what Bobby and John Cronin called "the most important lawsuit in the history of environmental jurisprudence." Back in the Sixties, the Consolidated Edison utility had announced a plan to build the biggest pump storage facility in the world on the Hudson beneath Storm King Mountain. Bob Boyle went to war against it, publishing graphic pictures in *Sports Illustrated* of grisly fish kills in the turbines of Con Ed's power plant at Indian Point. He enlisted scientists, who showed that up to 90 percent of the Hudson's striped bass laid their eggs near the mountain. It took fifteen years, but eventually the Storm King project bit the dust—simultaneously establishing the legal precedent that citizens have the right to sue corporations over threats to natural surroundings.

Just as they'd been for Boyle, striped bass were often at the center of Bobby's efforts to preserve the river. In fact, the first time he ever cross-examined a scientist on the witness stand, stripers were the main topic. The year was 1988, and Riverkeeper was suing to shut down the Chelsea pumping station. Constructed on the Hudson in the 1940s, about sixty miles north of Manhattan, it was designed for use in drought emergencies, capable of removing a hundred million gallons a day of river water, injecting it into the Delaware aqueduct, and sending it on to New York City. The city wanted to activate the station after a severe drought caused a substantial drop in its reservoir capacity. The trouble was the intake pipe off Wappingers Falls was located at a point in the river that recent studies revealed to be a locus for spawning of Hudson striped bass.

At the permit hearing, a crucial question became what would happen to the fish once they'd been whisked into the aqueduct, which ran for nine miles underground before dumping into a reservoir. The municipality's Environmental Impact Statement (EIS) relied on an argument used by power plants in the past: that the adult stripers would be able to sense the impending approach of a screen and find an existing escape hatch, thus carried by jets safely back to the river.

It sounded good on paper, but Bobby and Boyle's Fishermen's Association weren't convinced. Ian Fletcher, an expert biostatistician, came to Bobby's house to go through the EIS line by line. It took three days. "The point Ian made," Bobby recollected, "is that fish in the Hudson River, unless they have contact with the bottom, don't even *know* that they're moving with the tides. Not if they're in the middle of the water column. It's the concept of relative motion, like us when we're in an airplane and feel like we're not moving. Down there it's dark; the fish only have six inches of visibility. So once they'd hit that screen and turn around, they'd stay there just swimming against the current until they became completely exhausted. Ian had observed fish behavior in a darkened flume at his lab and by dropping marked carp into one of these intakes. Those were experiments never tried by the utility scientists, and he proved that all their assumptions about fish behavior were wrong."

The Chelsea pumping station's advocates had hired their own expert to defend the theory, who'd signed off on sworn testimony that was little more than a cut-and-paste job taken from old power plant applications. "We had him on the stand for a couple of days," Bobby continued, unable to conceal a gleeful smile. "I asked him: 'Well, if the fish could sense the approaching screen, how do they do that? Do you mean like a bat can sense things?' He said, 'Yeah, like a bat,' and started to become very uncomfortable. I said, 'Well, a bat has sonar; do fish have sonar?' At that point, I think he realized he'd gotten himself into a jam. He said, 'No, but they can sense the electrical stimulus in the water.' He just started making up stuff. And I said, 'Well then, why don't they avoid fish nets?' Right? Because a screen's no different—but they'll swim right into a gillnet on the river. Fish do have the ability to sense electrical activity. But if there's a structure that stands still in the water, they can't see ahead like a bat can. So we took the case to federal court—and stopped the pumping station."

Ian Fletcher, a native Scotsman who liked to wear a kilt and play the bagpipes, became Bobby's first scientist mentor. Fletcher had shown up a few years earlier—as did the striped bass—in a legal fight over New York's Westway highway project. *The Riverkeepers* described him as "a figure who struck fear in the hearts of industry scientists. A brilliant statistician and oceanographer, he not only published regularly in peer-reviewed journals, but also could translate arcane concepts into English so precise and eloquent that they seemed not just simple but obvious to the layman. Knowledgeable in math, physics, biology, engineering, oceanography, and fluid dynamics . . . Ian defended the truth above all else and was fearless about testing even his most cherished assumptions."

Fletcher's modus operandi would serve Bobby as a model for future contests against what he called industry "biostitutes," including Monsanto and the pharmaceutical giants. Westway, "a $2 billion construction project that would have added 242 riverfront acres to Manhattan's West Side and enriched the city's most powerful developers by billions of dollars," had

seemed impossible to stop. It had the backing of the unions as well as leading politicians and media. Indeed, the *New York Times* "forced the dismissal
of one of the country's most influential reporters, Sydney Schanberg (who
had previously risked his life in the killing fields of Cambodia), when he
wrote a series of hard-hitting articles questioning Westway's wisdom."

The proposed Westway site had been officially declared "biologically
impoverished" when the Hudson River Fishermen's Association learned
that striped bass loved to hang out under the piers that would be demolished during construction. Boyle's group sued after the Army Corps of
Engineers issued its fill permit in the spring of 1981. On the morning
that Ian Fletcher flew in to examine the file boxes provided in compliance
with discovery requests, the scientist saw a front-page *New York Times*
photo of President Reagan handing Mayor Ed Koch a poster-sized $85
million check to commence the building process and figured "we don't
have a chance." Mayor Koch quipped during the proceedings: "Let's hope
the striped bass will have the sense to move across to the Jersey shore." To
which Boyle retorted: "Striped bass don't go shopping for apartments!"

As Dr. Fletcher examined fish collection data from thousands of different points in the harbor, he concluded that 64 percent of the striped
bass habitat would be wiped out. "He then retraced the Corps formula-
tions and proved that their consultants had known the true impacts and
had deliberately massaged the numbers to mislead the court." Bobby still
remembers the scientist's moment of truth: "He essentially spent four days
locked in a room with all of their paper data files, and in the end he came
out and said 'I got' em.'" During the trial, after Dr. Fletcher explained his
findings, Judge Thomas Griesa said he had "sentenced people to prison in
securities fraud cases where the conduct was less blatant."

On January 31, 1982, striped bass made the front page of the *New
York Times*: "US Judge Blocks Westway Landfill as Threat to Bass." In
announcing his decision, Judge Griesa spoke of the "importance and
value of striped bass. . . . The striped bass fishery contributes to the economic well-being and enjoyment of literally millions of citizens. . . . The
proposed landfill would have the impact of destroying this habitat." The
Army Corps, in lockstep with Westway's promoters, didn't give up, and
a second trial ensued with Fletcher again exposing the latest fraudulent
study. Judge Griesa then blocked all federal funds for the project.

Boyle wrote in *Sports Illustrated*: "The striped bass has been the noble creature that has led all our fights in the Hudson, and so far he remains undefeated." It was also the bass, through Boyle, that brought the problem of PCB contamination into public consciousness, after General Electric's discharge of some 1.3 million tons of toxic chemicals into the Hudson between 1940 and 1975. This resulted in a ban on commercial fishing for Hudson stripers when samples showed 90 percent contained levels of the carcinogenic compound above the Food and Drug Administration's standard.

Now, after Bobby gassed up his boat the *Lickety Split*, heading back toward the island he pointed to the shoreline and said: "The resident population of stripers is still highly contaminated. They're outnumbered by the migratory fish, but there's no way to tell the two apart by looking at them." His voice intensified as he continued: "General Electric knew during the fifties and sixties that PCBs were dangerous—we have letters from Monsanto Corporation, which was the manufacturer, warning GE not to let any of it go into the environment. It would have been an easy thing for them at the time to collect the PCBs and recycle them. But they could get away with dumping them into the river, so that's what they did."

In 1976, Congress had finally banned the manufacture, sale, and distribution of PCBs. But it wasn't until December 2000, after a ten-year fight between GE and the federal government, that the EPA ordered the company to spend $500 million over five years to dredge PCBs embedded in the river bottom north of Albany. "The science indicates that if we don't dredge, then most of the fish in the Hudson will be inedible for over a hundred years," Bobby said, aiming his boat toward a cove about a quarter mile from the campsite. "If dredging happens, most fish will be edible within two years."

What about GE's argument that dredging would stir up the river sediments in a detrimental manner? "That's GE's propaganda," Bobby responded, his eyes flashing. "The sediments get stirred up anyway during storms, so that stuff is constantly recirculating in the river. Besides being a probable carcinogen, the real problem with PCBs is that they're very potent endocrine disrupters. That means they disrupt the formation of nervous tissue, so when fetuses are exposed, the babies are more likely to have cognitive impairment. Even small amounts of PCBs have the potential to produce a number of problems in our children. And every woman between Oswego and Albany has elevated levels of PCBs in her breast milk."

His voice trailed off momentarily. Then he continued: "I don't let my kids eat any fish from the Hudson. Frankly, I myself am not gonna let the polluters get between me and the fish. It's probably not a smart thing for me to do, but I eat fish anyway. I had myself tested. In my body, I've got two to three times the amount of PCBs as a normal human being my age. For a man of fifty, not intending to have any more children, the risks are more internalized. But we don't feed our kids tuna anymore either, or other freshwater fishes, because those are loaded with mercury."

Bobby's son Conor carried an Epipen on the trip, in case he had a severe asthma attack. A peanut allergy had been finally diagnosed by a physician at the Mayo Clinic, a year after Conor suffered a strong reaction in the proximity of peanut butter. Where this was coming from remained a mystery, but Bobby felt "there's a good chance it's due to chemicals or pesticides in our food." This seemed to be the origin of the concern that would eventually lead to his starting the World Mercury Project in 2016, two years later to become the nonprofit Children's Health Defense organization.

The fight against General Electric over PCBs marked the origin of something else—Bobby's disenchantment with corporate media. In 1986, GE had become the parent company of NBC-TV; and, fifteen years later, NBC's president Robert C. Wright—also vice-chairman of the GE corporate board—brought four other members to a private meeting with the New York City Council to argue against the Hudson cleanup. David Whiteside, then a young man who'd become an organizer of Riverkeeper campaigns in the South, would later observe that "Bobby got some bad press for being an aggressive environmentalist trying to sue the good old institution that Thomas Edison had invented. Portraying him as an extremist, they attacked his credibility and even questioned his sanity." It wouldn't be the last time. But time after time, Bobby would prove his detractors wrong.

CHAPTER ELEVEN

ECUADOR DEBACLE: A LESSON IN PETRO-POLITICS

Maybe it was his father's deep friendship with Cesar Chavez, who organized the Mexican farm workers. Maybe it was that less than two months before his death, RFK ignored the admonitions of his advisers that he bypass visiting the Pine Ridge Indian Reservation in South Dakota as a waste of time in the heat of primary season. "Indians don't vote," the candidate was advised. When staff aide Fred Dutton sent him a note that "we're in a campaign, and he should knock off the Injuns," Senator Kennedy replied in longhand: "Those of you who think you're running my campaign don't love Indians the way I do. You're a bunch of bastards."

In 2022, Bobby elaborated: "My father believed that America should be an exemplary nation, but we would never live up to our promise as a moral authority around the world unless we first went back and corrected or made meaningful amends not only for slavery but the genocide of the American Indian."

In *The Riverkeepers*, Bobby would remember: "Several of us would accompany my father on most of his campaign trips. Whenever he landed in a new city, he would ask to tour nearby Indian reservations. Many of us were given Indian names by tribal leaders on visits to various reservations.

Indian or Eskimo leaders often visited the house, bringing gifts from great chiefs, peace pipes and headdresses, objects my father treasured. One winter a contingent of Alaskan Indians arrived for lunch during an unusual Virginia snowstorm and spent the afternoon with my father in the backyard building a great igloo."

Bobby continued: "Viscerally he admired the Indians for their courage and toughness. But the conditions in which most Indians lived offended his sense of justice and his notion of America's mission in history. . . . Once he told us that on the Pine Ridge Reservation he had seen an entire Sioux family living in the burned-out hulk of an abandoned car. As he had said about the residents of Bedford-Stuyvesant in Brooklyn, New York, 'when you're older,' he told us, 'I hope you will help people like this.'"

Long after, Bobby would become a founding board member of *Indian Country Today*, the largest Native American newspaper. Traveling to Pine Ridge to meet with Tim Gallego, the editor and publisher, Bobby told him: "The night before he died, my father won South Dakota because of the reservation vote." Tim told Bobby, "We did 99 percent of the vote for your dad. And we're still looking for the people who voted against him."

Robert Kennedy had started a Senate subcommittee on Indian education to address the injustices of the tribal schools. After his father's death, Ted Kennedy had done his best to step into his shoes with disenfranchised groups that had looked to him for help. "I would run into people on every reservation who had known my father and my uncle and been touched by their work."

In 2000, Bobby had an epiphany in Arctic Village, the most northern native community in North America. He had traveled there to support their battle against prospective oil drilling amid the caribou herds in the Arctic National Wildlife Refuge. "I spoke to the whole village gathered in a communal wooden cabin following a drum ceremony," Bobby remembered. "Afterward one of the elders came up to me. He said, 'Your uncle came by here in April 1969. He'd flown in somewhere on a float plane and then taken a dogsled to our town. It was night, 24-hour darkness, and we didn't know he was coming.'" Ted Kennedy made the trip to complete some of his brother's unfinished business on the Indian Education Act. Bobby Jr. brought along his son Bobby III, seeking to prevent the invasion of America's premier sanctuary for virtually every major form of Arctic wildlife.

So it was part of the legacy and in the blood, and Bobby cued his environmental philosophy from something Seneca Chief Oren Lyons once said to him: "It's vanity to say we are protecting nature for the sake of the planet. The planet is four billion years old. Its crust is forty miles thick. It has survived freezing and warming and volcanoes and earthquakes. Nature will survive without us. But what will we be without nature?"

For Bobby, of course, there were the dramatic river adventures that punctuated his infatuation with Latin America. As his Hudson Riverkeeper partner John Cronin put it: "When he first started working, he had a natural interest in environmental issues and a competing interest in political issues of Latin America." Beginning simultaneously in 1985, Bobby had helped develop the NRDC's international program for environment, energy, and human rights. In that vein, he would soon take what he learned from the Hudson River wars and travel to Canada and Latin America to assist indigenous tribes in safeguarding their homelands and opposing large-scale energy and extractive projects in remote wilderness areas.

Bobby would write: "The Ecuadorian Amazon is among the most biologically diverse forests on the globe. Some scientists believe that the Oriente, Ecuador's great rain forest, was one of the few regions of the Amazon basin that remained humid during the Pleistocene ice ages and that areas like this one would probably function as 'safe houses' or speciation centers again, should major or minor climactic shifts occur."

Bobby had spent several weeks in 1978 traveling through Ecuador in the company of the progressive presidential candidate Jaime Roldós Aguilera, who, after being elected, died in a suspicious plane crash linked to the CIA. His successor resumed opening the country's coastline and Amazon basin to Texaco and other American oil companies. The industry went on to place some 330 wells and pumping stations in the jungle, extracting 1.5 billion barrels of crude—almost half of that ending up in the United States.

In the process, more than twelve million acres of forest had been consumed. Reckless development saw a single pipeline leaking seventeen million gallons of petroleum into the Amazon's sensitive headwaters - a spill 50 percent larger than what spewed from the *Exxon Valdez* in 1989. Texaco left in its wake a thousand or so waste pits that resulted in cancer and miscarriages soaring among native peoples who bathed in those waterways. And Texaco was not alone. A dozen foreign oil entities held contracts with the state oil company Petroecuador to drill on Indian lands and protected areas.

In July 1990, the *Oil & Gas Journal* reported that a group led by Conoco Ecuador Ltd. had plans for a $500 million, five-field project in the Oriente region that would tap reserves estimated at 200 million barrels. That same month, Bobby flew to Quito, Ecuador's capital, with two other NRDC environmental experts to meet with Judy Kimerling. Formerly an assistant attorney general in New York State, she had taken on Occidental Petroleum over its dumping of poisonous waste into New York's Love Canal before deciding to take a small grant from NRDC and look into drilling impacts in the Ecuadorian rainforest. What Kimerling saw enraged her.

"I had made fifteen previous trips to Latin America," Bobby recalled in the Preface to an NRDC-published book, *Amazon Crude*, in 1991, "and my work as an environmental lawyer during the past seven years had brought me to some of the worst toxic waste sites in New York. I did not expect to be surprised. However, nothing in my experience prepared me for the scenes that Judy Kimerling showed us in the Ecuadorian Amazon. Most people know that the Amazon rain forest is endangered. But the role of US oil companies in its destruction has largely gone unnoticed by the outside world. Because of the remoteness of the Amazon's oilfields, much of what we witnessed on that trip had never been recorded before."

From conquering never-rafted rivers, Bobby now was called to explore inner jungle reaches penetrated by petroleum giants whose deeds were devastating the eight indigenous tribes that relied on the Oriente's natural resources for their survival. It was the Hudson River's dilemmas magnified a hundredfold, and without the laws to do anything about it. Flying above an open production pit near a drilling platform inside a wildlife reserve, "its brimming surface of gleaming crude oil reflected the orange glow of the gas flare, and an adjacent stream wore an evanescent petroleum sheen." Bobby shouted over the engine noise: "They'd go to jail for this in New York." Judy Kimerling shouted back: "This isn't New York!"

Here, a witch's brew of toxic chemicals permeated the environment. "Each pit has an overflow pipe to a nearby body of water," Bobby wrote. "The toxic soups also percolate through soils into groundwater or flood into lakes or streams when the pits collapse. The only treatment occurs when the oil companies burn the pits to reduce their petroleum content."

Meeting with the chief clinician at a community health center, along with representatives of fourteen villages accounting for about forty thousand people from the Aguarico River basin, Bobby was stunned to hear the same tragic stories over and over. The local doctor informed the visitors: "We have studies from a nearby area without oil and there is not a single case of malnutrition. Today, we have 70 percent malnutrition in children six to twelve and 98 percent in the most contaminated areas. Because there are no animals left to hunt and no fish left in the streams, the major sources of protein have disappeared. The children have anemia. All during the dry season, they come in here with pus streaming from their eyes and rashes covering their bodies from bathing in the water. The parasitism rate for this area is now 98 per cent."

On day three, in the company of two local Carmelite priests, Bobby and his small contingent reached the jungle community of Shuara where Texaco developed its oil patch many years before. Beside a well, Bobby saw six men "mired in the gooey banks of a shimmering black lagoon, clothing and arms stained with oil. Some of them were shirtless and wore no gloves, their unprotected bodies tarred black. They cut the trunks of blackened trees and other vegetation and scooped up oil and contaminated sediments with shovels and their hands for burial in a series of open pits ten to fifteen yards from the lagoon. Two nearly naked men stood neck deep in oil in the center of the black pit. Periodically they dove under the oil in an effort to attach a hemp rope to a submerged tree stump. As they pulled the tree toward the shore their eyes gleamed white against the black ooze that coated their skin and hair. At the end of the day, the company hosed them down with gasoline to remove the crude oil."

It was surreal. A world populated by brilliant flocks of macaws, of spider and woolly monkeys, of palms and plantains and strangler figs - all depleted amid the scarred and oil-drenched earth and the recently burned river. When Bobby asked, in Spanish, how much the workers earned, one revealed that PetroEcuador paid 60,000 *sucres* a month, the equivalent of

two dollars a day. "When they are sick, the company fires them," Father
Jesús explained.

Then Jacob Scherr, a young NRDC attorney on his first trip with
Bobby, witnessed something he would never forget. Bobby suddenly
approached one of the workers to shake his oily hand. Rather than soil the
clean white palm of the foreigner, the man jerked his hand away. Bobby
then reached out a second time and grabbed onto the man's hand. "I still
get goose bumps," Scherr recalls. "That was the moment I knew, this is a
really special guy."

* * *

What ensued after the fact-finding mission would provide painful lessons
for Bobby on a number of fronts and presage later confrontations with
not only corporate powers, but the media and competing environmental
groups. It all began in early October 1990 at an NRDC press conference in New York where Bobby and Scherr told reporters of the ecologi-
cal disaster they'd witnessed. Now the Ecuadorian government was about
to open up the last refuge of the Huaorani people to Conoco Ecuador.
Exploratory blocks 16 and 22 covered 900,000 acres of rainforest, much
of it within the boundaries of Yasuni National Park.

The NRDC made public their letter addressed to E.S. Woolard, Jr.,
Chairman of DuPont, the parent company of Conoco, strongly urging
that those plans be dropped. Conoco officials responded by requesting a
meeting, which took place at the end of November in NRDC's New York
office. Bobby recalled in 2022: "DuPont had just bought Conoco and was
trying to polish its environmental bona fides. And they said, let's see if we
can all come up with a model for what oil development ought to be, and
we don't even mind taking a loss on this."

The company argued that development of these oil deposits was inevitable and that, even if they backed out, its partners might proceed anyway or PetroEcuador might simply take over the operation. But Conoco
pledged to exercise the highest standards regarding both the environment
and the rights of indigenous peoples. Their operations would be much
different than what started in the 1970s and led to widespread exploitation and pollution. And they'd be willing to explore having independent
NGO oversight.

The company promised they would not only clean up Texaco's mess, but they would serve as barriers against colonialist settlers claiming the land for logging purposes. Bobby remembers: "The settlers would come in and cut down a couple of trees and go get the deed as a squatter's right - and displace the Indians who'd lived there for ten thousand years. Conoco offered to hire lawyers to work with us and the CONFENIA [Confederation of Indian Nations of the Ecuadorian Amazon] to show proof of ownership by the tribe. There were no roads and the only way to get across into Huaorani land was by river, but the settlers were already flooding across it. Conoco proposed setting up an Indian police force, where badges were required and only given to the people who live there. The Indians with CONFENIA wanted to do it, and they asked us to represent them at the table to make sure all the safeguards were legal and enforceable."

No environmental group had ever explored the possibility of working with a major oil company in creating a kind of regulatory structure beyond what the government did. Bobby called the Conoco lawyers and recalled saying: "Look, in addition to doing the independent environmental oversight and management plan, would you also be willing to give a portion of your profits to the Indian community for rainforest preservation and for development projects?" The attorneys gave a resounding yes.

Bobby and Scherr twice returned to Ecuador over the next few months to consult with their CONFENIAE counterparts. But it was clear from the outset that their colleague Judy Kimerling wasn't comfortable with working in any way to find middle ground with Conoco. As the *New York Times* opened its exploration of the situation (February 26, 1991): "Citing lifeless streams, deforested wildlife preserves and the degraded health of Indians and settlers, American and Ecuadorian environmental groups are demanding that oil companies abandon plans to expand drilling into the pristine half of Ecuador's Amazon River basin. . . . On the other side, Conoco Ecuador Ltd. and the Ecuadorian Government argue that new techniques and elaborate precautions and regulations will permit clean production in the rain forest."

The NRDC discussed the situation with other concerned US groups, including National Wildlife Federation and the Sierra Club, and also briefed CORDAVI [Corporation for the Defense of Life], which consisted of two part-time Ecuadorian lawyers working closely with Kimerling. She refused to participate in any negotiations unless CORDAVI gave the go-ahead. Instead, in April, the group leaked Conoco's confidential minutes of the latest meeting with NRDC to the media, with CORDAVI accusing the NRDC of conducting "secret negotiations" with the oil company. Kimerling, given two month's severance pay, asserted that she'd been fired; Bobby and Scherr said she'd been dismissed for "unprofessional performance."

The controversy quickly spread. The Rainforest Action Network spearheaded forging a consortium of opposing environmental groups, most of whom sought the banning of Conoco's drilling rights altogether, with no need for the Indians at the negotiating table. The Sierra Club Legal Defense Fund filed a complaint with the Inter-American Commission on Human Rights, arguing that drilling on native lands violated their human rights. Articles appeared as far away as the Australian press maintaining that RFK Jr. was going too easy on Conoco.

But Bobby genuinely believed that it was possible to achieve better results through compromise, that a middle ground between indigenous rights, environmental consensus, and corporate profits was not only possible, but preferable.

Conoco, which years earlier had been granted oil leases by the Ecuadorian government, seemed to have expressed a sincere desire to be green. Indeed, Scherr admits feeling more comfortable with the company's environmental representatives than some of the customarily allied American nonprofits. Because of his willingness to try and forge a pioneering agreement, Bobby's announcement that "NRDC is committed to following the lead of the indigenous people" fell on many deaf ears. This was despite the NRDC Board at its June 1991 meeting declaring that the organization's "no drilling" policy would ultimately change only if a settlement—with Conoco or anyone else—was consistent with "the self-initiated and self-determined needs of the indigenous people, maximum protection of environment of the Oriente, and NRDC's global concerns about sustainable development and energy."

But it was too late to stop the fallout. While Bobby and Scherr were in Ecuador, a "constitutional court" (in truth a meeting of the Ecuadorian legislature) decided that Conoco had unconstitutionally been granted its licenses. Conoco went to the government and said this needed to be straightened out before it would make a planned $200 million investment in protecting the rain forest. The government ignored the plea. Scherr read the decision and it made no sense. The next thing they knew, a major US environmental group sent a letter with its two-member Ecuadorian counterpart to the US Treasury Department, charging Conoco with having violated the Foreign Corrupt Practices Act.

On October 11, Conoco announced it was pulling out of oil development in the region. That very same day, a Dallas-based company called Maxus moved in to take its place. According to Bobby, "Conoco said basically, if you can't control the rest [of the environmental groups] and we're going to get screwed by this anyway, why should [we] do it? And they left. Then we left. And exactly what we said was going to happen happened."

The NRDC pulled away that autumn of 1991, because there was clearly no way to make further progress. After promising to follow through on Conoco's proposed measures, Maxus came in almost a billion dollars in debt and without any oversight from NRDC or a similar environmental group. The new company paid lip service to negotiating with the Huaorani people, an agreement labeled "trinkets and beads" by the daughter of the Ecuadorian president. The Rainforest Action Network called for an economic boycott of the new company. In 1995, an Argentinian firm bought Maxus and began extracting oil from the Oriente with little apparent concern for what any outsiders thought.

In retrospect, it was clear that Bobby was right, that compromise often leads to a better overall outcome. As Lawrence Leamer wrote: "Whereas up until now activist organizations such as the NRDC had played the role of principled, passionate critics of oil exploration in remote pristine regions, Bobby and his colleagues [wanted to] help set up righteous monitors of a company attempting to profit from the responsible exploitation of one of the world's crucial resources."

Or, as his uncle JFK once described himself, Bobby was learning to be an idealist "without illusions."

* * *

Throughout the Ecuadorian struggle, Bobby remained close with indigenous leaders. The Shuar chief even visited him in Mt. Kisco. "They were very direct and had very strong leadership, and when they shook your hand they'd shake it like a Westerner and look you in the eye. I went falconing with them. And when they walked through the woods near my house, they would be looking up at the bark of every tree and chattering about it in their own language. I wish I knew what they were saying, but they were comparing it to something. And they had tremendous interest in every plant and every tree. They'd never seen a peacock before, and I had my daughter Kick catch one so they could take their photos with it. Then we gathered tail feathers, which they were anxious to take back and use for fashion headdresses. We went fishing in my lake. We caught a bunch of yellow perch, and we made a fire on the side and cooked them."

Bobby concluded that "if we'd just been dealing with the Shuar, we would have been able to hold it together—because they were also people who weren't afraid to confront the environmental groups and would have said to them, 'Why are you trying to interfere with our territory? We've been here for ten thousand years and we're not telling *you* what to do.' They were really savvy, and even just the way they would walk, they had this kind of confidence and fearlessness about them. They were real warriors."

So in the end, the Conoco debacle did not dampen Bobby's commitment to native peoples. If anything, it was strengthened. During this same period in 1992, with Pinochet no longer ruling the country but still in command of the Chilean army, Bobby would return leading a global campaign to stop the damming of the Bio-Bio River some five hundred miles south of Santiago. Pinochet had privatized all the country's rivers and turned them over to the Endesa state utility. The Bio-Bio, however, remained paradisiacal, and Bobby accompanied the local Pehuenche Indians and the largest contingent of kayaks and rafts ever to go down the river (some thirty people in all, including his brothers Michael and Max) to call attention to the situation. By this time, such an outing became known as "adventure advocacy."

Bobby later wrote of the expedition: "Although we were able to stop three of the four proposed dams, the single dam the utility company built inundated the most beautiful stretches of the Bío-Bío. I feel lucky to have seen the river before its destruction, which was a tragedy for all humanity."

CHAPTER TWELVE

FAMILY MATTERS

The 1990s marked, in many ways, a turning point for Bobby, both personally and politically. In 1991, he went on a two-day, two-hundred-mile bike race with his cousin, President Kennedy's son John Jr., to benefit the Special Olympics. The race originated near Bobby's home in Westchester County and ended at the summit of Mt. Snow in Vermont. Bobby was almost seven years older than John, and, until now, they hadn't spent more than casual time together. "Being with him was like being with one of my brothers," Bobby wrote. "Like his mother [Jackie], he was so thoughtful and protective of our family. He was brilliant, funny, and candid, as always. He entertained me with stories that highlighted his extraordinary ear for mimicry. He could do a French nanny, a Hispanic drug dealer, an Irish-American politician or a Boston cabbie ('I knew your Faththa'), a thrifty Scotsman, an Irish bog worker, or a Brit from Monty Python, and a pitch-perfect Dominican patois of his childhood nanny, Provi Paredes."

Ending the first day of the race, they tossed around a football and a Frisbee and engaged in some "deep talk."

"John and I spoke about the extraordinary goodwill toward our family," Bobby recalled. "It was an almost daily occurrence for both of us—in airports, restaurants, and on the street—that strangers would approach to hug us and thank us, and it was not uncommon for people to see our faces and burst into tears. So many people were touched by our fathers and those memories were a place for the repose of their most idealistic hopes for our nation. We had a kind of trustee's obligation to respect that currency and use it to make our country a better place." Eventually, Bobby often said a sense of obligation would lead from political and environmental activism, to running for public office, to becoming a public servant.

On May 20, 1992, at the age of thirty-eight, Bobby recorded in his journal: "Franz Leichter [a New York State Senator] called me to ask me to run for [Republican] Dee Dee Goodhue's seat for the Senate. I said no and made a commitment to myself to go for ten years without considering political office." It wasn't the right time, but the fact that he had to "commit" to waiting showed that it was a persistent thought in his mind. At some point, the time would be right.

That Bobby was increasingly a "player" in such circles became obvious that election year. In June, Senator Al Gore told him that "he was hoping to be picked as vice president and was on the short list, but [presidential candidate Bill] Clinton was wary of having two Southerners on the ticket. I told him he should change his accent and he said he'd tried that last time. I saw his famous private sense of humor that is so absent in public. Gore said that Clinton really wants Cuomo, his number one favorite, and joked that if he did that, he better also hire a food taster."

Gore was by far the strongest environmentalist in the field, and Bobby immediately went to work on his behalf, hosting a breakfast for him with two hundred political and eco-leaders through the New York League of Conservation Voters that Bobby cofounded. He went to the Democratic Convention in July, where he brought his namesake son Bobby along and gave a speech during a tribute to his father. "Joe Biden told me that Robert Kennedy was the reason he is in politics," Bobby wrote in his journal. During the ensuing campaign, Bobby met with Gore to discuss water issues and the need for a carbon tax.

The Clinton-Gore ticket's defeat of incumbent George H.W. Bush marked the end of twelve years of Republican rule of the White House. On November 3, Bobby watched the presidential election on TV and wrote in his journal: "so happy about this election that I couldn't sleep."

His mother felt similarly. Clinton called Ethel the night of his victory to thank her for the heartening letter she'd sent him when he was under media scrutiny for an alleged affair earlier in the year. The new president expressed hope that some of her kids would come to work for the administration. On December 7, Bobby received a call asking if he'd consider taking the job of Deputy Administrator of EPA for Legislation. Bobby struggled with making the decision and wrote down his concerns—less time for his family, withdrawing from the Riverkeeper movement, leaving a critical vacuum in the alliance for watershed protection,—"and finally, losing my independence and flexibility." Although he received many calls of support, Bobby declined. He would also turn down an offer to run the Region 2 EPA office.

Right before the Inaugural, his mother called to say that Clinton wanted to visit Robert Kennedy's grave. The Secret Service closed the Arlington Cemetery to the public, when Ethel, Bobby, and eight more relatives and close friends arrived. John, Max, and Teddy were already waiting at the gate. Clinton's advance men told everyone that he'd been talking about making this pilgrimage since the election. Army trucks drove them to the gravesite, followed soon after by Clinton, Hillary, and a sizable entourage.

Clinton went to the grave alone, before the others joined him and all knelt in prayer. The president took his time reading all the inscribed quotes at the site, and afterward they all talked informally. Clinton told the Kennedys this wasn't actually his first visit. While in Washington during the campaign, he'd run to escape the press corps and spent dawn up here by himself. Then he paused to say quietly, "He met me, you know."

Clinton related the story of how, when he was Legal Forum president at Georgetown Law School, he'd scheduled Teddy to speak. But with an overflow crowd on hand, the senator's staff could not locate him. In desperation, Clinton called Hickory Hill and reached Robert Kennedy on the tennis court. "Ten minutes later," Bobby later recounted, his father "showed up with khaki pants over his tennis shorts and a button-down shirt over his wet T-shirt. Two kids were in tow, whose heads were wet from the swimming pool. I was one of those kids. He gave a riveting speech."

On February 19, 1993, Bobby drove up to Hyde Park to see the president unveil his new economic plan at the FDR house. He penned in his journal: "Clinton gave a wonderful speech. I was proud of him. He knows how to persuade people, and he's interested in solving human problems.

He has an intellectual curiosity and an open mind, and that's something we haven't had for many years."

As Bobby was leaving, Clinton saw him at the back of the crowd and signaled him. The president asked Bobby to join him for a tour. They walked around the Roosevelt Museum and examined the objects on FDR's desk. Then they went outside to the grounds. "Clinton put his arm around my shoulder and walked to FDR's grave, where he put down a rose. Then he took my arm again and escorted me back to the Museum. He interrogated me about my life and the Hudson Valley. He was like a sponge. He was greatly interested in history and the role of the region in the American Revolution. He told me about his first visit to West Point, and I told him about Benedict Arnold's efforts to hand West Point to the British and the British strategy for dividing the Army of Virginia from Boston. I observed that Arnold was perhaps our best US General, citing Quebec and Ticonderoga and Saratoga. He asked about the betrayal. I said it was because Arnold was resentful at being passed over by Congress for promotion in favor of European generals with accents but no combat experience. He was very interested and said he hadn't realized that."

Clinton asked Bobby to visit FDR's house with him. According to Bobby, "he promised to whip through his media interviews. Two hours later he finished and we walked over to the house and studied FDR's bird collection, with which I was very familiar. We stayed for a long time, and he had a barrage of questions. Roosevelt, a great naturalist and ornithologist, in 1943 took part in the annual Dutchess County bird count, logging 108 species, many by calls alone."

Clinton told Bobby about his own childhood and described growing up in a 1,200 square-foot house. Bobby wrote: "I told him that I was overcome with emotion hearing his recent speech at the Lincoln Memorial and that I'd not seen such leadership from a president in my life. And I thought he would be one of the best presidents in our history. I meant all of this, and I'd wanted to tell him that for a long time. After we got back to the driveway, he came over to say good-bye, and I said, 'You're going to change this country. You already have.'"

Ethel had been considering selling Hickory Hill but after Clinton's election decided to remain, and the homestead "had a brief Renaissance as an alternate White House" with officials assembling there for weekends and dinner. On Easter Sunday, she hosted an egg hunt for the Cabinet.

That summer, Sunday June 6, 1993, marked the 25th anniversary of Robert Kennedy's assassination. "It was like Tennyson's *Ulysses* with all the old heroes gathered—Cesar Chavez's family, Jim Whitaker, Rafer Johnson, Marion Wright Edelman and Peter Edelman, John Lewis, Andy Young. At six o'clock we went on buses to Arlington," Bobby wrote. Twenty thousand people gathered at his father's grave, including the president and Hillary, who got a huge standing ovation.

* * *

A year earlier, after Bobby argued a Riverkeeper case before the Second Circuit where the judges seemed to favor him, he went next door to observe John Kennedy argue an arson trial. Afterward, Bobby recalled, "We talked about his future and we agreed that happiness came from being an effective human being and that usually included an ingredient of being in service. John had an uneasiness about accepting the political role that everyone expected of him. He thought maybe he would be more effective if he took a different course."

Bobby wrote in his journal: "The great thing about talking with John is that he was absolutely ruthless about examining himself and his motives in all things. We had that in common. John was very grateful for his job at the DA's office, which does not allow him to participate in political campaigns. And this has rescued him on many occasions, including with our family." The next day they continued their annual bike ride together, "staying together the whole time talking about how to stay in stable loving relationships, which was what both of us wanted."

It was Sunday and neither of them had been to mass, so they stopped at a Black Baptist Church, standing in the back listening to the service and the singing. When some of the congregants recognized them and began turning their heads to stare, the two ducked out. To their surprise, word had spread to the pulpit. The preacher called a temporary halt to the service and led the whole congregation out to say hello. John and Bobby shook hands, received kisses, took photos, and signed autographs.

Growing up and through his twenties, John's looks were thought to far outmatch his brains and depth. Beyond the famous salute at his father's funeral, he was known as a sweet little kid with a Beatles' haircut and an amused smile. He and Bobby shared many of the same outdoor interests, skiing together from Lake Placid to Sun Valley. They traveled to Florida, chasing tarpon and fly-fishing. John swam for several hours, training for a swim around Manhattan. He had also kayaked across Greenland and Norway, nearly drowning or freezing to death along the Scandinavian coast.

John coached Bobby in how to deal with the tabloid press, which ambushed them on a regular basis. "You've just got to do the next right thing and be able to look at yourself in the mirror." John also approached the constant assault on privacy with a sense of humor, saying "that the press were a nuisance but if they ever disappeared, he'd probably be saying to himself, 'Hey, what's wrong with me? Where did they go?'" Ultimately, he told Bobby, you had to ignore their sound and fury and not allow it to distract you from your inner journey. Which was really all that counted.

Bobby and John also "talked about Catholicism and the television preachers. [John] said he had doubts about the whole thing but was particularly cynical about people who claimed to have all the answers and to know the path. He quoted the *Tao Te Ching*: 'The path that can be narrated is not the Divine Path.' He said he felt Grace when he walked away from temptation or bit his tongue when someone did something obnoxious, or took time to do something for someone he didn't like. Sometimes he could walk away through discipline, other times it just took Grace. But either way he felt Divine Power and Providence when he walked away from self will and did the right thing."

There was also Bobby's relationship with John's mother. When he quarreled with his own mother as a kid, Bobby would sometimes seek shelter with Jackie. "She wrote me over forty letters when I was at school the year after my father died, which I treasured. On October 2, 1969, Jackie sent me Konrad Lorenz's book, *King Solomon's Rings*, with a note about how he wrote so many fascinating things about animals."

On February 12, 1992, Bobby wrote in his journal: "Jackie said some very nice things about me which . . . made me feel good. She said that she felt God was guiding my life and that no matter what suffering I went through, that the miracles are with me all the time and that the important things work out in the end."

After she was diagnosed with non-Hodgkin's lymphoma that came on quickly, Bobby visited Jackie before her death at 64, on May 19, 1994. "I felt devastated," Bobby wrote in his journal, "because I love her and she was such a source of stability to the family." John planned the funeral, saying he wanted it to reflect his mother's "love of words, the bonds of home and family, and her spirit of adventure." While outside the Park Avenue church some four thousand people stood vigil, Bobby was one of the pallbearers. Jackie's companion for many years, Maurice Tempelsman, read aloud *Ithaca* by the Greek poet Constantine Peter Cavafy and brought the seven hundred attendees to tears. Teddy kept his composure and delivered an extraordinary eulogy.

CHAPTER THIRTEEN

POWER PLAY ON NATIVE GROUND

On Earth Day 1990, shortly before he traveled to Ecuador's Amazon, Bobby was on hand in New York when a contingent of some forty men from the Cree and Inuit tribes in Canada arrived in the city's harbor. They had paddled an odeyak (a hybrid craft with the bow of a Cree canoe and the stern of an Inuit kayak) all the way from Montreal, a distance of 375 miles. They had journeyed along the St. Lawrence and Hudson rivers, holding public education events along the way that were widely covered by the media. Their goal was "to dramatize the potential havoc [a] massive hydroelectric project would wreak on their lands in northern Quebec."

The Cree were among the world's last remaining hunter-trapper societies, and the expedition had been organized by Matthew Coon Come, Grand Chief of Grand Council of the Crees, which governed nine communities comprising the Cree Nation. Coon Come was assuredly a man after Bobby's heart. Born two years after Bobby, his mother had gone into labor on a dog sled, and his dad made a shelter on the ice where he was born. Coon Come's name meant "He who wakes dawn." Raised in a tent on his father's trap line like many aboriginal youth of his generation in Canada, he'd been sent south at sixteen to a residential Indian school several hundred miles from his home.

"The federal government wanted to take the Indian out of me," Coon Come would recall. "It did not succeed. They told me that my culture and my people's ways of life would never sustain me. They lied. I am a son of a hunter, fisherman, and a trapper. My father taught me how to walk the land, and to love and respect the animals and all of creation."

After suffering physical abuse but completing high school, Coon Come returned to his people in the north, learning the Cree ways with his father and grandfather. What might be called his "political career" began when he attended a meeting of his community's elders to discuss Phase One of the planned James Bay Project. This was to include multiple dams in several phases. Realizing that the lands on which the Cree had hunted and trapped for five thousand years were to be flooded, the teenager was fully supportive of a lawsuit launched by the Cree in 1971 against the powerful government-owned Hydro-Quebec utility corporation.

The approximately twelve thousand Cree and six thousand Inuit of the remote region lost their battle in the courts and were forced into negotiation. In 1975, they signed the James Bay and Northern Quebec Agreement allowing the erection of a series of eight hydroelectric power stations known as the LaGrande River complex, covering an area the size of New York State. In return, the tribes were to receive over $170 million in land-use compensation as well as health care and education services and protection of fishing and hunting sources. The accord also called for tribal involvement on environmental review and consultation around any future development in the region. But while the pact was touted as the first to recognize a level of indigenous rights to self-determination, it also institutionalized the government's power to develop their sovereign lands.

And now a second phase was on the drawing board that would make Hydro-Quebec the most important source of cheap and abundant energy in North America. Robert Bourassa, the Quebec Premier who oversaw the first James Bay Project, returned from political exile to announce James Bay II in 1988. He had bemoaned in his recent book, *Power from the North*, that "everyday millions of potential kilowatt hours flow downhill and out to the sea. What a waste!" As *New York Times Magazine* writer Sam Howe Verhovek later described the Premier's vision: "Bourassa has advocated transforming James Bay into an enormous freshwater lake by erecting a dike across its mouth—a distance of one hundred miles. From there the water would be pumped through a series of enormous channels to the arid stretches of the Great Plains, the American Southwest, and California. Bourassa has referred to his plan simply as the 'Grand Canal.'"

The first client was to be the New York Power Authority, which in 1988 signed a $5 billion contract to import the hydroelectricity. It seemed, at first glance to outsiders, like a perfect match of Quebec's supply and New York's future demand, slated for readiness by early in the twenty-first century.

Coon Come and his paddling followers begged to differ. He had gone on to study political science at Trent University and then law at McGill University. As he outlined in a speech culminating their long river voyage to New York, the first James Bay Project had resulted in mercury leaching into the waters from flooded areas and making fish inedible upon which the Cree diet depended. Other unwelcome impacts included alteration of natural patterns of river flow that forced the Cree to adapt to new ways of traveling and harvesting wildlife resources. In 1984, ten thousand migratory caribou had perished in raging waters, the first year that the reservoirs were full enough to require controlled releases down the riverway.

The impacts of James Bay II threatened to be far worse. Some 620 miles north of Montreal, the project proposed to push north and dam the 150-mile-long Great Whale and ten other rivers that flowed through the transitional area between Quebec's tundra and taiga before emptying into the Hudson Bay. Altogether Hydro-Quebec planned to build several dozen huge dams and hundreds of dikes, reshaping terrain the size of France into one of the world's largest hydroelectric complexes, whose turbines possessed a dozen times the force of Niagara Falls. Reservoirs would transform more than 1,300 square miles of land and aquatic environments. Three underground generating stations to tap the water would have an installed capacity exceeding 3,200 Megawatts.

Bobby, who was among Coon Come's large audience that night, came to realize there was more at play than generating power. He recalled in 2022: "Quebec [was] essentially financing its secession from Canada. It was controlled by a French Canadian Quebecois party, originally led by René Lévesque, and because the amount of money going to Hydro-Quebec

POWER PLAY ON NATIVE GROUND

annually would be enormous, this would have given them a source of permanent foreign revenue with which to secede or, better, bargain for a deal to stay part of Canada. Hydro-Quebec *is* the government up there, it's very incestuous. But nobody ever consulted the Cree."

Matthew Coon Come was both eloquent and charismatic. Now amid the publicity accompanying his expedition, he managed to meet with New York Mayor David Dinkins to present his case against the new export agreement. Coon Come also met Bobby after his speech, where they struck an instant rapport, and the tribal leader invited him to visit the region.

Fortuitously, Eric Hertz had also listened to Coon Come's talk. Hertz was a former playwright and avid environmentalist who'd started an adventure travel company, Earth River Expeditions, a couple of years earlier. He realized that something was missing from the presentation—the Cree weren't able to show the rivers except via aerial photos. He approached one of the tribal representatives and suggested that his company could perhaps bring policy makers and media to see the area firsthand and take a rafting trip.

The Cree agreed to finance Hertz's spending several weeks flying over what he witnessed as "rivers with horrendous rapids that were unmanageable and unportageable" before he found a stretch of the Great Whale where a float plane could drop people off to camp. Hertz advertised a commercial trip and set about connecting to a couple of well-connected state legislators for a maiden voyage. He also wrote Bobby, whom he'd yet to meet, extending an invitation. Bobby responded that he'd like to come, provided he could bring along his young son Bobby III as well as his brother Michael and his son. Hertz said he'd need written agreements that the boys would have to walk every rapid along the banks. Over the centuries, the Cree had taken their canoes downriver but portaged every rapid. And the Great Whale had four major rapids that had never been run, the size of those in the Grand Canyon. Like many of Bobby's trips in Latin America, this would be a first descent.

So it came to pass. Earth River Expeditions journeyed to the sparsely inhabited James Bay region eight hundred miles from the nearest pavement and described by an earlier adventurer as racked by "wind, snow and natural violence of every sort." In a subarctic climate, winters lasted on average from late October into early May. The float plane carrying Bobby and his companions arrived in summer. "It was difficult to get in," Bobby III recalls, "and we were out there with probably eight rafts and fifty people for close to a month." The contingent included a dozen members of the New York State Legislature, which had agreed to buy power from Hydro Quebec.

Bobby remembers: "We slept in teepees every night. As soon as we landed, the Cree women would go cut down saplings, about thirty trees. It would take them a half hour to 45 minutes to set up a teepee that twenty people could sleep in. There was a fire in the middle of our campground, and we watched the Northern lights drop like curtains over the taiga. Meteor storms made the sight even more dramatic."

This part of northern Quebec was the traditional home of the Cree, many of whom didn't speak either French or English, spending summers in small towns and wintering in wigwams on their traplines. It included the section of the Great Whale that was to be fully dammed if Hydro-Quebec went ahead. And the pristine rivers were all Class-5 whitewater, potentially more treacherous than Bobby had encountered in Peru or Venezuela. His team had already named one of the rapids The Pearly Gates.

Bobby went on multiple adventures with Matthew Mukash, a local chief at the forefront of the opposition to the hydroelectric project (later elected grand chief of the Cree Nation in 2005). As he had done so many times in South America, he was able to connect to tribal leaders through his sense of adventure and his bravery. Bobby soon became close friends with both Mukash and Matthew. These sorts of personal connections, like the connection between JFK and the Soviet Premier Nikita Khrushchev, can lead to unlikely settlements or compromises but, as we've seen, can also take humanity back from the brink of annihilation.

* * *

Adding another element to the Great Whale mix, once Hydro-Quebec caught wind that Bobby Kennedy was in their territory, they had helicopters follow the group downstream, monitoring their every move. Bobby Kennedy was not, and never will be, a friend to the corrupt and greedy, and they know it.

As a result of its earlier incursion for James Bay I into the LaGrande River Complex, a 1984 study of the downstream Cree village of Chisasibi revealed that 64 percent of its residents contained levels of methyl-mercury in their bodies deemed unsafe by the World Health Organization. Methyl-mercury buildup is known to cause birth defects, including neurological injury and cerebral palsy, along with stillbirths and convulsions.

They demanded a massive area, Bobby explained, and "it was too large to remove trees and so they buried them all." Huge amounts of mercury from the soil, "when all of that organic material began to break down and ferment," was released into the water. When that happened, Bobby continued, "the fish became hideously toxic." As a result, virtually all the Indians were poisoned. "Mercury is a thousand times more neurotoxic than lead," according to Bobby. "I heard horrific stories of villages with birth defects and loss of IQ. If you really want to damage a population in every way, give them mercury—and that's what happened."

It was déjà vu for Bobby with what he'd been up against in the Hudson. And it was a mercurial precursor to what he'd find himself exposing a little more than a decade later.

Bobby continued: "Matthew Coon Come knew that Hydro-Quebec's expansion plan would devastate many more villages. They were offered a billion dollars at one point, to ten thousand poverty-stricken Cree. Matthew said no. And that is why he is an amazing leader." Bobby knew all too well that real leaders don't choose profits over people.

Coon Come had a flair for turning his cause into an international cause célèbre. While overseeing the Cree's role in crafting a Universal Declaration on the Right of Indigenous Peoples, he told a United Nations panel in 1991: "Who would consider the total elimination of the Hudson River? Yet great rivers have been destroyed where my people live." That winter, Coon Come's words were echoed by a mass die-off of eider ducks during a freeze that closed floe edges and open water habitats, resulting in a 75 percent decline in the breeding population.

American environmental groups, including the NRDC, Sierra Club, Greenpeace, and Audubon Society, were enlisted into the fight. The tactical warfare came to include demonstrations at state capitals in New York and New England (where a contract had also been signed with

Hydro-Quebec), focusing particularly on utility offices and hearings; more traveling road shows with music and slides; divestment campaigns by students on some university campuses; and tribal delegations traveling to Washington, The Hague, and Geneva. At least one legal case went before the Canadian Supreme Court, which in September 1991 ordered Hydro-Quebec to submit to a stricter environmental review process.

All this coincided with Earth River Expeditions continuing to run eight "conservation awareness" trips to the region, six of them with Bobby at the forefront. Hydro-Quebec essentially tried to buy off media coverage with a propaganda tour, which backfired. *National Geographic*, which had published a positive article about James Bay I, took a second and highly critical look. The *New York Times Magazine* and *Condé Nast* published strong pieces questioning what Hydro-Quebec was up to. CNN followed suit. Hydro-Quebec's PR campaign tried another ploy, taking handpicked groups of mostly Inuit Elders on helicopter tours of the project, seeking to temper their anxiety.

The turning point came from New York. A State Assemblyman from Buffalo who had accompanied one of the first river sorties, William Hoyt, was himself an avid wilderness canoeist who had paddled Arctic and sub-Arctic Canadian rivers for more than 2,000 miles. As chairman of the Assembly Energy Committee, he became the Legislature's primary opponent of New York's proposal to buy hydropower from Hydro-Quebec and flood the area he'd visited.

On March 25, 1992, Hoyt had just delivered an impassioned speech when he tumbled to the Assembly floor. He had suffered a heart attack and died shortly thereafter in an Albany hospital. The *New York Times* reported in Hoyt's obituary that "Richard M. Flynn, the chairman of the New York Power Authority, who had his share of quarrels with Mr. Hoyt, called him a 'man of integrity who fought tirelessly for causes in which he believed.'"

The latest cause was that of the Cree. Hoyt had introduced a bill requiring Hydro-Quebec to pass the same environmental standards as any company desiring to build a power plant in the state. In memory of Hoyt, his fellow legislators quickly voted his bill into law. New York Mayor Dinkins had already taken a stand with the indigenous activists. And these two occurrences were all that Governor Mario Cuomo needed.

Bobby had been lobbying Cuomo for almost two years to do something about the Hydro-Quebec situation. "I had a good relationship with him because my sister Kerry was then married to his son Andrew," Bobby explained to me. "Mario was teetering and trying to figure out a way to reject the plan, when Assemblyman Hoyt gave his ringing speech against it and suddenly dropped dead."

Two days later, New York State canceled its $13 billion contract with Hydro-Quebec to provide a year-round power supply of a thousand megawatts. Governor Cuomo said he'd accept the recommendation of the Power Authority's Flynn that the state no longer needed the power. "Circumstances have changed dramatically since this contract originally was signed in 1988," Cuomo said. Flynn added that he'd tried but failed to negotiate a better deal with the company.

But a spokesman for Hydro-Quebec claimed this wouldn't affect the "ongoing process or the realization" of the project on the Great Whale, which was to undergo three phases of construction. Hydro-Quebec pushed ahead, commissioning a feasibility study that eventually filled thirty volumes and five thousand pages, costing $256 million Canadian dollars. Using federal funding, the Cree hired experts to pick holes in the report.

In May 1993, Bobby met with John Adams of NRDC about the situation and wrote in his journal: "They were talking about the beating we were taking on this and on Ecuador, but they wanted to go forward with the James Bay program. I was kind of surprised because this kind of beating felt good to me. It was insignificant, and it meant that we were doing something important."

That same month, Bobby brought Bobby III along to a conference of Indian leaders in South Dakota. The event fell on the 25th anniversary of his father's campaign trip there in 1968. "The Indians told heartbreaking stories of despair and rejection and suicide on the reservations and about how my father had given them hope," Bobby recalled. After the gathering, accompanied by his sister Kerry, they went to a sweat lodge to receive their Sioux names. The medicine man, Rick Two Dogs, explained the ceremony and gave them the last of the eagle plumes that his grandfather had once bequeathed to him.

Stripping down to their swimsuits, the three crawled into the small hut where red-hot stones were being loaded into the pit and the canvas door was sealed. The medicine man prayed and poured water on the stones "and we broiled like lobsters." Their guides seemed distracted by all the smoke, so Bobby dug an air hole to provide oxygen so his son wouldn't pass out. Afterward, Bobby used it himself.

At the close of the ceremony, several women sang and chanted, and on the third passing of the pipe the names were given: Kerry, Good Earth Woman; Bobby III, White Buffalo Boy; and Bobby Jr., Spotted Eagle Man. Bobby recorded: "They said that when I see an eagle, it's a sign that God is watching over me, and I had always believed that all my life."

* * *

"Bobby Kennedy gave us instant credibility for many people who had never heard about our cause," Cree chief Coon Come said in 1994. "And it also made it more difficult for our opponents to dismiss us as a bunch of primitive Indians." On his continuing sojourns to the US, Coon Come met with President Bill Clinton and UN Secretary General Boutros Boutros Ghali. And he stopped off several times to visit Bobby. Years later, Bobby III remained awestruck by having watched Coon Come's calling "something like fifty Canada geese out of the sky. I'd seen other people do the goose call before, but it was like he was hypnotizing them or something."

Early in April 1994, on his first day as the new president of the New York Power Authority, S. David Freeman told a legislative hearing that the state should reject a 20-year, $5-billion contract extending an agreement to buy electricity from the Canadians. Some estimates had been that the project would be completed no earlier than 2018 and perhaps not until 2031. "We don't need the power, the price is too high, and there are unresolved environmental questions in Quebec," Freeman said. Governor Cuomo had already signed the agreement, but, as the *New York Times* noted, there was "an escape clause." Consolidated Edison was in the process of renegotiating with Hydro-Quebec because the demand and price of power had fallen sharply over the past two years. The utility that reared its ugly head against the Hudson River's striped bass hadn't gone away.

The news reached Hydro-Quebec just as the company was celebrating its 50th anniversary. The *Washington Post* headlined that the utility "finds its high-voltage plans short-circuited." It quoted Reed Scowen, Quebec's official representative to the US: "To put it mildly, selling power to the US is not a matter of life or death for us, either for the Quebec economy or for the health of our public utility." The article also quoted Bobby: "Now it's in their interest to suggest with bravado that [energy exports] are insignificant. Their investors don't think it's insignificant." As investment banker and Hydro-Quebec critic Robert Blohm put it, the company "does not currently earn enough US dollars to service its massive current US dollar debt of $10 billion, the largest US dollar debt of any company in the industrialized world," which US contracts for Great Whale power were intended to repay. "There's no money left in the kitty for ego projects," allowed the managing director of the Canadian Bond Rating Service Inc.

Perhaps rubbing salt in the wound, a little more than a week later, Matthew Coon Come received the Goldman Environmental Prize awarded annually to six activists worldwide. The April 18, 1994, announcement quoted him: "I've traded in my bows and arrows for the electronic media, the judicial process, and a public awareness campaign. I'm not anti-development. But we have to question the whole notion of business as usual because that won't save the land or the animals." For Goldman's press release, Bobby was quoted: "Because of his leadership, the Cree have been able to assert themselves. It's a model. For the first time in the Americas, Indians have been able to assert control over their land."

A Canadian federal judge ruled for the Cree that a stricter environmental review was needed. The long fight was over. In 1995, Bobby joined Coon Come at the Toronto International Film Festival's premiere of a documentary, "Power: The James Bay Cree v. Hydro Quebec." That same year, Coon Come bartered a revenue-sharing arrangement with the provincial government that would apply to all natural resource projects on Cree land in the future.

* * *

Some two thousand miles away, along the wild Canadian Pacific Coast, during this same time frame a different kind of battle over use of tribal lands was in full swing. Bobby traveled for the first time to Clayoquot Sound on central Vancouver Island to lend his support in 1993. He wrote:

"Environmental and native groups were challenging MacMillan Bloedel, a large timber corporation that was illegally logging native lands, clearcutting the last of the island's untouched coastal rain forest. Some of the area's ancient trees—towering three hundred feet with a girth of thirty feet at the base—had been seedlings when Jesus walked the earth. MacMillan Bloedel was converting them to paper and boards. The Nuu-chah-nulth people who'd lived there for thousands of years banded with environmental organizations to raise international awareness about the logging."

Bobby and his son Bobby III met with the local chiefs and discussed a boycott, as Canadians had done in helping to end apartheid in South Africa. The next day, taking a dugout canoe to Opitsu, Bobby wrote, "the village men picked the canoe up out of the water on logs and brought us up onto the beach on their shoulders, where the Chiefs assembled wearing traditional garbs and gowns. This was only the second time in fifty years that this honor had been bestowed upon a white person. (It's usually used for visiting Chiefs who don't want to get their feet wet on the beach.) We walked to the Community House where they gave us a potlatch [a ceremonial feast], showering us with gifts, cedar headdresses and masks and bracelets."

Bobby had recently flown to Bakersfield, California, for Cesar Chavez's funeral on April 27, 1993. He spoke about his experience with Chavez and how forty thousand people, "many in rags, including the growers he antagonized, attended his funeral." They attended, not because Chavez "was rich or powerful, or had achieved a high political office, but because he had asked people to sacrifice and allowed them to participate in this battle for dignity," Bobby wrote. "And that even people thousands of miles away, who spoke no Spanish, had never seen a farm worker or been on a farm, stopped eating grapes, and made that personal sacrifice because he had called us to act out of something more noble than self-interest." That sense of purpose, of moral courage, was spreading, and, once again, Bobby was showing his leadership.

Nearly eleven thousand people came to engage in peaceful protests against the corporate invaders in Vancouver. Their series of blockades became known internationally as the War of the Woods. Every day for several months, protesters would blockade a remote logging road before being carried or dragged into a bus that would transport them to a police station to be charged and released. This made it the largest act of civil disobedience in Canadian history. "Government enforcers arrested eight hundred people, three hundred alone on one notorious August day in 1993," Bobby recalled.

Bobby was declared Persona Non Grata by the Prime Minister of British Columbia as the fight to save Vancouver Island's remaining old growth moved to boardrooms. The NRDC, along with the Sierra Club, Greenpeace, and other environmental groups, began pressuring high-profile customers of MacMillan Bloedel, including phone-book publishers and the *New York Times*, to stop buying its newsprint. Home Depot agreed to join the boycott.

That December, Vancouver Island's Nuu-Chah-Nulth Chief Francis Frank called Bobby to explain the deal he had negotiated with the government and the company, giving First Nations veto power over cutting. Although the government could still stack the deciding panel with lumber company representatives and be allowed to steal the power back from the Indians, this marked the first acknowledgment of the rights of First Nations to participate meaningfully in the management and resources of their traditional lands.

What precipitated the agreement was a remarkable trip that the tribal leaders from Vancouver Island took to visit the Menominee Indians in Wisconsin. The five Nuu-Chah-Nulth tribes that Bobby represented as a lawyer wanted to do sustainable logging. Research undertaken by the NRDC found that no one in the world was doing it better than the Menominee. "They had taken something like three million board feet out of their forest since the 1880s, and there were four million board feet standing," Bobby told me. "In other words, the forest has gotten healthier and healthier because of the ways they were harvesting it. The trees were getting bigger and bigger. And the maximum leader of the tribe was the chief forester. They had their own sawmill and were making high-quality finished products and the tribe was flourishing. Their forest was so well defined that NASA was calibrating satellite positions based upon its stretch line, because it was so distinct from the neighboring countryside." Bobby arranged for the Nuu-Chah-Nulth to spend a week in Menominee country before going on to Washington for meetings with the Senate Foreign Relations Committee.

On the fifth anniversary celebration in 1998 of the successful blockades in Clayoquot Sound, Bobby would bring along his daughter Kick. "From a stage ingeniously constructed of driftwood, I delivered a barn-burner speech. I told them . . . that their most profound accomplishment was in democratizing a process that had been the province of a few industry decision makers. Now, thanks to them, the decision over logging would be made by all the stakeholders—community members, fishermen, environmentalists, timer workers, local officials, Natives, and so on. Their fight for democracy made them heroes comparable to the freedom riders, the lunch counter integrationists, the Selma marchers, and free young people who thought the big issues were solved after Vietnam and civil rights. Let them come to British Columbia, I said. . . . We climbed the steep cliffs beneath the full moon and found the path through the dark woods to our cabin."

In 2000, UNESCO would designate large parts of Clayoquot Sound an international biosphere reserve, "making much of the area a protected haven for tourists and boaters wanting to experience the incomparable majesty of the Vancouver Island wilderness."

"We shall all be alike—brothers of one father and one mother, with one sky above us, and one country around us, and one government for all. For this day the Indian race is waiting and praying." There is very little I can add to those eloquent words, except to say that I believe the day of Chief Joseph's vision is not far off. I believe we will see the dawning of it in our own lifetime, and that its full light will come to shine on our children.

—Conclusion of remarks by Attorney General
Robert F. Kennedy, before the National Congress
of American Indians, September 13, 1963

CHAPTER FOURTEEN

MIDLIFE TRANSITION

Bobby had had several girlfriends and was still seeing the London socialite Rebecca Fraser when he met Emily Black in a law school class at the University of Virginia in September 1978. A petite, quiet, and attractive brunette, she was a Midwestern girl, daughter of a schoolteacher mother and a stepfather who managed a lumberyard in Bloomington, Indiana. Her biological father had died of a heart attack when Emily was two, and she had two siblings. In 1968, she'd been a teenager in the crowd when Robert Kennedy came to her hometown to campaign in the Indiana primary.

Asked what attracted him to Emily, Bobby told me: "She was super smart and super funny, and curious about everything. She'd come from very humble roots and through her own hard work and brains made it into one of the best law schools in the world. She could hold her own in any conversation with our circle of friends, people like Sheldon Whitehouse and Chris Bartle and lots of very impressive people."

Bobby wrote of Emily: "She had been raised in a home with an authoritarian stepfather who made arbitrary decisions to justify his power in her life by quoting the Bible. She devoted her life to justice and fighting arbitrary, self-righteous and moralistic authority."

Emily had graduated Phi Beta Kappa in political science from Indiana University and was in her second year of law school when she and Bobby started dating. Their April wedding, a year later in 1982, took place in her hometown church. TV cameramen and reporters came from all over the world. This was American royalty come to the Hoosier State, featuring a mingling from across the generations—not only Ethel and Uncle Ted, but Arthur Schlesinger Jr., Richard and Doris Kearns Goodwin, Joseph Alsop, and Bobby's other relatives, family, and personal friends, as well as the friends and relatives of Emily.

After they graduated from law school in June, the couple moved into the elegant uptown New York apartment that Lem Billings had willed to him. At the time, Bobby had been hired as an assistant district attorney in the office of Robert Morgenthau, a childhood friend of JFK's who'd appointed him US attorney for the Southern District. Emily was to commence her own legal career as a public defender with the New York Legal Aid Society, helping poor and indigent individuals facing criminal charges.

Emily had given birth to their first child, Robert Francis Kennedy III, in September 1984 and a second child, Kathleen, in April 1988 (nicknamed "Kick" after his father's sister, the baroness of Cavendish, who died in a plane crash forty years earlier). Since 1984, when Bobby started at Riverkeeper, the family had been living in a sprawling white colonial house on an 11-acre estate in Mount Kisco. But as Bobby traversed the jungles of the Ecuadorian Amazon and the frigid reaches of the Canadian wilderness, his personal life went through a major transition as he approached midlife.

Not long after Kick was born, the couple began drifting apart. Emily had been "his rock," as Bobby's friend Chris Bartle put it. "Bobby needed the kind of assurance and support only she could provide." But the house was often filled with weekend guests from many walks of life, and Emily simply didn't possess Bobby's enthusiasm for socializing or competitive outdoor games.

During the eighth year of their marriage in 1990, Emily decided to return to part-time lawyering after having been a stay-at-home mother since Bobby III's birth and worked mostly from home but also shared an office in Tribeca with her best friend from law school. Two years later, she and Bobby separated. They kept the house, but in April 1992, Emily took an apartment in the city and Bobby moved into a small cabin in the woods near the reservoir on Cross River.

Bobby recalled: "We had been working on the relationship for several years. But it just could not be saved." He recorded in his journal at the time: "Emily said that her principal objective was to get through the divorce with dignity and love and support one another and try to feel out the steps God has laid out for us, and trust the results would put us where He wants us to be. And to protect the children. And to get through this without anger and be good and supportive friends to each other."

On March 17, 1994, Bobby spoke at a black-tie dinner in Pennsylvania commemorating his father's return to public life there thirty years earlier. The next day Hamilton Fish Jr. announced his retirement from the US Congress because of a recurring prostate cancer. Politicians and press across New York as well as members of his family called Bobby urging him to run to fill the vacancy. "But I was going through a period that was difficult for my children, and I felt that I needed to give my attention to them." Once again, the timing wasn't right.

There was no dispute over assets, and Bobby and Emily signed a separation agreement in May of 1994. That summer, Emily wrote to her former mother-in-law after Ethel reached out to her in a letter: "I have a lot to thank you for Ethel. I have learned so much and grown through your example . . . [among other things, I have learned] to face difficulty with courage and to get beyond it with generosity, and an enduring sense of humor and appreciation for the joy in life. I feel lucky to have been a part of your family and to have had that sense of belonging for these last years. . . . I think you know that I will never love anyone as I have loved Bobby. I am grateful that we had each other to lean on while weathering the storms of young adulthood. I am proud of all he has accomplished and will always support him in heart and spirit." Emily and Bobby agreed to share joint custody of Bobby III and Kick.

* * *

Bobby had known Mary Richardson since she was fifteen. Like the Kennedys and the Skakels, she was raised in a large Irish American Catholic family, alongside six siblings. Her father, a professor and secretary of the faculty at Stevens Institute of Technology, had died of cancer when she was twelve. Mary engaged in rebellion against her schoolteacher mother much as the adolescent Bobby had with Ethel. "When you lose a parent young, you hold resentment against the surviving one," she once said. Around the same age that Bobby ran away from home, she'd gone off for six months to join a circus where all the performers slept together in a huge tent. Then, sent away to school in Europe, she moved in with an Italian prince.

Her family scraped the money together to send Mary to the progressive coed Putney boarding school in Vermont. There, on the first day, she met another fifteen-year-old named Kerry Kennedy. They bonded instantly. "The next weekend we hitch-hiked to Boston," Kerry later recounted, "to see my siblings Michael, Bobby, David, Courtney, and Kathleen," after which Mary "spent nearly every weekend and vacation with our family." Roommates all through high school, Mary and Kerry continued to live together at Brown University and afterward while working in New York and Boston. "We were inseparable," Kerry would recall. "We shared friendships, a closet, a cash card. People couldn't tell our voices apart."

Mary got straight As at Brown and remained a stellar student at the Rhode Island School of Design. She had also grown into a tall beauty with long brunette hair who dressed elegantly in black and craved adventure. As a kind of "honorary Kennedy," she fit right into their milieu. When Kerry's older brother Joe started a nonprofit organization called Citizens Energy in 1979—primarily aimed at aiding low-income and elderly households by creating projects for discounted and free home heating services—Mary designed, wrote, and produced its first annual report.

Launching a Manhattan career as an architectural designer, Mary was also becoming "a master at securing donations for good causes and considered the challenge a contact sport." Like Bobby, she felt a powerful kinship with the people of Latin America. When the military cut off supplies to the poorest regions of El Salvador in the mid-1980s, Kerry recalled, "she volunteered to help the RFK Center for Justice and Human Rights put together a truckload of food, clothes, and medical supplies for the Mothers of the Disappeared." With the help of Kerry's connections, Mary managed to ship ten precious containers to the CoMadres.

Kerry called her "the kindest person I've ever met. And she was that way to everybody, not just celebrities but fishermen, cabdrivers, and the guy who installed her floors." In 1986, Mary became active in Joe Kennedy's first successful run for the US Congress in Massachusetts. She took a job as an architect at Parish Hadley, a prestigious New York design firm, a few years later becoming part of a green team renovating then-Vice President Al Gore's official Washington residence at the Naval Observatory.

Mary also remained, in Kerry's eyes, a kind of Renaissance woman—"a rafting guide on the Kennebec, she could build a campfire and on it cook a gourmet meal for 100, throw on a pot, weave a blanket, knit a hat, plumb a john, sponge paint a ceiling, and have a successful run as part of a girl group of go-go dancers in the cages at Studio 54." They continued to share a residence together until Kerry married Andrew Cuomo in 1990.

* * *

Mary and Bobby became friends through Kerry, but there was no romantic relationship between them until the spring of 1993. Bobby would recall: "One day when I had invited Mary to come to an art exhibit, the scales fell off my eyes. I suddenly saw her for the first time. I began dating her cautiously and fell deeply in love." He wrote in 2015: "I felt like she had been veiled in smoke all those years because I was not ready for the relationship. Now the smoke was lifted, and I was continually surprised at the depth of her character and baffled that I hadn't noticed those things before. It was like exploring a new country, climbing mountains and discovering from each new rich top another beautiful valley below, filled with adventure and laughter and surprises.

"By August 1993, I knew that I wanted to spend the rest of my life with her. She had everything I ever wanted. She was smart, thoughtful, honest, curious, brave, and beautiful. And I felt I had so much to learn about life and love from her. She had struggled and prevailed and gained wisdom in the process. We began living together in a shack on the Cross River Reservoir."

Bobby also recalled that Emily "really liked Mary, and we were living very nearby. When Emily left, she would ask the two of us to stay with the kids, and she loved the relationship Mary had with our children." Bobby III remembers: "When they first started dating, Mary was literally running the whitewater rafting trips, as like an outfitter-chef; she knew where every tent pole was."

Bobby proposed to Mary on a trip to Ireland, in Dublin's Phoenix Park, and they began planning a wedding for nine months down the road.

On March 24, 1994, he flew to the Dominican Republic to obtain the divorce from Emily. Numerous foreigners were doing similarly, because the country's laws enabled someone to end a marriage in only a few hours. On April 15th, Lem Billings's birthday, Mary and Bobby were wed in a small ceremony, on the deck of the Riverkeeper's research boat, the *Shannon*. They went fishing and caught a number of perch and cod and after diving kept some oyster shells as souvenirs. Bobby had just turned forty, and Mary would be thirty-five in the fall. They would move into the estate in Mount Kisco, while Emily remained at the first Westchester home with the two children.

Kerry and Andrew Cuomo would move a mile down the street to be close to Bobby and Mary. Kerry reflected years later; "I have never seen two people more thoroughly enchanted with one another and more completely in love. They brought out the best in one another, and spoke about each other with wonder and awe." A healthy baby boy was born on July 24, 1994. They named him Conor.

CHAPTER FIFTEEN

LUCK OF THE IRISH

In the Summer of 1963, a few months before his death, President Kennedy had visited the Emerald Isle of his forebears and considered the trip "one of the most moving experiences" of his forty-two years. "This is not the land of my birth," he said, "but it is the land for which I hold the greatest affection." He traveled with his youngest sister, Jean Kennedy Smith, from Shannon to Dublin, speaking to record crowds gathered along his route.

In Dunganstown, JFK visited Mary Kennedy Ryan, his third cousin whom Bobby would also meet in later years. There the family's heritage dated back to the 1700s, when they paid rent to an English landlord. The Kennedys educated their children in a secret "hedge school" and were among rare residents of Dunganstown who could read and write.

The president met with his Irish counterpart, Eamon de Valera, to whom Jackie wrote following the assassination: "He never would have been president, had he not been Irish. All the history of your people is a long one of overcoming obstacles. He felt that burden on him as a young Irishman in Boston—and he had so many obstacles in his life, his religion, his health. . . . The people of Ireland had faced famine and disease, had fought against oppression and died for independence."

Ireland was the British Empire's first colony and its last. In 1690, King William of Orange defeated Britain's Catholic King James II and his Irish allies at the Battle of the Boyne, dooming the final hope for Irish freedom. Five years later, the Crown introduced retaliatory penal laws prohibiting

Catholics from owning land or animals, getting educated, holding public office, practicing a profession, serving on juries, voting, or practicing their religion. The law banished Catholic clergymen. For the next three centuries, the landless and bedraggled Irish Catholics lived lives of bullying and coercion on an island ruled by a foreign government, with a permanent military garrison. It's no wonder that Jack Kennedy, who knew his history well, nursed a lifelong antagonism to every variety of colonialism and imperialism, and a wariness toward involving America in colonial wars.

Following a 1916 rebellion, the Irish Republican Army (IRA) fought a bloody war against the British for two and a half years. And then, in 1920, the British negotiated a peace by splitting the island. The twenty-six counties to the south, with their Catholic majority, became an independent republic ruled by its own parliament. The northern six counties, industrialized and wealthy, with Protestant majorities far outnumbering Catholics, remained as part of Great Britain but also had their own parliament. The British gerrymandered the county lines to guarantee a permanent Protestant majority. The major police forces, the Royal Ulster Constabulary (RUC) and the Ulster Special Constabulary, were all Protestant and Unionist.

A young generation of Catholics, inspired by the Kennedy presidency and Martin Luther King's peace marches, took to the streets of Northern Ireland to protest the centuries-old system of apartheid that kept them in impoverished ghettos. One of these was Paul Hill. He had grown up in West Belfast, where his merchant seaman father was a Protestant and his mother a Catholic. Hawking newspapers at age nine, he never forgot the sight of women weeping in the street when JFK was killed. Or the photograph of the small barelegged boy in his winter coat, offering a salute, that his mother had him purchase from a local shopkeeper. It was John F. Kennedy Jr. Thirty years later, Paul Hill would marry John's cousin.

In response to the peaceful protests in the Sixties, the Ulster volunteer force, a secretive paramilitary death squad, allied with the police, detonated a series of bombs in Catholic neighborhoods. The Protestant RUC police force attacked civil rights marchers in Derry and beat two Catholic men to death. These deaths precipitated the worst rioting in the North since the 1930s. The Union government asked the British Army to come in and put down the rebellion, which they did.

From Washington, Senator Ted Kennedy began voicing strong concerns and expressing support for the goals of the Northern Ireland Civil Rights Association. His initial posture was confrontational to the British position. As the violence escalated, so did Teddy's rhetoric. In 1970, he delivered the bicentennial address at Trinity College in Dublin, condemning the British policy of internment without trial in Northern Ireland. To the exasperation of the British leadership, he called Northern Ireland "Britain's Vietnam."

Meanwhile, the ghetto fought back. Hundreds of women risked their lives by marching into the street to confront the British troops. The women's march broke the siege, but the violent response to the peaceful protests by army police and loyalists radicalized Catholics across the North and sent many young Irish flocking to join the IRA. As the British militarized the North, the IRA transitioned from a defensive police force to a guerilla movement. The IRA became a terrorist organization, and it enjoyed broad support from Catholics in the United States. During Bobby's years at Harvard, every Irish pub in Boston and Cambridge had a jar on the bar collecting funds for an IRA surrogate.

A state of virtual civil war existed as British troops patrolled Belfast streets and a Protestant paramilitary movement (the Ulster Defense Association) arose to hold onto Northern Ireland's place within the UK. In 1974, an elite IRA squadron unleashed a murderous bombing of two pubs near London that catered to British soldiers. Twenty-year-old Paul Hill, on the run from his parents, was asleep in his North London flat at the time of the bombings in nearby Guildford. Overnight, though, Hill found himself falsely implicated by an older member of what became known as the "Guildford Four." After several days of deprivation at the police station, Hill signed a confession with a cocked revolver in his mouth. He would receive a life sentence.

Bobby recalled: "He was really just a kid who wanted to sit in the pub and watch rugby. All of a sudden he becomes the scapegoat for this IRA squad. He would've had some protection if they'd put him in the Maze, the big prison where you're allowed to wear your own clothes and are treated more like a POW. But instead, they put him in British prisons and moved him thirty-eight times. All the British criminals hated him because they looked at him as a terrorist. So Paul lived a nightmarish life for fifteen years. If you can survive that, you come away with some wisdom."

A personal incident cemented Senator Kennedy's resolve to do something about the escalating violence. His niece, JFK's daughter, Caroline, had secured a year-long internship at Sotheby's in London and was living with her friend Rebecca Fraser. She was the daughter of Hugh Fraser, a boyhood chum of JFK and now a Conservative Member of Parliament and vocal IRA critic. Bobby would soon meet Rebecca while enrolled at the London School of Economics. They started dating, and she would for a time become his fiancée.

Every morning, Rebecca would drive Caroline to work in her father's red Jaguar.

On October 23, 1975, the two young ladies were already on her front stoop when Fraser dashed back into the house to answer a ringing phone. Caroline waited in the hall. Minutes later, a time-bomb planted under Fraser's car by the IRA detonated and blew the car to tree level. The powerful blast knocked down Fraser and Caroline and cut them with flying glass. The phone call had saved their lives.

In 1977, when Bobby was a junior at Harvard living off-campus in a small rental, Uncle Ted asked if John Hume could stay in his spare bedroom when he came to teach at the Kennedy School of Government in the spring. Senator Kennedy considered Hume, a brilliant Northern Ireland Parliamentarian, the best hope for bringing eventual unity to the troubled region. His advocacy of peaceful change and his condemnation of violence earned Hume the antipathy of both the IRA and the Ulster death squads. Both sides occasionally made attempts to murder him.

Bobby first visited Ireland in 1978, where Hume showed him the neat bullet hole that an IRA sniper had fired and missed in the living room of his modest home in Derry. Bobby also traveled to Belfast with Rebecca Fraser. "The city was racked with poverty," he remembered. "A maze of narrow streets intersected Belfast's impoverished neighborhoods. Most of the homes were century-old tiny 'two-ups' with outside toilets. There were pubs on every corner and blocks of idle, abandoned mills and engineering factories. The 'Peace Line,' a thirty-foot high barricade, divided the Catholic Falls Road from the Protestant Shankill Road. Parts of the city looked like Germany after World War II, with piles of rubble, bomb debris, and entire blocks bombed out."

Bobby's uncle persisted in his peace effort, cofounding Friends of Ireland in Congress in 1981. Four years later came the Anglo-Irish Agreement, where the British Government acknowledged for the first time the role of the Irish government in Northern Ireland, including advising on human rights and policing. Initially, Britain's Prime Minister Margaret Thatcher had opposed the agreement after the IRA tried to bomb her cabinet meeting. But Ted Kennedy reached out with House Speaker Tip O'Neill and Bill Clark, President Reagan's Secretary of the Interior and a friend of Thatcher's, and persuaded her to endorse the arrangement.

At the same time, a broad-based campaign including clergy, politicians, and judges from Britain and Ireland mounted support for the wrongly convicted Guildford Four. In 1989, an appeals court determined that detectives had manufactured the confessions, and three of the men gained their release. Paul Hill, however, was not among them. After serving fifteen years, he was taken back to Belfast, having previously signed another coerced confession to the 1974 murder there of a British soldier. Refusing a deal for immediate parole if he would accept guilt, Hill vowed to fight the conviction.

Freed anyway after posting a small bail of some $6,000, he set to work fighting on behalf of the "Birmingham Six," another group incarcerated under bogus charges. Hill helped win their release and, in a book called *Stolen Years*, recounted how all these cases had been trumped up by British intelligence as part of their covert war against the IRA. (The book would become the basis for the award-winning 1993 film, "In the Name of the Father.")

As a US Representative from Massachusetts, Bobby's older brother Joe had been one of the Guildford Four's most vocal supporters in Congress. In 1990, he arranged for Hill and another ex-prisoner to address the Human Rights Caucus in Washington. Ethel Kennedy attended and was impressed enough to tell Hill afterward that she wished her daughter Courtney had been able to attend. Courtney had traveled with her sister Kerry to Ireland and "felt like I was home" and later briefly enrolled at Dublin's Trinity College. Returning to New York in 1980, she married and later divorced a TV executive, while working for the foundation named after her father.

Hill was en route to address a law society in New York where Courtney was then recovering from a broken bone suffered in a skiing accident. At Ethel's impetus, the two arranged to meet for lunch. Courtney expected to meet someone "shriveled up" physically and mentally. "I couldn't understand a word he said," she recalled, hearing a Belfast accent delivered with such rapidity. "But I thought, 'He's gorgeous.'" She attended the lecture— "he seemed to have this energy, and this integrity"—and they began a correspondence, then met up again in London. In March 1992, Courtney brought Paul up to Pace, where Bobby was teaching, and he gave a lecture about his jail experience in the British legal system.

After a three-year courtship, they married aboard a ship in the Aegean Sea in the summer of 1993, approaching Hill's 39th birthday, and soon moved to County Clare in Ireland. Earlier that year, at Ted Kennedy's request, newly elected President Bill Clinton had appointed his sister Jean as Ambassador to Ireland. It was another violent year, with the Unionists killing IRA members and innocent Catholics with planted bombs.

Hill's appeal of his conviction for the murder of British soldier Brian Shaw was scheduled for early March in 1994. Led by Courtney and their mother, Ethel, a Kennedy contingent including Bobby, Joe, Kathleen, Kerry, and Rory flew to Belfast to lend their support. Bobby arrived on the morning of the 4th. "The airline had lost my bag, and I needed to buy a shirt for court," he remembers. "A feud ensued because my driver, a Catholic, wanted to go to a Catholic mall, but the RUC [Royal Ulster Constabulary] insisted we ride in the car with them to a Protestant center. We were attacked by the British press. When we entered the courtroom, Paul and Courtney were surrounded by IRA people. A belligerent British soldier shouldered Joe, saying, 'Why don't you go back to your home?' Joe replied, 'Why don't you go back to yours?'"

Before the trial began, Hill explained to Bobby that it was British government strategy to encourage the Unionist terror organizations into sectarian violence, a false-flag operation to make it appear to the world as if it were a Catholic versus Protestant fray. The Kennedy clan was about to be on the receiving end. They were being shadowed by the terror boss of the Ulster Defense Association, Johnny Adair. Once asked by a journalist if he'd ever had a Catholic in his car, Adair replied: "Only dead ones."

He called himself "Mad Dog," and he had the chops to prove it. Standing five-foot-three, his body was covered with tattoos. Since 1990, he had orchestrated a wave of terror in West Belfast that involved the murder of upward of forty Catholics. An unpaid police informant, Mad Dog operated in collaboration with the British security forces and became a hero of the Loyalist movement.

In 1989, Adair had driven the getaway car for a Loyalist death squad in the murder of Pat Finucane, an internationally revered human rights lawyer who often defended IRA prisoners. Finucane had just been returning from dinner in Dublin with Bobby's visiting sister Kerry, when gunmen opened fire at his home in the presence of his family.

Most recently, one of Mad Dog's men had thrown a grenade into the home of Gerry Adams. He was the leader of the Sinn Fein party in Parliament and a key player in the peace process who had forged an alliance with John Hume in 1993 seeking to persuade the IRA to put down its guns. Adams had stayed with Courtney and Hill at their London flat and was now the prime target of Mad Dog's Company C death squad. The bomb bounced off the hardened glass that saved his family, destroying his porch and front door. Adams had been sleeping in safe houses for security.

So, as had often happened, the Kennedys were in the thick of it. Adair and his close associate, Winkie Dodds, closely monitored the movements of the Kennedy clan as they began their thirteen-day courtroom vigil. An article in the *Belfast Telegraph* (July 24, 2008) would reveal Mad Dog's plan to ambush the Kennedy car as it made the daily trip from a Catholic Church retreat in North Belfast to the Royal Courts of Justice. In an eerie déjà vu of Dallas, Adair selected Downview Gardens, near the Shore Road end of Donegall Park Avenue, as the ideal point to attack when the vehicle had to slow down at a security ramp, providing an almost stationary target. He intended to fire a rocket-propelled grenade from a Russian PRG-7 launcher. Hill's wife, Courtney, was the primary target, but he knew that assassinating the other Kennedys with her would create the international incident he desired.

Then Mad Dog made a serious mistake. He'd been attending a bail hearing for fellow loyalists at the nearby Magistrate's Court, then crossed the road to the High Court building where the third day of Hill's hearing was in progress. Police spotted him lingering in the courthouse for about thirty minutes, sparking an alert within the RUC Special Branch that led to extra security on the Kennedy family's route.

The Belfast paper noted that details of the plot to blast the Kennedys had emerged only now, fourteen years later, after members of Adair's notorious C company had rebelled against his leadership earlier that year and been debriefed by the new UDA leadership. In a memoir titled *Mad Dog* published in 2009, Adair had described a planned ambush of an IRA group of "sitting ducks" along a planned route using a rocket launcher but said nothing about a scenario involving the Kennedys.

The verdict for Paul Hill came back on April 21, 1994. The Crown had argued that Hill couldn't have been threatened with a pistol for his testimony because the police station contained no weapons. But the Crown's case against him collapsed when two police officers admitted there having been guns in the precinct the night Paul had confessed. So the three Court of Appeals judges, wigged and in red robes, found that Hill's confession was coerced and exonerated him.

Outside, Joe Kennedy made a statement on behalf of the family: "We're here to support my brother-in-law in his struggle for justice. Given the history of my family, I also want to let the Shaw family know that we know what it's like to be the victim of political violence. But one wrongful act should not condemn an innocent man for the rest of his life."

Bobby remembers: "We drove to the monastery where everybody was staying for safety's sake after our hotel had a bomb threat. So now we drove on to a small town in Donegal. When our priest friend Father Michael tried to pull up for gas, Paul rolled down his window and said, 'Not here.' It was Loyalist territory."

That night, John Hume came to the hotel and Bobby went out to dinner with him and his wife. "John had lost twenty pounds in the past year from worrying. He said he'd told Gerry Adams that the world had changed and the old arguments for their violence were no longer valid. I asked Hume what [Adair's] UDL would do if the IRA put away its guns. He told me: 'They would kill for a while, and then the British forces would be here solely to fight the UDL. That's why the UDL doesn't want the IRA to put down its arms. That is why they are threatening to kill me.' John is now their number one assassination target. We were followed everywhere by the Irish police, who had intelligence of a death threat."

But that night in the wonderful fish house restaurant off the fishing wharf, Catholics and Protestants all dined happily and conversed, including a Protestant leader who came over to talk with Hume. As Bobby put it, "that was John's vision for the future of the North."

A few weeks later, in May 1994, Mad Dog Adair was arrested and charged with directing terrorism. He ultimately received a sixteen-year jail sentence. That same year, after the IRA declared a "complete cessation of military operations," Adair's UDA and a sister movement called their own cease-fires.

Senator Kennedy visited Northern Ireland for the first time in January 1998, meeting with both Unionists and Nationalists in Belfast and Derry. The inclusive peace talks, chaired by former Senator George Mitchell, led to the Good Friday Peace Agreement that April, the most promising opportunity for lasting peace in the three-decade-old conflict in Northern Ireland.

For the most part, it's worked. After serving his time, Mad Dog Adair came out of prison and settled in Scotland, where early in 2021 he spoke in an interview about John Hume coming to visit him in Maze prison and convincing Mad Dog and his followers to maintain the cease-fire that they'd been considering abandoning. Hume had died in the summer of 2020. Mad Dog said: "John Hume wanted peace and he was prepared to speak to anyone or go anywhere to bring it about. There's no doubt, John Hume persuaded us. He won us over and we decided to give it a go. John Hume was a good man and he deserves his place in history."

Things seemed to have come full circle. Bobby tells a story, courtesy of Paul Hill: Two Irishmen were walking down the road after the Good Friday Treaty was signed. And one said to the other, "Do you realize this is the first day of peace in seven hundred years? We've been at war for seven centuries." And the other replied wistfully, "Ah yes, ah yes, but weren't it great while it lasted?"

CHAPTER SIXTEEN

WATERSHED MOMENTS

While his older sister Kathleen was elected lieutenant governor of Maryland in 1994, and his older brother Joe was by then serving his fifth term as a US Representative from Massachusetts, a *New York Magazine* cover story headlined toward the end of 1995 that Bobby was "The Kennedy Who Matters." Water remained the focus of his professional life, but the issue now assumed dimensions on his home turf that propelled him into a whole new realm politically. It required every ounce of the negotiating skills and personal approach that he'd recently learned (and lost) in Ecuador, prepped for (and won) in Northern Quebec. Now he'd no longer be crossing tribal lines, but party lines. He was not simply a Riverkeeper, but a guardian for the water quality of the entire state. And his adept maneuvering at the highest levels of New York politics was about to thrust him into the limelight toward assuming unsought high office himself.

When Lieutenant Ron Gatto arrived at the Riverkeeper office, he considered it a last resort. Gatto was one of twenty-eight police officers employed by New York's Department of Environmental Protection (DEP) to patrol and protect the city's upstate reservoirs and the two-thousand-square-mile watershed that feeds them. That winter of 1989–90, he'd fielded a complaint about one of the prisons in Westchester County dumping raw sewage into a trout brook that ran through the facility and caught them in the act. Gatto also uncovered medical and human waste being emptied directly into the Croton Falls Reservoir by the Putnam County Hospital.

But when he went to his superiors at the DEP, he remembers, "they ripped up the tickets I'd issued and said to stay away from things like this." Gatto didn't know what to do. A friend told him about the Riverkeepers, and Gatto spoke to Bobby over the phone, who invited him over.

Bobby and his Riverkeeper partner John Cronin had been concerned for several years that the quality of New York City's vaunted clean drinking water had entered a steady decline. The water emanated, Bobby wrote, from "a masterpiece of ingenuity and design—nineteen reservoirs and controlled lakes deliver 1.5 billion gallons of gravity-propelled water per day from Westchester and Putnam Counties and from the Catskill Mountains . . . [But] the water supply was unfiltered. No pollutant removal technology existed between the reservoirs and the water consumer." Bobby had sued the city to obtain its water distribution maps and found the Croton Reservoir's water to be the dirtiest in the system, while being delivered to families in the most impoverished neighborhoods of the city. Now Riverkeeper was helping organize minority communities to do something about it.

And here came Lieutenant Gatto, "a tough Italian weightlifter," as Bobby would later describe him. Three years younger than Bobby, he'd been born and raised in Yorktown when the Westchester County village still contained bears and porcupines in the woods where his father took him hunting and fishing. Gatto told the Riverkeepers: "I don't get it. If a kid steals a car or throws a brick through your window, any DA in this country will prosecute them—even though their crime only injures a single person. But if you drop a toxic chemical or disease organisms into the drinking water for nine-and-a-half million people, no one is interested in the case." Gatto recalled in 2022: "From that day on, Bobby and I became the best of friends."

As *The Riverkeepers* book would chronicle: "Gatto had a special talent for discovering illegal pollution. He could smell a leaking septic tank from a moving vehicle. To the shock of his passengers, he once brought his patrol car to a screeching halt on Catherine Street in Yorktown and began sniffing the air like a bloodhound. In a short time he discovered an illegal septic bypass at the Holy Comforter Nursing Home. . . . Gatto executed the first search warrant in the history of the agency and sent the first violator to jail. . . . Within one year Gatto made fifty arrests."

Still, the entrenched bureaucracy at the DEP continued to undermine their zealous patrolman's efforts. Gatto was disciplined repeatedly and four times investigated (and cleared) after anonymous sources brought scurrilous charges against him. So Bobby arranged for Gatto and a fellow officer to testify at a hearing chaired by the city council's president. Bobby also made sure that all the major New York newspapers and TV stations were on hand to learn about how the DEP's police captain and the agency's Sources Division "aimed to facilitate development around the reservoirs to ingratiate themselves with local politicians and developers." Gatto and his associate revealed how they were provided inadequate equipment and training to deter enforcement and received direct orders to drop investigations. As a result, incoming DEP commissioner Al Appleton transferred the compromised captain and promoted Gatto to second in command of the department.

* * *

Just as Gatto's relentless and gutsy work "broke the ice," as he puts it, a complex situation arose with no easy solution. In 1989, Congress had strengthened the Safe Drinking Water Act with a new provision, the Surface Water Treatment Rule, which mandated that any of the nation's systems taking their water from reservoirs, streams, lakes, or rivers needed to build a water filtration plant. "The city immediately agreed to filter the Croton waters, which received most of the watershed's sewage and were undeniably degraded," but which provided only 10 percent of the total supply.

The other 90 percent came from the upstate rural Catskill/Delaware system, propelled by the world's biggest pipes along the New York City Aqueduct from as far as 120 miles away that served the five boroughs and surrounding counties. Booming development, however, had witnessed twenty-five sewage treatment plants constructed from the 1950s through the 1970s. Altogether 112 such plants discharged into the watershed of the city's water supply, the only unfiltered supply in America allowed to occur. So the EPA now wanted to require filtering of the Catskill/Delaware water, but the cost would be enormous—an estimated $8 billion for construction and another $300 million annually for operation. "That would have bankrupted the city," Bobby reflected years later. "Water rates would at least double, rental and real estate prices would soar, with the worst impacts landing on low-income people who already have leaky pipes. But in fact, conventional filtration would not remove many of the pollutants and organisms associated with watershed development."

On March 9, 1990, the renowned columnist Murray Kempton wrote in New York's *Newsday* of Bobby's legal fight for better water: "The Kennedy history grew great on the urge to direct public affairs as commanders do, and now . . . we come upon a Kennedy dedicated to quarreling with the management of public affairs as guerrillas do."

Kempton concluded his piece: "He tried being the prosecutor his father had earlier been. It did not suit his nature, and he went away to struggle with private afflictions and disappeared to find his strength in the hills and rivers where all strength began. Now he has emerged and committed himself to being the healer that his father finally learned was the highest of callings. On Friday, he rose up as the healer who does not fear to abrade, and the sight of him would have made his father as full of pride as of love."

In June 1990, Bobby went to the bachelor party for his sister Kerry's soon-to-be-husband, Andrew Cuomo. Sitting alone on a bar stool was the groom's father, then the governor of New York. "Rather than just exchange pleasantries," the *New York Times* later reported, "Mr. Kennedy was seized by an urge to buttonhole Mario Cuomo about his pet concern, New York City's giant watershed. And so he did, for the better part of an hour. 'Nobody else was talking to him,' Mr. Kennedy recalled. 'So I made him sit and listen to me.'"

New York's US Senator Daniel Moynihan managed to slip a temporary exemption into the Surface Water Treatment Rule that let the EPA waive the filtration requirement for cities whose population was over a million and "could demonstrate a water protection program capable of protecting their supplies indefinitely." That's precisely what Riverkeeper thought was needed; the trouble was, with the exception of an occasional event like the one where Gatto testified, urban dwellers were clueless about the situation. Not a single article about watershed pollution or the prospect of a filtration order had appeared in the city's major papers.

Riverkeeper set to work. Students from the Environmental Law Clinic at Pace analyzed permit and discharge records and found that 30 percent of the reservoir's sewage plants violated their permits, and lawsuits against each were initiated. A report, *The Legend of City Water*, was compiled and distributed widely to government officials. Bobby wrote op-eds and met with editorial boards and took film crews on pollution tours of the watershed. He lobbied EPA administrator William Reilly to extend a temporary filtration waiver, and, on his last day in office, Reilly gave the city a year to develop a program to protect its reservoirs.

Riverkeeper engaged a pro bono PR firm that came up with an ad campaign planning to hang posters in bus stations with graphic pictures of sewage in the water. Bobby and his compadres addressed high school students who'd figured their water simply came from a faucet, completely unaware of their city's hydraulic connection to upstate New York. City Council members and county government officials were presented with coffee mugs spelling out the problems, including one denoting that "two percent of every cup of Croton water is human sewage."

In the process, Bobby antagonized the Establishment, who feared this could cost New York's Democratic Mayor David Dinkins the next election. Warning calls came in from uncle Teddy's office and Bobby's brother-in-law, Andrew Cuomo. His NRDC colleagues asked him to back off. But on an airplane Bobby had met Dennis Rivera, the powerful head of the city's Service Employees International Union 1199 (the largest such Local in the world). They had bonded immediately. Rivera told Bobby not to worry about all the flack and eagerly joined his campaign.

Another key ally was Bill Tatum, publisher of the African American *Amsterdam News*, with whom Bobby met often and who did editorials about the water supply. As noted earlier, Bobby's research team had discovered that the worst water was being sent to the poorest communities. Tatum helped Bobby recruit the Black Police Officer's Association, the group One Hundred Black Men, and other organizations from Harlem into the campaign.

The DEP's Al Appleton told Bobby that Mayor Dinkins was dismissing the need for land acquisition in the watershed as "the Kennedy strategy." Toward the end of May, Bobby received a call from the mayor. "I just want you to know that I'm really pissed off at you," Bobby remembered him saying. "I've known you and your family for many years. Your mother's been my tennis partner. We've always been close friends. The fact that you would launch a campaign like this without even picking up the phone and calling me. . . . is beyond my comprehension."

Bobby calmly explained that he appreciated Dinkins's strong stance on environmental issues and that he had refused to meet with his Republican rival Rudy Giuliani. In June, there would be a Walk for Water, an eight-mile march around the Croton Reservoir, to educate New Yorkers about where their drinking water came from. "I suggest that you come to our event," Bobby said. "Allow me to introduce you and say glowing things about your record and then announce that you'll commit another $300 million to watershed land protection."

After the Mayor and his aides went ballistic about what this would do to the budget, Bobby added: "If you don't spend it, you're going to have to spend $8 billion on a filtration plant and $300 million every year to operate it," a ruling that the EPA was scheduled to make in December. If he did, Bobby would bring a host of environmentalists to Washington "to persuade them that New York should be given a break." This was apparently the first time Dinkins learned about the EPA's looming decision.

On the eve of the Walk on Water, Bobby got a call from fellow attorney Eric Goldstein at the NRDC. "The city blinked," Goldstein said. "Your Irish strategy worked." The city agreed to put $250 million in the next year's capital budget. Mayor Dinkins showed up for the walk. So did Bobby's celebrity friends Glenn Close and Eddie Olmos. Bobby appeared every fifteen minutes on the *Today Show*, which had continuous coverage as more than a thousand people turned out in a torrential downpour. "Instead of our critical campaign," Bobby remembers, "we unraveled a giant poster three hundred feet high on July 4th weekend, with a picture of the Statue of Liberty holding a pitcher of water in her hand and a caption that said, 'Best Water on Earth.'"

Bobby and Dennis Rivera also kept the pressure on. They published an op-ed together on environmental racism in the *New York Times* (August 15, 1992), pointing out that "inexorably, society's wastes flow toward communities debilitated by social unrest, high illiteracy and unemployment, and low voter registration. They have become toxic dumping grounds while receiving few of the safeguards that prudence and decency demand but only political power can obtain."

They cited specific proposals to put the North River sewage treatment facility in West Harlem, where "the chronic stench of the poorly designed

plant is one more demoralizing obstacle to revitalization of the neighborhood," and to another plan for placing the state's largest medical-waste
incinerator in a densely populated South Bronx neighborhood.

Rivera recalls it was at the dining table of his apartment on the Upper
West Side where Mayor Dinkins then agreed to put forward a multi-
million-dollar bond issue for expanding enforcement and purchasing
properties that ensured they wouldn't contaminate nearby water supply
and for expanding enforcement.

The mayor also opened a series of negotiations that resulted in raising
the city's commitment for a water protection program to $750 million.
Thousands crowded the South Street Seaport at a rally in October 1993
to celebrate the mayor's announcement, followed by the EPA granting the
city another three-year grace period to implement its solution. Bobby had
worked with key US senators to make sure that happened.

* * *

Then everything changed overnight. After a bitter electoral race, Rudy
Giuliani defeated Dinkins in January 1994. The new administration opened
private negotiations with the Coalition of Watershed Towns and refused to
let environmental groups like Riverkeeper participate. The regulations proposed by Dinkins disappeared. "The Giuliani version eliminated all hazardous waste controls, pesticide restrictions, septic controls, and enforcement initiatives," and the new mayor "fired the DEP's press officers who
had developed relationships with the public interest community." According
to *The Riverkeepers*, "We were back at war with the city. But this time we
were ready."

Enter Douglas Kennedy, Bobby's youngest brother, who had just
started a job with the *New York Post* as an investigative reporter. His editors gave the go-ahead, and, says Bobby, "we began funneling tips to him
from our friendly DEP moles." Douglas's first article exposed that the
Croton reservoir had been shut down because of sewage pollution. A DEP
spokesman told the *Post* that it was actually "organic material" forcing the
closure. That night, September 8, 1994, David Letterman informed his
national TV audience that the *Post* article "scared the organic material out
of me." The country took notice.

Douglas's second story revealed the DEP's secret discovery of *Vibrio cholerae* in Croton water, a grim reminder of the cholera epidemic of 1832, "which killed one-fifth of the city's population [and] prompted the construction of the Croton Reservoir system in the first place." Bobby said later, "I was proud of Douglas. He got more attention for the issue in four days than I had gotten in four years." The Riverkeeper phone rang off the hook with DEP informants coming forward, including the detection of cryptosporidium cysts for the first time in the water supply coinciding with rises in hospital admissions of patients with the diarrheal disease, some of whom had died.

Now dozens of stories in other media criticized Giuliani for discarding the tough water protection regulations, and Riverkeeper mailed a thirteen-page report card on the mayor's actions to every member of the state legislature. A hearing that the DEP expected to be over quickly found hundreds of attendees in a standing-room-only crowd lambasting a panel of recalcitrant city officials.

After the hearings, with Giuliani upset by the stories portraying his administration as uncommitted to water-quality protection, DEP commissioner Marilyn Gelber asked for a truce. Riverkeeper agreed to suspend its public attacks so long as the city would redraft its regulations and reform the agency's upstate watershed division, and the DEP chief said yes.

Now the politics got strange indeed. Bobby also knew George Pataki, the Republican candidate running against three-term incumbent Mario Cuomo in 1994. Pataki, who lived in Garrison not far from Riverkeeper's office, had made contributions and frequently attended fundraisers for the organization on Bobby's boat. Once he became a first-term state senator, however, his donor base shifted from environmental constituents to the big developers and farmers of upstate New York. In April 1993, Pataki had cosponsored a bill that would have limited New York City's authority for regulating pollution in the watershed. He made it clear that, if elected governor, he would fight against what Riverkeeper advocated. So, beyond the familial ties to Cuomo, Bobby campaigned vigorously against Pataki and his otherwise decent record on environmental issues, to widespread press coverage during the campaign.

Then an odd switcheroo transpired. Giuliani, who'd started his political life volunteering for Robert Kennedy's presidential campaign in 1968, suddenly entered the water fray two weeks before the election. Standing on the steps of City Hall, Giuliani shocked the city by crossing party lines and endorsing Cuomo—calling the watershed controversy the main issue driving his decision.

Bobby recalled for this book: "Giuliani brought me in because I gave him the imprimatur of legitimacy. I attended a number of meetings with him and his staff. He turned things over to us and pretended to help us. What he's really doing is trying to sabotage his main rival in the Republican Party and keep the way open for himself to become governor in the future. So he needed an issue to hate Pataki, that Pataki was going to screw New York by forcing a filtration system down its throat. There's no possible rationale other than raw political ambition."

The mayor's surprise announcement did bring a sudden surge in Cuomo's poll numbers, but it didn't last. By the weekend before the election, upstate backlash began to build against Governor Cuomo. On election day, Pataki prevailed by four points. "Then Giuliani switched sides again, and I went back to suing him," Bobby recalls. With a simultaneous Republican takeover of Congress in Washington in 1994, the ever-calculating mayor believed that reducing federal mandates like the EPA's water filtration plan was more than likely. In January 1995, new water regulations released by the city were as bad as, and in some cases worse than, what Giuliani had presented before.

This only strengthened Riverkeeper's resolve. Douglas Kennedy continued to break stories about the DEP's malfeasance, reporting on proof found by the Riverkeepers that high-level technicians had been ordered to skew results in their drinking water labs to conceal contaminated water in the city pipes. Another article revealed that all nine DEP headquarters buildings supplied their staff with either bottled or filtered water. Commissioner Gelber responded by firing the deputy charged with managing the labs. The tide shifted as dozens more citizens attended public hearings and demanded strong regulations to protect their community water supplies while rejecting the filtration idea.

Still, things seemed at a stalemate. And Pataki wasn't quick to forgive Bobby for attacking his positions on the environment during the campaign. "Someone very high up" in the Pataki administration, Bobby recounted to *New York Magazine*, made an unfriendly phone call shortly after the election. "He made it very clear," Bobby said, "anything I wanted for the next four years, I could forget."

Yet by the spring of 1995, Pataki offered to mediate the watershed controversy between New York City and the upstate towns. Tension between the two constituencies had only escalated, as developers desiring an open door banded together with citizens still fuming over the city's earlier takeover of their land. Federal law didn't force the watershed communities to cooperate with the city's effort to avoid filtration. Finding any common ground would need to be voluntary on the part of all stakeholders. And environmental groups were told they wouldn't be welcome at the table.

Then, in April, Michael Finnegan, the governor's chief counsel and lead negotiator, placed a call to John Cronin. Pataki had come to realize, Finnegan reported, that the issue couldn't be resolved without input from the environmental community. Following an initial meeting in Albany, a week later the governor showed up at the Riverkeeper headquarters in shirt sleeves and jeans for a private meeting. "All right, how do we protect the water supply?" he asked a group of environmentalists, including the NRDC that Bobby served as an attorney. An hour later, Pataki concluded: "I want you in the negotiations. The state won't stay involved if the environmentalists aren't involved as well." Just so long as nobody leaked this news to the media.

Bobby realized Pataki was burying the hatchet. Later, the governor would tell Bobby what affirmed his decision. The actor Christopher Reeve ("Superman") lived not far from Bobby and had been one of Riverkeeper's closest allies on the issue, testifying at the public hearings and attending the press conferences. Reeve had spoken with the newly elected governor at Rockefeller Center's yearly Christmas tree lighting, telling Pataki that a politician with courage could become a hero on the watershed matter. Then, on May 27, 1995, Reeve was thrown from a horse during an equestrian competition and broke his neck, paralyzing him from the shoulders down. Watching a TV report describing Reeve's heroic efforts to survive and recover, Pataki felt inspired to make protection of New York City water the highest priority of his first year in office.

* * *

But there was another hurdle. Up in the Catskills region, residents had been badly burned by earlier water projects that benefited New York City. As *The Riverkeepers* chronicled, people were "understandably alert and angry. The history of family and community upheavals associated with the creation of the reservoirs remained fresh in their minds. The city completed the Cannonsville Reservoir in 1964. Towns had been flooded, churches ruined, property condemned, barns burned by city workers, and cemeteries relocated. Altogether, thirty-two communities were moved and the best bottom land in Delaware County permanently flooded. . . . The new talk of aggressive regulatory programs, controlled growth, law enforcement, and large-scale property acquisition rekindled bitter memories of how a distant city, with the approval of the state, once took control of their lands. Upstate developers financed an intensive direct-mail campaign that helped fan worry into fear. They delivered the message to every Catskill household that the city was coming back to take over the rest of the property and pay no recompense to local communities. A coalition of watershed towns formed to coordinate opposition."

Bobby was persona non grata with the upstate farmers, and it had seemed an irreconcilable situation. Toward the end of 1990, inside the ornate old courthouse in the city, the deputy mayor, DEP representatives, and other advocates had gathered alongside agricultural interests to discuss the proposed direct regulation of farming practices in the watershed. Bobby didn't attend, but a contingent from Riverkeeper sat at the conference table. When one of them suggested that big corporate farms dominating the region stood in the way of progress, Richard Coombe pounded his fist on the table and stood up. Coombe ran about three hundred head of beef cattle on a 1,500-acre farm a few miles from the Roundout Reservoir, through which more than half of the city's water supply from the Catskill basin flowed. For the past ten years, Coombe had also represented his district in the New York State Assembly and been a vigorous defender of farmers' interests.

He was a strong speaker, ready for a fight, and Bobby was his target. Coombe rightly pointed out that about 85 percent of the Catskill watershed was in agriculture and, above all, forestry. Coombe remembered saying: "And Bobby Kennedy Jr. lives on the east side of the Croton system, where 10 percent of the water goes. This is about land use! There are three hundred-some farms in the Catskills, and we're obviously doing good for the water because it's so pure you don't *have* to filter. Bobby Kennedy's uncle and father would be ashamed of him!"

Coombe conceded in 2022 that he was "pretty brutal" in his remarks, "but it must have resonated because they took it back to him and he called my wife." Now it was Bobby's turn to be angry. He planned to sue Coombe over what he'd said and demanded that the legislator come see him. Finally, Coombe took the phone and told Bobby, "No, you come here." And Bobby did.

"He'd been my sworn enemy, calling me names in the paper," Bobby recalls. "Then I went up and spent a day with him on his farm." Coombe remembers: "We argued it out. Bobby all of a sudden realized, holy cow, we're not a big corporate farm and really not the bad guy. At the end of the meeting, we went to the bottom of a hill. He couldn't believe all the native brown trout in the stream. I bent over and drank the water out of it and I said, 'Bobby, that's how pure this hill is for water with a lot of cattle on it.'"

A dialogue was underway, and for both men it would mark a major turning point.

Bobby explained that his objective wasn't to put the land out of production, but to figure out how to keep the farmers alive while simultaneously cutting down on the pollution. "A lot of it just meant incentivizing them to put riparian zones around their streams. The vegetation absorbs the nitrogen and phosphorus and the stuff that's going to destroy the water. If you put even a short riparian zone on either side, it prevents 90 percent of the cow shit going into the stream. But it's a pain for the dairymen because if they're to plow, it's better not to have any vegetation there. So you do have to offer incentives, but in the end it's good for everybody."

Coombe reflected back: "We should have remained mortal enemies and instead we found out we had the same common desire. I wanted to farm, but I also wanted the environment to be clean. He wanted the environment clean and really didn't care if I farmed if I did it right. It became a partnership rather than an adversary position."

It was certainly true that outdated farming practices at some two hundred Delaware County dairy farms endangered the waters with contaminated runoff. So Bobby began making more trips into what he'd considered enemy territory, where the Catskill groups were a conservative Republican bastion and particularly distrustful of the environmental interlopers. "I spent a lot of time in the Catskills, sitting at people's kitchen tables," Bobby remembers. DEP cop Ron Gatto sometimes accompanied Bobby and remembers: "He had a real beehive to commit to. We were up in Delaware County one time, and 'Motor Mouth' Miele as we called him was the new DEP Commissioner and giving a talk blaming the farmers and all these geese for polluting the water. They sent a bunch of boys out into the parking lot and overturned his car."

Amid the process, Coombe's distrusting constituents voted him out of office in the State Assembly, but a new path awaited. He was appointed the first chairman of a Watershed Agricultural Council. More than thirty people sat at the same table for meeting after meeting over the course of a year. According to Bobby, "it was an amazing group of stakeholders, because the dairy men were there and the governor and the hotel groups and people from the water supply and the EPA." Riverkeeper was the only one out of five involved environmental groups allowed direct input. Bobby and Cronin hired Dave Gordon, who'd studied environmental law at Pace, to be their day-to-day point person. "I was trying to get the underlying structure straightened out," Gordon remembers. "Development is controlled by local town planning boards. How do you fix the process and get them to enforce the rules, or get the city to weigh in if local towns aren't?"

Early in October 1995, at a meeting in the Capitol Building in Albany, Bobby surprised the negotiators with a strong statement supporting Delaware County's demand from the city for economic development money. This marked the first time environmentalists had taken the Catskill community's side in a dispute. Bobby would tell *New York Magazine* reporter Pat Wechsler: "I saw all these upstate people—farmers who had lost their lands, those who had seen their graveyards flooded in the name of cleaner water—put aside their resentments. It seemed easy enough for me to put aside my party affiliation."

In response, as Bobby and Cronin wrote, "they dropped their knee-jerk opposition to our every demand and even began supporting us in some areas. A feeling emerged that we were all part of one community and shared a responsibility to find a solution. The rhetoric diminished and the horse-trading began." As Gordon observed, "everybody exercised their power and raised their concerns in a manner geared toward preserving the process and getting the best solution without upsetting the apple cart. Anybody could have at any time, simply by lifting it."

Coombe recalls: "We had to get 90 percent of the farmers signed up willing to put in a whole new management plan. We did ten demonstration farms, different types in all corners of the watershed, and included a couple who were really anti-city. Then the city gave us a hundred thousand dollars for each farm to put in Best Management Practices, and it worked." Ultimately, the negotiations would replace existing regulations with a cooperative program that financed helping farmers retrofit on a voluntary basis.

Bobby estimates he attended just about all of the 250 meetings. He sat next to Cronin writing notes back and forth. "Negotiating among that many parties is tough," Cronin says. "There were four or five of us representing environmental interests, and we didn't always agree. Probably about ten times, it looked like the whole thing had blown up. But Bobby was the one who really discovered and then articulated the issue, and did it brilliantly. He'd seen the surface water treatment rule coming, and that New York City was going to have to do something about it. They wouldn't be able to get an avoidance from EPA on spending billions on a water treatment plant by continuing business as usual, with the lack of protection for water supplies in the Catskills. On the other hand, Bobby understood that the communities up there deserved compensation and that millions of dollars needed to be invested in protecting what he considered the world's best water supply."

With the Watershed Agricultural Council as the prototype, a sister group called the Catskill Watershed Corporation formed to bring not just farmers but local communities into the mix. According to Cronin, "Bobby defined the issue and had he not, both New York City and EPA were in an untenable position, because the city hadn't given EPA any reason to give them an exemption. New York's position had been, our water's always been good and always will be. Which is the same thing as saying, it's so big that we're diluting anything bad, so let us keep on with what we're doing. EPA knew that wasn't good enough. They basically monitored all the meetings but did not take a position, saying they'd do that when something landed on their desk."

Michael Finnegan, Governor Pataki's right-hand man, recalled Bobby's role like this: "The upstaters thought that if the monitoring [program data] showed that more development was possible, regulations should be changed to allow it. Many of the environmentalists thought only more regulation and not less should be considered. But Bobby said that if the data tell us that the regulations should be changed for more growth, then so be it. I remember he said, 'We're not anti-growth, we're for a clean environment.' The upstaters never looked at him the same after that."

After Dick Coombe's Delaware County jurisdiction came to terms with the city in mid-October 1995, the other Catskill communities soon followed suit. Here's what they all came to by Halloween: "The upstate communities agreed to drop nearly a dozen lawsuits that could have delayed the city's regulations and land acquisition program for decades, and to allow new restrictions on development within the watershed. In return, the city would spend $1.5 billion purchasing critical watershed lands (approximately 120,000 acres), building environmental infrastructures (storm sewers, septic systems, stream bank stabilization, and agricultural runoff controls), refitting all sewer plants with state-of-the-art microfiltration and nutrient removal technology, and creating eco-friendly jobs in the watershed. The agreement represented a historic transfer of wealth from city water consumers to fund economic vitality and environmental protection in the Catskill Mountains."

Coombe wasn't the only participant to change his mind about Bobby. Anthony Bucca, who was vice-chair of the Board of Supervisors in the Catskills town of Hunter, had disparagingly called Bobby "a Fifth Avenue environmentalist." "I regret that phrase now," Bucca said. "He's not the rich dilettante I made him out to be with that. We all got kind of caught up in our own hyperbole. I've got to give Bobby credit—with all the crises facing the city, he helped to get money out of them for us."

But it wasn't over yet; no agreement had publicly been announced. Bobby wasn't satisfied about the too-narrow protective buffer zones around the reservoirs and the new sewage discharges permitted in some parts of the system. Riverkeeper demanded a $100 million-plus water-quality monitoring fund and enforcement power against any violators of the accord. The state Department of Health as well as the watershed towns balked. On the eve of the signing ceremony, the city removed

Riverkeeper's clause granting citizens the right to enforce the agreement's key provisions in court. At which point, "we said we would not be at the next day's news conference."

After the situation appeared beyond salvaging at 9 p.m., a final draft got hammered out by all parties shortly after 3 a.m. Some of the participants, like Giuliani's chief deputy Peter Powers, were called in the middle of the night to an emergency meeting in the governor's office. Some last-minute compromises had to be made. But when midmorning came, Bobby joined Governor Pataki, Mayor Giuliani, and the EPA Regional Administrator as speakers unveiling the watershed agreement, whose 1,500 pages would take until January 1997 to bring to finalization.

Bobby and Cronin represented the environmental signatories. Cronin still marveled in 2022: "You have to step back and take a look at existing policies and law to understand what a remarkable achievement this was. There is no place in the Safe Drinking Water Act and all its regulations, or in EPA practice, that recognized what we did with the agreement. It was completely outside the law—not illegal, but outside statutory and regulatory law. It was an ad hoc effort that turned into a legally binding document. And while nothing was ostensibly sanctioned by statute, the watershed agreement set up a whole new set of practices and rules that the EPA ended up adopting."

Michael Finnegan wrote immediately thereafter: "Every once in a while something happens to demonstrate the powers of firm political leadership coupled with the goodwill of ordinary citizens." The governor's top aide heralded the "unorthodox process" as representing "a model approach to environmental dispute resolution."

When all was said and done, Finnegan also took Bobby aside. "If you ever run for governor," he said, "let me know and I'll switch parties."

As for his farmer friend Dick Coombe, besides running the Watershed Agricultural Council for a decade, in 1992 he was appointed by President George H. W. Bush to run the USDA's Natural Resources Conservation Service for the ensuing five years. In that capacity, Coombe traveled the globe, attending a world summit in Johannesburg. "Over the years, Bobby called me off and on asking me questions about agriculture," Coombe says. In 2012, his son having taken over the farming, Coombe opened a museum commemorating the watershed effort in his hometown.

CHAPTER SEVENTEEN

RIVERKEEPER MEETS THE STYX

The year 1997 began auspiciously, and celebratorily, with the finalization of the landmark New York watershed agreement. In its wake, Riverkeeper hosted its first annual fundraising dinner dance, which within two years would bring in almost $800,000 (nearly half of Riverkeeper's total budget). In September, Bobby and Cronin's coauthored book *The Riverkeepers* made it onto the *New York Times Best-Seller List.* The two headed "all over the country hawking our book." In November, Bobby and Mary's second child was born "in the midst of a fierce three-day storm with rain and windstorms that knocked out power across the country." They named the baby boy Finbar, which would become Finn for short.

At the same time, Bobby remained the watershed watchdog. On November 14, 1998, the *New York Times* published his op-ed headlined: "Why New York's Water Supply Still Isn't Safe." The city's own engineers at the DEP, Bobby warned, were ignoring the watershed pact. Mayor Giuliani's promises to reform the agency had turned hollow. "Instead, the department seems daily to be devising new concessions to spur development in the watershed," Bobby wrote, citing several examples and concluding that "development is gobbling up the open land of Putnam and Westchester Counties, raising taxes and destroying the drinking water. Eleven of the city's 19 reservoirs have dangerously high levels of nutrients because of overfertilization associated with careless development." He concluded with cautionary words that a top official with the EPA had recently told him. "You can sign an agreement and get all the favorable publicity, but if your hearts are not in it, you can just chip it away piece by piece."

On November 16, 1998, at a meeting with the Gannett Editorial Board, Bobby withdrew his name from the senate race to replace the retiring Daniel Moynihan, which many had been urging him to go for. "I want to watch my children grow and I decided to announce early rather than fueling speculation," he wrote. In another journal entry a month later, Bobby wrote: "Maura Moynihan called today to express sadness that I was not running for her dad's seat, which she said her family wished they could hand to me. It's too bad, and many days I regret my decision and the finality of it. The reasons I gave are valid and true but boxed me out of running for many years."

In 1999, Hillary Clinton called Bobby to tell him she wouldn't run for Moynihan's seat if he might still plan to do so. "I thought that was very gracious of her and told her I wasn't," Bobby recalls. That June Hillary announced her candidacy, and she and President Clinton purchased a home in Chappaqua, just down the road from Bobby.

Then tragedy struck. On July 16, 1999, John F. Kennedy Jr. was flying a single-engine plane with his wife, Carolyn Bessette, and her sister from New Jersey to Martha's Vineyard, to attend Bobby's sister Rory's wedding. Bobby and Mary were waiting to meet them on the island. "It was a foggy night," Bobby later wrote, "and when John's housekeeper notified me that the plane was an hour late, I knew that John was dead. All night the Shakespeare line revolved in my head on a loop: 'And to the gods, we are as flies to wanton boys; they kill us for sport.' I felt shattered, unable to move—a weight compressed my chest preventing me from breathing. My feet could not move. Even that despair was made bearable by my faith and the warm embrace of my family and my church. The repetitive rituals, so familiar now, wake and funeral, love, laughter and faith which grounds the strength to carry on."

Bobby told the *New York Times* (November 10, 1999) that his decision not to pursue the New York senate seat "was for this time around" and "if I wanted to run for something, there's always something to run for."

Late that year, the *Times* headlined in its Metro section: "Robert Kennedy Says Mayor Plays Politics with Water." Riverkeeper released a new report cataloguing existing concerns, with Bobby charting that the city had allowed developers to build new septic systems on steep slopes around reservoirs risking contamination. Mayor Giuliani was planning his own run for Moynihan's seat as the Republican Party's leading candidate.

Bobby met privately with Hillary at his Pace University office. "She asked me to cut an ad for her because she was being tagged as a carpetbagger. I agreed and reminded people that my father was also tagged as a carpetbagger when he ran. Her campaign told me that had been their most impactful ad." In May 2000, Giuliani withdrew from consideration after receiving a diagnosis of prostate cancer and the announcement that he and his wife were separating. Hillary went on to win handily in November 2000.

* * *

As a boy, like his father before him, Bobby became an avid reader of Greek literature. It might certainly be said that, as he grew older, his life took on many aspects of the ancient epics: "royal blood" with a wide variety of siblings, addiction and redemption, ongoing near-death experiences in the face of powerful forces, and in future years, what many would see as a transition from hero to outcast. The falcons Bobby trained to land on his arm were, in a mythological sense, emblematic. Falcons had been worshipped since earliest times as cosmic deities; the Egyptian god Horus is depicted as a falcon wearing a cobra-headed crown. The falcon came to represent the fight against our fears as well as being a metaphor for clear vision.

In Greek mythology, the river Styx formed the boundary crossing between Earth and the Underworld. Some versions describe the Styx's miraculous powers, making invulnerable someone who bathed in its waters. This was said to have happened to Achilles after his mother dipped him in the river as a boy. As he came of age, two of Achilles's mentors met untimely deaths: Achilles accidentally killed Chiron with a poison arrow and murdered his music teacher after Linus corrected him on his studies. During the Trojan War, Achilles would be struck and killed by an arrow shot into his heel—the appendage by which his mother had held him at the river Styx. Thus, our modern expression "Achilles' heel" is a metaphor for one's vulnerable spot.

So there was something archetypal about the painful parting of ways between Bob Boyle and his two proteges, with the pride of the Riverkeeper's patriarch proving to be his Achilles' heel. At this point, Riverkeeper had filed more than a hundred successful suits against polluters resulting in more than a billion dollars in fines—and resurrecting the once-comatose Hudson River as Bobby's father had pledged to do before his life was cut short. The Pace Environmental Law Clinic was providing over a million dollars annually in pro bono legal services for Riverkeeper. Bobby enlisted celebrity friends like Jerry Garcia, whose Grateful Dead had been his favorite band as a teenager, to do a painting for Riverkeeper, then made into a T-shirt that the rock group sold on its tours in the nineties and raised around $400,000 in one year alone. Promoting *The Riverkeepers* book, Bobby made about 150 speeches a year encouraging his audiences to get involved, and the movement grew phenomenally. Beyond New York, thirty-eight River, Bay, and Sound Keepers sprang up across the country, in Canada, and in Costa Rica, adopting the Hudson River model.

At a ceremony at the Garrison headquarters honoring Riverkeeper's founder and President Robert H. Boyle, Bobby described his great admiration for the man who brought him onboard and gave him purpose and support in the mid-1980s. If he had to be stuck on a desert island with one person, other than his wife, he said he would choose Boyle. But Boyle was having second thoughts about their time-honored relationship.

Disputes threatened their relationship over the ensuing year. "In memos, Kennedy kept demanding more 'democracy and openness,'" according to *Talk Magazine*. A much more widely read periodical, *Time Magazine*, named Bobby and John Cronin as "Heroes for the Planet" in August 1999. Writer Roger Rosenblatt described the two as "serious and good-humored men in their late 40s who look like kids, think like politicians and talk like poets," who had formed a "partnership based on vigilance and the law." That notoriety, specifically for Bobby and Cronin, could have been responsible for some of the strain with Boyle and others.

Around the same time, actor Alec Baldwin, an advisory board member at Riverkeeper, wrote to Boyle and the organization's treasurer, John

Fry. "I have kept tabs through other people on the personality-related clashes between various individuals in the organization and I want all of you to understand clearly that I don't believe Kennedy is responsible for any of this. I think some people to whom I'm sending this letter may be responsible for this and want to pin it on [him]. . . . If you allow whatever valid disagreements people have with each other to become the fodder for driving Kennedy out of the organization, you are making an irreversible mistake." In the end, Boyle and several other board members resigned.

As the years went by, Bobby never uttered an ill word about Boyle. Indeed, he continued to regard him as one of his primary environmental heroes. Once, when asked to define the situation by a reporter who'd covered Riverkeeper for years, Bobby replied: "Part of Bob's charm is that he is a curmudgeon and it was part of his strength. He has become more charming and stronger with age."

Could it have gone another way? Without Boyle's impetus, there would never have been a cleaned-up Hudson. His book, *The Hudson River*, published in 1965, had been the Bible of the movement until *The Riverkeepers* thirty-two years later. Boyle had helped mold two leaders from the next generation who turned the organization into a national movement. They'd fought together against hostile forces until, some believed, Boyle's jealousy over having to pass the torch overcame him. He died in 2017 at the age of eighty-eight.

Riverkeeper continued and vastly expanded Boyle's mission. Among other things, the local Hudson group provided information to FBI and EPA criminal investigations about the city DEP's mishandling of mercury in the water supply system, leading to guilty pleas for felony violations. "My vision is the same as that of Bob Boyle," Bobby said, "that we have a riverkeeper on every major waterway across the globe and every minor one as well—at least here in North America. That we have patrol boats on every one of those water bodies—fighting for local communities and for the rights of those communities to use and enjoy their local waterways."

Indeed, this was far from the end of Riverkeeper's impact. Case in point: In 2001, on behalf of the Pensacola Coastkeepers, Florida attorney Mike Papantonio filed with Bobby two suits against ConocoPhillips and its predecessor, Agrico Chemical Company. Dozens of Pensacola residents had found their properties contaminated by a toxic plume spreading from Agrico's fertilizer plant, which had closed in 1975 and been added to the EPA's Superfund list in 1989. In April 2004, ConocoPhillips—the third largest integrated oil and gas company in the US with a net income of $4.7 billion—agreed to a $70 million settlement with Papantonio and Riverkeeper. The money would be divided in varying amounts among a maximum of seven thousand people who had owned the polluted land. Today, there are more than 350 riverkeepers under the umbrella of the Waterkeeper Alliance around the world.

CHAPTER EIGHTEEN

UP AGAINST THE NAVY
IN PUERTO RICO

*As I sit by the sink in my cell, I've resolved to use the final week of my
confinement to consider, with my contraband ballpoint, the events that
led me to spend my summer vacation in a Puerto Rican prison.*
                                        —Robert F. Kennedy Jr., July 2001

The saga of Bobby's incarceration for an act of civil disobedience has
its origin a year earlier, when Dennis Rivera called him with an urgent
request. The two men had become good friends after working together on
the New York watershed agreement. Rivera joined the board of Bobby's
new Waterkeeper Alliance, and the health care workers' union that he led
launched and funded the Puerto Rican Coastkeeper.

Born in the mountains of Puerto Rico, a US territory subject to our
laws, Rivera was an American citizen by birth but without voting repre-
sentation in Congress. He'd been involved since his early 1970s college
days in local campaigns aimed at stopping the US Navy from conducting
training exercises and shelling two small islands. The first demonstrations
successfully prevented the planned evacuation of Culebra, when President
Nixon ordered a halt to further bombing, and relocation of the US base.

So the Navy then turned full attention to the larger nearby island
of Vieques. Some 9,300 people lived there, predominantly of African

and Taino Indian heritage and thus looked upon by many other Puerto Ricans as second- or third-class citizens. "That's why the US Navy had put its installation there," according to Bobby. "It would have been much more strategic for them to build it on one of the barrier islands off North Carolina, but those people had Congressmen." So, dating back to the start of World War II, the Navy had held live target practice on Vieques and rained down thousands of pounds of explosive ordnance—quantities that eventually exceeded the power of the Hiroshima bomb, with residents sandwiched between the bombing area and munitions depots.

Resistance jelled in 1999 after a fighter jet's missed target killed a Viequan civilian security guard. In response, nearly 70 percent of the island's residents voted to ask the Navy to stop the bombing and immediately disengage from Vieques. A month later, hundreds of civilians and religious and political leaders moved into the impact zone, constructing an encampment from plywood and canvas, complete with a church. The Navy temporarily stopped bombing, but it still owned two-thirds of the twenty-by-five-mile island, calling this the Atlantic Fleet's most important training ground.

Bobby agreed to accompany Rivera and see the island firsthand in April of 2000. His first impression was not unlike the landscapes he'd encountered in Latin America. "Vieques has the sleepy, almost timeless ambiance of a Gabriel García Márquez novel," Bobby later wrote. "Broad expanses of unspoiled tropical hardwood forests carpet the island, which is home to fourteen threatened and endangered species. Ranchers drive cattle across scrub savannas, while droves of sea turtles nest on some of the most beautiful white sand beaches in the Caribbean. Vieques's coast is spangled with brown pelican rookeries, vibrant coral reefs, and mangrove estuaries where rare Antillean manatees calve. Mosquito Bay, just east of the town of Esperanza, is one of the brightest bioluminescent bays on Earth. On moonless nights, you can read a book by its phosphorescent light."

The other reality loomed as an enemy within. Bobby would recall how he and Rivera "visited protesters in their makeshift village in the impact zone, and a group of fishermen took me across the naval embargo line so I could scuba dive to inspect the reefs. The coral had been shattered by bombs and crushed beneath a graveyard of sunken decoy ships, and the reef bristled with unexploded bombs. I found evidence of clear civil and criminal violations of three federal environmental statutes—the Clean Water Act, the Resource Conservation and Recovery Act, and the Endangered Species Act—and agreed to represent a group of Vieques fishing and community organizations in their legal fight against the Navy."

When he departed Vieques, Bobby told the Associated Press: "We've got to get the Navy out of here." If such activities were carried out on the US mainland, he said, "there would be a revolution." He and Rivera then developed a plan. "Bobby and Pace University would do the research and technical work," Rivera remembers, "and I would raise the money to pay for this. We hired people and went to work,"

Their investigation quickly revealed that the US Navy's activities had had a devastating effect on the island's inhabitants and their environment. Various government studies showed that Viequans suffered the highest infant mortality rate, the highest cancer rate, and highest overall mortality rate in Puerto Rico; medical tests on residents also indicated dangerous levels of contaminants in their blood including cadmium, arsenic, mercury, lead, and uranium, all associated with the detonation of naval ordnance. Many Viequans also suffered from vibroacoustic disease, a potentially lethal thickening of the membrane around the heart caused by persistent exposure to sonic booms.

While this deadly impact on human health was upsetting, Bobby felt the best legal case involved the Navy's violations of the Endangered Species Act. The ESA requires that a federal agency, including the military, complete a biological assessment before disturbing an area where endangered species are known to exist. There were many of these in the Vieques naval maneuver area, including four species of turtles and four species of whales. One of the largest rookeries of endangered brown pelicans in the Caribbean occupied a rocky outcropping off the southern coast of the impact zone. Yet the Navy had never completed a single biological assessment, despite being ordered to do so by the US Fish and Wildlife Service and National Marine Fisheries Service. Vieques's appeared to be an airtight case to enjoin continuing activities until such assessment was performed.

In August 2000, Bobby filed actions against the Navy on behalf of the NRDC and two dozen other groups for violating environmental law and

Viequan civil rights. Two months later came a motion in a San Juan federal court requesting a preliminary injunction to cease further bombing until a trial could be held. But any hopes for a quick victory were dashed when the case was assigned to Chief US District Court Judge Hector Lafitte, who had already been hard on protesters and extremely hostile to any legal challenges to naval activities in Puerto Rico. Bobby recalled: "Judge Lafitte sat on our injunction request for eight months and did nothing. I filed a series of motions asking the judge to act, which he ignored. I filed a motion to the Court of Appeals in Boston, which ordered Lafitte to act, an order he also ignored."

On the political front, the fateful presidential election of 2000 was approaching. Bobby had already thrust himself more into the political arena than he'd been willing to do in the past. At a fundraising lunch in October 1999, he'd introduced Democratic candidate Al Gore, as "our last best chance to save the planet." But inexplicably, despite his stellar record on environment, Gore largely ignored the issue during the campaign. "That was a great disappointment to me," Bobby said later. "I urged him to do it. And I believe he would have become president if he had."

An even more disappointing moment would transpire before the Clinton administration left office. Both Bobby and Dennis Rivera had seemingly strong relationships with the president and conducted negotiations with him around the Vieques situation. "Dennis had pulled out the stops to get him elected," Bobby told me. "He was the most powerful union leader in America, with a polling operation that could bring out twenty thousand people." And after Bobby helped Hillary in her successful Senate campaign, "she gave me full access to her office and helped me on a number of environmental issues." So why not this one?

In December 2000, Bobby recorded in his journal spending an hour on the phone with the outgoing president, "begging him to stop the bombing, which he could do with an executive order. He decided at the end not to do it. What the hell, he could have done something really good for the people of Puerto Rico and instead he used the last hours of his presidency to pardon Marc Rich and a bunch of other scoundrels. That left us very disillusioned." Rich, a big donor to the Clinton library and Hillary's Senate campaign, had fled the US to evade $48 million in taxes when Clinton pardoned him for tax evasion.

For this book, Bobby reflected; "We were pleading the case of the most vulnerable and alienated people in Puerto Rico, and he refused to act on it. I suppose he was calculating that those people couldn't help him after he left office. But it seemed inexplicable, and, for the first time, I had an uneasy feeling about the Clintons."

Not surprisingly, the incoming George W. Bush administration ruled that the Navy could resume full-scale bombing. In April 2001, Rivera and New York Governor Pataki toured Vieques by helicopter with the American naval commander. It was simply impossible, the commander said, to simulate all the necessary components of classic amphibious assault anywhere but here. "We agreed to disagree," Rivera remembers. "And Bobby said: 'We have only one way to go, and that's civil disobedience.'"

Bobby later reflected: "I was faced with a difficult decision. Federal law notwithstanding, the Viequans were convinced that naval power could trump every legal and political institution that was relevant to their daily lives. They had rights, and I had pursued them, without achieving results. I hadn't even been able to get them their day in court. Once again, the law had failed them. With my hands tied and the Navy poised to open fire, I felt all I could offer my clients was to join them. So I decided to venture into the impact zone and join the protesters trying to force the Navy to stop its exercises."

* * *

Bobby reached out to his actor friend Edward James Olmos, the first American Hispanic to receive an Academy Award nomination (for the film *Stand and Deliver*) and at the time the star of TV's popular *Miami Vice* series. Olmos had met then-teenage Bobby in 1968, when he went on tour with Robert Kennedy's presidential campaign as the singer for the rock band Pacific Ocean (scheduled to play at the victory party the night the senator was assassinated). Living along the Hudson River in recent years, Olmos became a spokesman and documentary chronicler for Riverkeeper. Well known as an activist within the Hispanic community, he'd often been to Puerto Rico, knew Dennis Rivera, and opposed the Navy's assault on Vieques.

"Bobby called me in desperation to say he needed my help," Olmos remembers. "I told him I was in Argentina, marching with the mothers of the 'disappeared ones.'" Bobby remembers telling his friend: "I was calling to invite you to get arrested on Vieques this weekend, but you're off the hook. But Eddie insisted on coming. When he got to San Juan, he'd had only three hours sleep in the past thirty-six."

Bobby and Rivera flew together to Puerto Rico, where they rendez-voused with an exhausted Olmos who'd taken four flights to make it in time. They were supposed to receive a message from the fishermen some-time after midnight that the coast was clear. "When the call came, it was to tell us that we'd have to wait till daylight," Bobby remembers. "Somebody must have tipped the police/Coast Guard because there were twenty-plus patrol boats circling at the mouth of the harbor like hungry hammerheads."

As the sun rose on April 28, 2001, Bobby stood on the patio of the Casa Cielo, a small hotel overlooking a lush green valley unfolding down to the tiny port of Esperanza. Two miles away he could see a dozen fish-ing boats racing out of the harbor to try to divert the US Navy and Coast Guard vessels that patrol the waters around the live impact area of the naval bombing range on the east end of the island. "Those are our decoys," Wilda Rodriguez, a leading Puerto Rican activist, explained to him.

About ten miles to the southeast, Bobby glimpsed a group of dark silhouettes—cruisers and destroyers from the Navy's USS *Enterprise* bat-tle group preparing to begin the day's exercises. Closer to shore, a Coast Guard cutter blocked access to the firing range from Esperanza's harbor. As the fishermen's flotilla approached, a pair of Zodiac pursuit craft low-ered from the cutter and gave chase. The decoy action was under way.

Bobby and his companions jumped into a waiting van and sped down-hill toward the harbor. Olmos was amazed to see hundreds of people cheer-ing the flotilla as friendly police guided them through the crowd. They clambered aboard an iron-bottomed thirteen-foot *panga* with a fresh coat of marine paint that obscured the registration numbers. The local fisher-men at the consoles, as well as two young teenagers, "had purple shirts over their heads with eyeholes like Hamas" out of which came their stares.

As Bobby recounted in an article for *Outside Magazine* (October 2001): "Outside the harbor, the waves were high enough that we had to grip the bowlines to keep our feet. We were spending as much time in the air as the flying fish we watched in huge schools skirting the water on every side. Between bone-jarring bumps, I could hear the naked screw whining as the prop went airborne. About two miles out, we saw the decoy fishing fleet running back toward the harbor, hounded by the Coast Guard Zodiacs, but when they realized what we were up to, the speedy inflatable boats peeled away to intercept us. The masked man at the wheel of our panga buried the throttle, and we began a ten-mile race to the live fire zone. When the Zodiacs pulled in front of us, we skirted them and raced onward. In each Zodiac a heavily armed team wearing flak jackets and helmets pointed their rifles at us and shouted fiercely for us to heave to and allow them to board. Instead we went faster, skipping across water like the flying fish."

Olmos recalls Bobby "standing up looking like George Washington" as their captain twirled a rag above his head, signaling the decoy boats to turn around. He soon marveled at one of the boys emerging from a crawl space to sneakily fill their gas tank and give a thumbs-up, as the Zodiacs closed in yet again and their panga veered toward shore and crossed a line of breakers. Ahead of them lay a shallow boneyard of rocks and coral heads. Olmos remembers: "I'm going, holy shit!, because we're maneuvering right into the reef where their rubber ducky boats couldn't follow. If they hit the coral, they'd explode." As the pursuers pulled up and stopped, only their lone small craft made it through the uncharted hazards into the opening of the bay.

Bobby wrote: "A Navy SH-3 Sea King helicopter took up the chase, however, swooping low in hopes of turning us, and watching to see where we would land. A half-dozen Navy spotters with high-power binoculars also followed our movements from the military's hilltop observation post, perched on the edge of the live impact zone. As our panga approached the beach, we scrambled onto the bow."

Once onshore, Olmos remembers asking what the game plan was. He says Bobby and Rivera were now ready to turn themselves in. Bobby wrote that Olmos presented a different approach, turning to him and Rivera saying: "Let's try to hide from them, and see if we can stop the bombing for one day." They would need to spread out.

According to Bobby, with the helicopter hovering, "we sprinted across the beach to a muddy, crater-pocked moonscape that had once been a mangrove estuary. Dodging discarded equipment, twisted metal, unexploded shells, while breathing dust from the chopper's downdraft, we scurried up a ridgeline, figuring that once we reached the other side we would at least be out of sight of the observation post. Then we took off in different directions. The naval police, we knew, were already on their way."

What ensued remained vivid when I spoke to Olmos in 2022. He first sat down atop a tank, on which the Navy's advanced bombs had left behind holes larger than a foot in diameter. In the distance he could see a line of trucks heading down a road in their direction. As long as the trio "occupied" the island, no bombing could take place. Olmos calculated he was about two miles from the observation tower. He counted out the ninety-or-so seconds between the helicopter looking for his companions and circling back to where he'd parked himself. Then he made his move. He jumped off the tank, fell to the ground, and began to crawl. He worked his way about twenty-five feet down a cliff, dug a space under some bushes, and went into hiding as the chopper soared over again. Stripping off his white T-shirt, he quickly fell asleep.

Bobby described his own escape: "I hid under a partially demolished decoy half-track. As I lay there, considering the possibility that I might take a direct hit if bombing resumed, I noticed daylight gleaming through the clean, telltale holes made by depleted-uranium bullets. I decided to hunt for a less radioactive hiding place. With the helicopter shadowing me, I hiked southeast, found a few acres of brush crowding a mangrove swamp, crawled into the thickets, squeezed under a mangrove root, and covered myself with branches and twigs. For two hours the helicopter hovered nearby, trying to flush me out.

"Then I heard soldiers cursing the thorns as they searched for me, some of them passing just a few feet away. Finally, a soldier with a dog spotted me and blew a high-pitched whistle and the sound surprised me and I came out. Petty Officer Larry Roberts handcuffed me with zip-tie Flex Cuffs and walked me through the swamp to a military access road. There I saw Dennis, handcuffed and squatting in the hot sun behind a camouflaged truck."

Olmos was still at large, getting some much-needed rest. At one point he heard a voice nearby telling someone via a walkie-talkie: "There's no one here, we cannot find him." He dozed off again, waking up as it was getting dark. The helicopter and the trucks had vanished. He looked at his watch. It had been ten hours since they hit the island. Because some of the shells were depleted uranium, he wondered if he might be lying in a radioactive area. Or if his adversaries might think he'd left the island and commence some nighttime target practice.

So Olmos emerged and put himself in the line of sight from the observation tower. Then he sat down again on an abandoned toilet seat. About an hour later, a Navy patrol approached him. "Damn, it *is* you!" he remembers the leader exclaiming. "Lieutenant Castillo from *Miami Vice*! Can I take a photograph?" (It turned out Olmos's show aired five times a week in Puerto Rico.) After which the actor was handcuffed and placed in a truck.

Bobby and Rivera had already been taken to the naval compound at Camp Garcia, locked in dog cages for several hours to await the capture of their partner. Otherwise, they hadn't been mistreated, and, once Olmos arrived, disposable cameras materialized and many of the soldiers had a photo taken with them in their cuffs. It was an odd feeling, Bobby would recall. These weren't fans snapping a shot with a Kennedy, but hunters posing with their trophies. He smiled for the camera.

Their guards then searched them, took their shoes, socks, and belts, and marched them toward a chain-link enclosure containing several dozen protesters. As they approached, the prisoners cheered and chanted, "¡Kennedy! ¡Olmos! ¡Vieques, *sí! ¡Marina, no!*" According to Bobby, a guard ordered them to shut up, and when they didn't, soldiers opened up with pepper spray. Prisoners writhed on the ground in their handcuffs, fighting to escape the agonizing burns, pissing themselves, and one man crapped his pants. Then Bobby, Olmos, and Rivera were thrown in among them.

Later that day, protest-weary MPs would beat other prisoners, including US congressman Luis Gutiérrez. By then, the three "celebrity prisoners" had finally been provided some water and a sandwich. Their hands still cuffed behind them, they were loaded onto a barge full of cargo containers. Crammed against the sides, they were soaked by the rough seas—"if something were to happen, we'd be dead," Olmos remembers—on the long wet ride to Roosevelt Roads Naval Base on the Puerto Rican mainland.

During the boat trip, Petty Officer Roberts, the sailor who captured Bobby, told him that he aspired to become a Navy SEAL. He asked about President Kennedy, knowing of his commitment to Special Forces and the SEALs. Looking westward from their slow-moving transport, Bobby could see El Yunque, the mountainous Puerto Rican rain forest that had once been home to an American Special Ops jungle warfare school that he had visited when his uncle was in the White House.

Bobby later wrote: "Listening to Roberts, I experienced a familiar sting of ambiguity about opposing the military service that was such an important icon of my childhood. But I was also thinking that those who love an institution most should be the first to criticize it when it does wrong. Every nation has a right to ask its citizens to sacrifice their lives during time of war, but on Vieques, the Navy was endangering the lives of children, women, and men without their consent, all for the sake of a dubious military exercise."

Reaching the holding tank at Guaynabo, Bobby recalls, "Eddie, Dennis, and I were ordered to stand abreast, facing three prison guards. They told us to strip naked, to lift our private parts for visual inspection, and then to turn, bend, and spread our cheeks. It was in this position that Eddie seemed to experience second thoughts about having answered my call. He turned to Dennis and me and said, 'Hey guys, lose my number.'" Olmos recalls Bobby almost keeling over with laughter.

They were to be tried separately; Olmos expected they'd get a slap on the wrist and be released. But two days later, a magistrate freed each of the men on $3,000 bail pending trial. When Bobby arrived home, the reaction of his friends and family was generally supportive. "My mother counts among her close friends enough admirals to sail the Atlantic Fleet, but she was nevertheless as proud of my civil disobedience as if I'd been elected to the Senate."

At a fundraiser for the RFK Memorial on Hickory Hill hosted by the cast of the *West Wing* TV series, Bobby smiled greeting Secretary of Defense Donald Rumsfeld. "Thanks for putting me up in Puerto Rico last weekend," he said. Rumsfeld replied with a laugh: "I heard you couldn't find a hotel down there, so I thought I'd help you out," Ethel Kennedy, sitting beside the secretary during dinner, goaded him in earnest good humor to abandon Vieques.

Shortly after returning home, Bobby was briefly hospitalized with an irregular heartbeat, diagnosed as atrial fibrillation. In early June, he and Rivera were notified that their trial date would be July 6. Bobby's brother-in-law Andrew Cuomo helped persuade his father, Mario Cuomo, to act as cocounsel with their Puerto Rican lawyer. The former New York governor also set to work lobbying the Bush White House to end the bombing, pointing out the flaws in the Navy's argument that Vieques was a military necessity. On June 14, Bush announced a compromise—the Navy would stop using Vieques as a target area by May 2003, a solution that pleased neither the Navy nor its opponents who demanded bombing stop immediately.

Three days before the trial, Mario Cuomo called Rivera and Bobby to inform them that he'd worked out a deal to delay sentencing, so Bobby could be with his wife, Mary, when she had their new baby due toward mid-July. The catch was that they had to plead guilty and waive their right to appeal. According to Cuomo, Judge Lafitte favored the deal and had indicated that an agreement to sign on might incline him toward a more lenient sentence. Bobby asked Mary what she thought. "Don't take any deal," she told him. Rivera agreed.

So Bobby flew to San Juan on July 5. At the Caribe Hilton, Reverend Jesse Jackson signaled him from a beach chair. His wife, Jacqueline, had just been released from ten days in jail for protesting at Camp Garcia. Jackson had returned to Vieques with five congresspeople to attend the trial. "Suffering is often the most powerful tool against injustice and oppression," Bobby recalled Jackson telling him. "If Jesus had plea-bargained the crucifixion, we wouldn't have the faith."

Bobby's defense was based on the doctrine of necessity, that a defendant can't be convicted of trespassing if he shows that he entered the land to prevent a greater crime from being committed. "We intended to prove that we had engaged in civil disobedience for a single purpose: to prevent a criminal violation of the Endangered Species Act by the Navy that the federal court had refused to redress."

But Judge Lafitte didn't wish to hear such evidence, responding that he was "not going to allow political views, philosophical views, none of that." What the *New York Times* described as "caustic exchanges" went back and forth between the judge and Bobby's defense lawyers. After the

Navy presented its case, Lafitte announced that he found the defendants guilty and would allow statements prior to sentencing. The trial lasted seven hours.

Mario Cuomo delivered a closing statement that the *Times* wrote "mesmerized the courtroom audience with his characteristic eloquence." The former governor invoked the celebration of Independence Day and continued: "We ask the court to recall that this nation was conceived in the civil disobedience that preceded the Revolutionary War, the acts of civil disobedience that were precipitated by the Fugitive Slave Act of 1793, in the famous Sit-down Strikes of 1936 and 1937, all through the valiant struggle for civil rights in the 1960s and the movement against the Vietnam War. Always they were treated by the courts one way: not like crimes committed for personal gain or out of pure malice, but as technical violations, designed to achieve a good purpose."

After pointing out that President Bush had recently shifted policy and ordered the bombing tests stopped by 2003, Cuomo went on: "It is clear that the die is cast, in significant part because of the role of Mr. Kennedy, Mr. Rivera, and many others who chose to participate in acts of civil disobedience. Because there appears no urgent need to deter them or others from the kind of massive protest that occurred in the past, we ask that the defendants, having been found guilty of civil disobedience, be sentenced to time served."

Then Judge Lafitte, following a fellow judge's having meted out a 90-day sentence in May to Reverend Al Sharpton for similar action, turned to Bobby and said: "It hurts me to sentence you, but I have to promote respect for the law." Bobby received thirty days for trespassing on Navy property. He smiled and wished "good luck" to Dennis Rivera, who was on his way to the bench for sentencing. "See you soon," Rivera replied. They were, indeed, about to become cell partners for a time inside a maximum-security prison in Guaynabo. Olmos, separately sentenced, would serve twenty days on a different cellblock.

Bobby recalled: "That night we arrived on our cellblock, where roughly half of the prisoners had been arrested on Vieques, cheering *disobedientes* and treating us to an ovation of clapping and congas. Many of the more than seven hundred Vieques protesters imprisoned at Guaynabo suffered severely for acting according to their consciences. They lost their jobs, and some of their families were going hungry. By comparison, this was a vacation for me, although perhaps not one I would have chosen under ordinary circumstances. The food was tolerable and accommodations simple, if tight; but we were strip-searched whenever we left the cellblock, and guards regularly searched our rooms for contraband.

"At night, however, we could see through a tiny darkened window that crowds had gathered outside the prison to cheer us and wave banners of support. A strong sense of community infected the entire cellblock, including those who were there for rape, drug smuggling, gang activities, and murder. We played basketball and dominoes, engaged in heated discussions during communal meals, and every night the prisoners made popcorn and played merengue, bolero, and salsa on guitars and congas. As it happened, there were several well-known Caribbean musicians serving time at Guaynabo, so some nights the music was magical."

Not surprisingly, Bobby became the unofficial "leader" of the whole cellblock. When Ethel Kennedy came to visit, one of the guards walked over to her with tears in his eyes and apologized for her son's detainment. "He came to help us and our children and our children's children," the guard said. Other prominent visitors included Senator Hillary Clinton and the actor Benecio del Toro. Bobby's sister Kerry and his wife, Mary, also came to call, accompanied by Bobby's then-six children. They introduced Bobby to his two-week-old son, named Aidan Caohman (after two Irish saints) Vieques Kennedy. Dennis Rivera and Edward James Olmos were designated the newborn's cogodparents.

On July 29, three days before Bobby's release, almost 70 percent of the island's residents voted in a nonbinding referendum to force the Navy to stop the bombing and leave immediately. The Navy, predictably, stated that the vote had no bearing on its plans and that it would go ahead with more maneuvers on August 2. And so, freed from prison, almost immediately Bobby picked up his seven-year-old son Conor and returned to Vieques. He called the Navy's resumption "an exercise in bullying" and encouraged protesters to invade the range again.

It took almost another two years but ultimately, the *New York Times* headlined, "Navy Leaves a Battered Island, and Puerto Ricans cheer."

Bobby was quoted that the withdrawal was a mixed blessing. "The problem is, they're leaving the poison behind," he said. "There are tens of thousands of unexploded bombs. Fish are contaminated, crabs are contaminated, seagrass is contaminated. The soils are contaminated with toxins. The fact that they're leaving the island would be great, if they would only clean it up."

The Navy responded with a statement that the property had been transferred to the Department of the Interior, which "is required to develop the land for use as a wildlife refuge, with the area used for exercises with live bombs to be designated a wilderness area and closed to the public."

The 14,500-acre training range was later designated an EPA Superfund site believed to contain mercury, lead, napalm, depleted uranium, and other contaminants. An $800-million cleanup effort in Vieques and Culebra led by the Defense Department and the Army Corps of Engineers remains ongoing, scheduled for completion by 2032. According to s Government Accounting Office report in 2021, so far crews have removed munition including 32,000 bombs, 12,000 grenades, and 1,300 rockets from Vieques. "Substantial work remains," the report stated. "The Navy also faces challenges on Vieques with community distrust of the military handling cleanup efforts."

Looking back, Bobby stated: "I couldn't help but think that if the Navy had only done the right thing years ago, and committed itself to ameliorating the health, environmental, and economic effects of its maneuvers, it might have found the patriotic citizens of Vieques welcoming it as a good neighbor. My most poignant moment in prison was a reminder of that patriotism. One night a group of Viequans apologized to me tearfully after hearing that an American flag had been burned on the island. This was not why they had gone to prison, they hastened to assure me; they loved the United States but could not stand idly by while their beloved island was being wantonly destroyed. Like so many battles in history fought by great powers, the US Navy's fight to keep Vieques was lost through its own arrogance."

CHAPTER NINETEEN

JUSTICE FOR
MICHAEL SKAKEL

When Bobby first got sober in 1983, his twenty-three-year-old first cousin, Michael Skakel, had been regularly attending Alcoholics Anonymous for a year. "I was hard-wired to drink myself into oblivion every day," Skakel told Bobby when they met for the first time early that winter and soon began going to twelve-step meetings together. Bobby found himself strongly impressed by his cousin's honesty and generosity with other struggling addicts.

Until then, the two families had been like oil and water. Bobby's mother had long felt estranged from her Skakel clan. The Skakels lived in a sprawling mansion in Greenwich, Connecticut, with an eight-car garage and a ten-acre lawn surrounded by hundreds of acres of woodlands. As Bobby later wrote: "They had a distinctive family history with their own iconoclastic gestalt and greater wealth than the Kennedys. My mother's generation of Skakels prided themselves on being the anti-Kennedys. They were rough-and-ready carbon [oil and coal] Republicans with seasoned contempt for the nanny-state, regulating, soak-the-rich sort of government they imagined the Kennedys promoting. Among their crowd, any association with the Kennedys was a kind of social demotion."

The instant friendship between Bobby and the younger Michael ran in the face of all that. "We lived close by and loved to do a lot of the same things," Bobby recalls, "fishing, scuba diving, hunting with the hawks, and skiing. He was an athlete who was named to the US National Speed Ski team. We traveled together often with my wife [Emily] and our kids, who we taught how to ski at the Skakel's winter place in the Catskills. And we attended literally hundreds of [AA] meetings. In this context, and others, we shared our deepest feelings."

Michael had had a rough upbringing. His mother died of cancer when he was twelve, leaving Ethel Kennedy's brother Rushton to raise seven children with the help of numerous servants. "In a highly presentable family, he was never very attractive: dumpy, blowzy, red-faced," as one journalist described him. The runt of the clan, Michael became a target for his father's anger. He was dyslexic and with low self-esteem, a teenage alcoholic starting a year after his mother's death. His own family had compelled him to attend a reform school in Maine called Elan, where Michael was often subjected to hazing, to supposedly speed his recovery. Bobby, having gone through his own tormented youth, not only related to, but loved him.

It was in the early part of their relationship when Michael told Bobby about what had transpired on the terrible night of October 30, 1975. His school friend and next-door neighbor, fifteen-year-old Martha Moxley, had been bludgeoned to death with a golf club, her body found in a wooded area on the back side of the Skakel property late the next morning. During that time, Bobby had moved down to Alabama to research his senior thesis for Harvard and had never even heard about the tragedy. Now he pressed Michael for details.

Michael and several other teenage boys had been drinking beer, playing backgammon, and listening to eight-track tapes in his father's Lincoln Continental outside the Moxley mansion in the gated community of Belle Haven on Long Island Sound. When Martha, a pretty blonde-haired sophomore, walked over with some girlfriends out for harmless pre-Halloween pranks (they called it "Mischief Night"), Michael invited her to come along to a friend's back-country estate some eleven miles away. After she declined, citing her approaching 9:30 p.m. curfew, Michael and Martha made plans to go trick-or-treating the following night. With his three friends, he drove to the stone gothic fortress down the road, where they watched *Monty Python's Flying Circus* on TV, drank more beer, and smoked some pot before arriving back home between 11:30 and 11:45 p.m. At which point, though they wouldn't know it until the first round of police questioning the next afternoon, Martha Moxley was already dead.

There was an embarrassing part of Michael's recollection that he also shared with Bobby. He'd been quite inebriated upon returning home that night and didn't stay in his house long. After a typically loud departure overheard by his sister Julie, he'd ended up climbing a tree at the nearby Moxley house overlooking the front bedroom he guessed as being Martha's. There the teenager made a half-hearted attempt to masturbate before worrying that someone might spot him and scurrying back down the tree. His father had, after all, beaten him silly when he caught Michael in the same act six months earlier.

Michael told Bobby that, on the way home again, he sensed a presence in some dark bushes near the Moxleys' driveway. "Come out of there, and I'll kick your ass!" he yelled, throwing stones in that direction. He explained to Bobby in 1984: "I was always scared of the dark, and something that night made me scared shitless. I ran home from streetlight to streetlight." Finding his downstairs doors now bolted, he climbed through his bedroom window at about fifteen minutes after midnight. He'd been out again for forty minutes, and his sister, still awake, was surprised to hear him back so soon.

Michael wasn't alone in feeling something weird going on. Not long after he'd left the area with his buddies, Martha's mother, Dorothy Moxley, was painting in the master bedroom when she heard a loud "commotion" in the yard, on the side of the house where Martha's body was later discovered. That was somewhere between 9:30 and 10 p.m., she told police, a ruckus consisting of "excited voices," incessant barking, and what she thought were her daughter's screams. The racket was so strange that she ventured to the window to look outside. Unable to penetrate the darkness, she then turned on an outside porch light—but turned it off again after a few seconds fearing that whoever was there might see Martha's bike on the porch and steal it.

By the following midmorning, the entire community was searching for Martha. After her body was found at around 11:30 a.m., hordes of police, media, and curiosity-seekers descended. The last person seen with Martha on the night of the murder was Michael's seventeen-year-old brother, Thomas Skakel. Friends had seen her flirting with and then kissing Tommy before "falling together behind the fence" with him near the backyard pool at around 9:30 p.m. Greenwich police made him the prime initial suspect, along with a just-hired live-in tutor, twenty-four-year-old Kenneth Littleton. He'd begun working for the Skakel family only hours before the killing, gone drinking with his new charges earlier that night, and repeatedly changed his story about his whereabouts at the time of the murder. Other suspects included a live-in gardener with a well-known voyeuristic appetite for young girls, and even Martha's brother John, who'd already been accused of raping a neighborhood girl that same week.

Michael Skakel passed a polygraph and was never among the suspects.

No one was ever charged, and the case languished for almost two decades. In the early 1990s, the writer Dominick Dunne went to see Martha's mother. Dunne later recounted in *Vanity Fair*: "I had just written three bestselling novels in a row, and they had all been made into TV miniseries. I told Mrs. Moxley that I thought I could write another based loosely on her daughter's murder, since no facts were known publicly at the time, and it might turn a spotlight on the long-dormant case. She said she wasn't sure. Then I told her that I too was the parent of a murdered daughter. Our daughters had been born a year apart, and each was viciously attacked by a man she knew on October 30, although in different years. That moment marked the beginning of our friendship. She said okay, I could write the book."

In 1993, Dunne published a bestselling novel titled *A Season In Purgatory*. Its protagonist is the patriarch of a large, wealthy, and politically well-connected Irish Catholic family in Connecticut whose neighboring teenage daughter is bludgeoned to death with a baseball bat. A successful true crime writer comes forward two decades later to accuse the patriarch's son, then being groomed as a presidential candidate, and a nasty investigation ensues.

As far as Bobby was concerned, Dunne's roman à clef intentionally pointed a finger at John Kennedy Jr., a charge that the writer denied. Dunne did reveal that his interest had first been piqued when he covered the 1991 rape trial in which William Kennedy Smith was ultimately acquitted, "after a bogus courthouse rumor had it that Smith had been in the Skakel house in Greenwich on the night of the murder in 1975." On his national book tour, Dunne loudly blamed Tommy Skakel for the crime.

Whichever Kennedy family member Dunne had chosen to feast upon, CBS Evening News produced a long segment on how his book revived interest in the case. "Martha Moxley was soon back in the news," Dunne wrote, "and I was on television quite often talking about her murder. I learned that in certain houses in Greenwich the subject was being discussed again for the first time in years."

Michael continued to represent America on the World Cup Speed Skiing circuit; he ranked third in the US at age twenty-six. And then, seemingly out of nowhere, the roof began to fall in. In May 1996, Dunne's miniseries *A Season In Purgatory* aired on CBS. The writer escorted Mrs. Moxley to a press conference to announce that she was doubling the reward for information about her daughter's killer to $100,000. To various media outlets, Dunne allowed that Tommy Skakel ought to come clean about his involvement.

And Dunne passed along news to Mrs. Moxley about a potentially game-changing development. A young man working for a private detective agency had stolen and provided to Dunne a copy of the Sutton report, the result of a several-years-long investigation allegedly funded by Rushton Skakel in an effort to clear his son Tommy's name. Dunne swore Dorothy Moxley to secrecy, but it now appeared that a different Skakel scion had committed the terrible deed—and his name was Michael.

Mrs. Moxley couldn't contain herself and relayed this to detective Frank Garr, who had retired from the Greenwich police force to work full-time for the state attorney's office on the case, and who now asked Dunne to give him a copy of the report. "There was nothing in it," Garr told Leonard Levitt of *Long Island Newsday*. "It was all theories and speculation," a sentiment the detective also voiced to Mrs. Moxley.

The report suggested that Tommy may have helped his brother move the body, but "in all probability" Michael had been the perpetrator. The private detectives, who had signed confidentiality agreements, interviewed all seven children in the Skakel family. And Michael "confessed," for the first time, about the private sexual activity in which he'd engaged in the tree outside the Moxley home. He'd stayed silent about this earlier, for fear of retribution from his father. Since the tree's location was a football field away from the pine tree under which the girl's body was found, Michael's revelation was innocuous at best.

On October 3, 1996, out in Los Angeles, homicide detective Mark Fuhrman avoided going to jail by pleading no contest to a felony perjury charge during the O. J. Simpson trial. A known racist, Fuhrman had denied using the N-word to attorney F. Lee Bailey, which he'd actually uttered ten times. The court sentenced Fuhrman to three years of monitored probation, ending his law enforcement career.

But Fuhrman had made an influential friend in Dominick Dunne, who had been present to cover the 1995 O. J. proceeding. In 1997, Dunne read Fuhrman's subsequent book about the O. J. case, *Murder in Brentwood*, and found it compelling enough to meet with him in New York and share the Sutton report. Dunne then hosted a cocktail party for Fuhrman to introduce him to Connecticut law enforcement officials. According to Dunne, Fuhrman "immediately became interested in writing about the case . . . because I knew he would get the most attention on talk shows and in newspapers." Fuhrman used the unsubstantiated report as the basis for cranking out (with ghostwriting help) *Murder in Greenwich: Who Killed Martha Moxley?*, published in February 1998.

Promoted as revealing "explosive new information," the book named Michael as the likely killer. "There were all kinds of inaccuracies Fuhrman used to make Michael look guilty," Bobby realized. "The tree Michael climbed was not the tree the body was found under, which Fuhrman conflated. I knew from Michael what the real story was, and that Fuhrman had tried to fit it into a particular box."

Dunne wrote the foreword, shifting his earlier pronouncement of guilt from Tommy to Michael. Detective Garr soon followed suit by changing his focus from the Skakel's tutor to Michael, claiming to uncover key witnesses who'd been with Michael at the Elan drug treatment center after the murder.

Bobby would write in his 2016 book *Framed*: "Michael [Skakel]'s teenage ordeal at the brutal drug treatment program in Maine left him severely afflicted with post-traumatic stress disorder [PTSD]. The relentless public attacks naming Michael a murder suspect darkened his worldview and aggravated his PTSD. His paranoid suspicions about our family were a symptom. This paranoid impulse caused Michael to record a fateful series of interviews with author Richard Hoffman in 1998."

Michael had decided to publish a memoir, and Hoffman packaged a proposal tentatively titled *Dead Man Talking: A Kennedy Cousin Comes Clean.* The ghostwriter touted this as "the first account by an insider of the avarice, perversion, and gangsterism of America's royal family." According to Bobby, "Michael told Hoffman he believed the Kennedy family had a hand in his misfortune, including the growing clamor—triggered by Fuhrman—to make him the scapegoat in the Moxley murder."

In January 1999, Frank Garr illegally seized the book proposal, interview tapes, and personal items belonging to Michael from Hoffman's home in Cambridge, Massachusetts. Garr then leaked transcripts to the tabloids. (Garr had his own book deal in the offing.) Dominick Dunne was also privileged to receive a copy of the proposal, a second, toned-down version titled *The Obvious* being submitted to five publishers. The book would never see the light of day, but several sections ended up posted on the Internet. As we shall see, one of the interview tapes would be used by the prosecution in seeking to prove Michael's guilt.

Before long, Michael had come under grand jury investigation. On January 19, 2000, after a warrant was issued for his arrest, he surrendered to Greenwich Police and was released on $50,000 bail. That June, a three-day Probable Cause hearing took place in Stamford, Connecticut, to determine if he could be tried as an adult for a crime he allegedly committed as a minor. Bobby and his brother Douglas attended a morning session and sat with the Skakel family. Bobby afterward gave a strong statement to the media, saying: "It is a horrible, unspeakable tragedy, and it compounds it to blame Michael, who is innocent."

But Michael wouldn't speak to him, and none of Bobby's other siblings wanted anything more to do with their cousin. Nor did they like what Bobby was doing. This marked the second time he'd gone against the grain of what his family considered acceptable. The first had happened in 1993 around the Conoco oil deal in Ecuador, when he stood up for the local Indians against environmental groups that wanted the rain forest emptied of humans. "A lot of them had a 'look but don't touch' attitude, and I'd come out as a hunter and fisherman who believed that nature enriches us, that it's where we go to sense the divine."

In an article for *Vanity Fair*, Dunne wrote of observing Bobby give Michael "an in-full-view bear hug at the lunch break" during the hearing, adding that Bobby's "defense of his cousin surprised me." On May 7, 2002, testimony began in *State of Connecticut v. Michael Skakel*. It was a media circus, with fifty-five reporters and eighteen satellite trucks in attendance for six weeks. Mickey Sherman, the flamboyant attorney hired by the Skakel family to represent Michael, proved an unmitigated disaster. One month later and twenty-seven years after the event, a jury declared Michael guilty of the first-degree murder of Martha Moxley. The forty-one-year-old was sentenced to a term of twenty years to life in prison.

Bobby would write: "Michael had an airtight alibi—eleven miles away with five eyewitnesses when the murder was committed, which his lawyer failed to turn up. The State offered no physical or forensic evidence, no fingerprints or DNA, no eyewitness testimony linking Michael to the murder. Indeed, bungling police investigators had lost many items of physical evidence that might have exculpated Michael. . . . With no evidence linking Michael to the killing, the State tried him based on the perjured testimony of three confession witnesses suborned by a crooked and malevolent cop obsessed with winning his career case."

Those witnesses had all attended the Elan reform school in Maine, where rumors about Martha Moxley's death had followed Michael. He'd been compelled at one point to carry a large sign reading: "Ask Me How I Killed My Neighbor." The most incriminating "testimony" alleging a confession from Michael was read aloud by actors, because the fellow who'd come forward in response to Mrs. Moxley's $100,000 reward offer and claimed this to a grand jury had died of a heroin overdose before the trial. A shady, menacing drug addict was effectively replaced by a charming, credible surrogate to sway the jury.

Bobby had been in the gallery for the closing arguments on June 2. Asked some years later in a TV interview how such a verdict could possibly have been reached, he replied: "The prosecutor used a trick, which he pioneered at the time, a video that was very, very deceptive. He took tape recordings of Michael talking about what he'd been doing up in the tree when he was fourteen, saying, 'I hope nobody saw me do that.' The state played that comment over gruesome pictures of Martha's body, with flashing red blood dripping down the screen. They paid $60,000 to make it. Mrs. Moxley, who had never seen pictures of her daughter in that state—and had left the courthouse when the murder was discussed—was in the audience. She gasped and fell against her chair. That video was used to gloss over the fact that there was no evidence connecting him to the crime. By the time it was over, Michael was cooked."

Bobby sat down to work on a 14,000-word article for *The Atlantic*, headlined "A Miscarriage of Justice" when it appeared in the January/February 2003 issue. He opened by saying: "Until I recently visited [Michael] in prison, the two of us had been estranged for several years. . . . On the two days I attended his court proceedings last year in Norwalk, Connecticut, he was cold and distant. Many people asked me why I would publicly defend him—a cause unlikely to enhance my own credibility. I support him not out of misguided family loyalty but because I am certain he is innocent. . . . Michael's conviction shocked his six siblings into talking about the case with one another, and with me. For the first time, they shared their memories of the night when Martha Moxley was killed."

Bobby wrote about the distortions of Dunne and Fuhrman, the cooperation of the Skakels with the authorities, and the unlikelihood that 120-pound Michael "could have wielded a golf club with the savagery or strength needed to shatter the shaft and then drive it through Martha's body" and then dragged a girl of matching weight over a hundred yards into the woods.

Bobby went on to describe much more compelling evidence implicating the family's briefly hired tutor, Kenneth Littleton, who ended up a vagrant arrested for a number of crimes in Florida in the early 1980s. Bobby recounted a bizarre incident where Littleton "climbed a sixteen-story structure and gave President John F. Kennedy's *'Ich bin ein Berliner'* speech. When he was arrested, he told the police that he was 'Kenny Kennedy,' the black sheep of the Kennedy family. . . . By August of 1991, when Connecticut law enforcement authorities reopened the Moxley case, Littleton, still a prime suspect, had again been institutionalized."

Even if no other suspects had surfaced, in 1975 the statute of limitations for murder in Connecticut was five years. Prosecutors were barred from bringing murder charges for crimes committed beyond that period, no matter how solid the evidence. It was a settled bedrock precedent. In order to prosecute Michael, they had to overrule three Connecticut Supreme Court decisions ruling that the five-year statute of limitations applied to any murder committed in 1975, '76 or '77. But Michael's lawyer failed to raise the issue when he was charged—"presumably because he wanted a high-profile trial," Bobby concluded.

Michael's hapless lawyer, paid almost $2 million by the Skakels, did not disguise his friendship with Dominick Dunne. Bobby wrote: "The day after the conviction, Sherman told me that he was going to a Court TV party for Dunne. When I questioned the propriety of his attending, he said, 'We're friends. What can I say, I'm a kiss-ass.'"

In May of 2014, Bobby appeared on CNN talking about his cousin Michael Skakel's murder case. Michael's inept first lawyer had gone to prison for tax evasion in 2011, and, in October 2013, a judge had granted Michael a new trial based on his habeas corpus appeal, ruling that Mickey Sherman's defense of him in 2002 was "constitutionally deficient" and so lacking that "the state procured a judgment of conviction that lacks reliability." Michael was released from prison after his family posted a $1.2 million bail.

Bobby had come across a new key witness. After his article "A Miscarriage of Justice" appeared in *The Atlantic* following Michael's conviction, he had received a three-page, single-spaced fax at the magazine's office early in 2003. It was from someone named Crawford Mills, who said he was a former classmate of Michael's at the private Brunswick School. He opened by saying: "Unless I've been lied to, the jury got it all wrong."

Mills went on to recount how, two years earlier after Michael's arrest and when his trial was still pending, his friend Tony Bryant confided in him that he'd been in Belle Haven on the night Martha was killed—in the presence of two friends from a Manhattan public high school who ended up murdering her. "Tony's full name is Gitano Bryant," Mills wrote. "More than one Bryant has had successful fields in the NBA. Tony's brother Wallace played for the Mavericks. His cousin Kobe plays for the LA Lakers."

Celebrity seemed to stalk this case in the strangest ways. As Bobby summarized how Mills came into this knowledge: "During his Brunswick years, Mills had met the men whom Tony fingered as the killers. He knew them only as 'Adolph' and 'Burr.' Both of them subsequently confessed to Tony their roles in killing Martha. Except for his mother, Tony told no one the secret for twenty-seven years. But Michael's indictment prompted Tony to recount his tale to Crawford. Tony wanted Crawford to take the information about Adolph and Burr to the police, but he begged Crawford not to identify him as the source. Tony was married with four young children and was living in Florida, where he was president of a tobacco company. Tony felt he needed to keep a low profile to protect his family and his business. Having brought the two murderers to Greenwich in 1975, Tony, who is African American, feared becoming a suspect himself. He had shared what he knew of the events of October 1975 with his mother soon after Martha's murder. She was particularly fearful that Tony's skin color would make him an attractive target for law enforcement."

After reading Mills's letter, Bobby called him. Crawford revealed that he'd told this story before the trial to, among others, Michael's lawyer, Mickey Sherman, and to Dorothy Moxley, but no one seemed interested. The *New York Times* wouldn't return his calls. Crawford was then employed as an audio technician at CBS, where, a few days following Michael's conviction, he found himself miking up Mrs. Moxley in preparation for her appearance on Bryant Gumbel's *The Early Show*. Mills introduced himself as the guy who'd been sending her letters about Tony Bryant, but she just stared at him blankly. After he provided a thumbnail sketch of Bryant's allegation, he added: "Mrs. Moxley, I don't think Michael did it." She remained stone-faced and simply replied: "I know Michael did it." As Crawford then unsuccessfully sought to interest attorney Sherman, slated to appear on the same program, Mrs. Moxley angrily reported Mills to the show's producer. He was abruptly out of a job, ejected from the building by security guards.

Mills called Bryant, who was sorry about his friend's dismissal but still refused to get involved. "I'm going to out you," Crawford told Bobby he'd said to Bryant. Mills continued his crusade in the aftermath of the Skakel trial, even writing a letter to the judge. A reporter from the *New York*

Times who'd covered the proceedings responded with a "very polite" yawn. Six months later, Crawford's sister sent him a copy of Bobby's article.

Bobby still had no idea if Crawford's story was true, but he managed to track down Tony Bryant. "I've been waiting twenty-seven years for this call," Bryant said when Bobby identified himself. Although part of him seemed relieved, Bryant was initially reluctant to talk. "I have certain relatives," he explained, "that would not like any type of publicity concerning this thing that happened in Greenwich and any connection to it." He knew Michael somewhat and didn't like him. "Nevertheless, I want to do what I can to help. Because I know what happened to your cousin. He got screwed. He really did. He is innocent." Bobby engaged Bryant in five separate tape-recorded phone conversations between the end of February and beginning of March 2003, where he went into much more detail than he'd revealed to Mills. Bobby turned the tapes over to a private investigator retained by Michael's new attorneys, and the PI then went to meet with Bryant in Florida.

His story, as it unfolded, was a grim one indeed. He had attended the Brunswick school in Greenwich for two years, where he was the only black kid, "It wasn't the worst thing in the world, but it wasn't easy," Bryant said. There he'd met Michael and befriended Mills. After Bryant then moved to New York with his mother, an Academy Award-winning producer of educational films, he continued to socialize with many of the young people around the Belle Haven neighborhood, occasionally including Martha Moxley.

At Charles Evans Hughes High, a "rough school" in Manhattan, Bryant became fast friends with two new classmates of unusually large physical stature. One he identified only as "Adolph," an African American from the South Bronx who was "not somebody to mess with." The other, who Bryant called "Burr," was from the Pacific Northwest and of mixed Asian, Native American, and Caucasian heritage. Burr was the "gasoline" and Adolph the "engine . . . always trying to outdo each other." At the age of fourteen, Bryant was attracted to their rebellious and seemingly fearless natures.

The pair had met Martha at a Greenwich street fair in mid-September of 1975 (an event mentioned by the girl in her private diary). Adolph was immediately "infatuated" and soon "obsessed" with Martha. Encountering her again at a couple of church mixers, he expressed to Bryant his fierce jealousy of some other boys. He commented to Bryant and others that someday he was going to have her. On several occasions he said he wanted "to go caveman on her," which Bryant took to mean Adolph would drag Martha away by the hair and sexually assault her.

On the night of Thursday, October 30, 1975, Bryant, Adolph, and Burr took the train from Manhattan to Greenwich to take part in "hell night," when older children would commit petty pre-Halloween vandalism. Wandering around Belle Haven that night, the teens smoked marijuana and drank a couple of six packs of beer lifted from someone's garage refrigerator. They marauded the neighborhood, occasionally ducking into the woods or behind stone walls to dodge security guards. They ended up in an undeveloped meadow that extended north from the Skakels' back porch, where young people commonly congregated out of sight of security.

Bryant observed Martha, Michael, and his brother Tom coming in and out. Continuing to get drunk and high on pot, they picked up some golf clubs that were lying around the property where Rushton Skakel had recently sponsored a chipping tournament for a hundred Mitsubishi executives. Bryant's two friends walked around Belle Haven carrying the clubs, telling him they had their "caveman stick." The pair embarrassed some of the girls coming into their circle with sexual overtures. "They were like, where are the bitches?" Bryant remembered. Adolph said, "We just gotta get into something. I'm not getting out of here unsatisfied." Bryant, sensing that things were getting out of control, decided he'd had enough. He told Adolph and Burr he had to catch the last train due to his mother's curfew. He asked if they wanted to come, too. They said no, they were going to stay the night with a local boy they knew. As Bryant departed, Adolph repeated that he was going to "get caveman tonight" on a girl.

When Martha's body was found the next day, her pants and underwear were pulled down, but she had not been sexually assaulted. Pieces of a broken six-iron golf club were found close by. An autopsy showed that she had been both bludgeoned and stabbed with the club, which was traced back to the Skakel residence.

Two days later, Bryant's mother confronted him after reading an article in the Saturday *New York Times* reporting the discovery of Martha's body. "Don't you know this girl?" she asked. "Yes, I do," Bryant said. "There is no way you are going back to Greenwich," his mother added, instructing her son not to speak with anyone about being in Belle Haven that night.

When Bryant next encountered Adolph and Burr at school the following Monday, they made no effort to hide what they'd done. "Well, I got mine," Adolph boasted. "We did it, we achieved the cave man," Burr said. They never mentioned the girl by name, but it was clear who and what they took pleasure in boasting about. "They made a joke of it. They were proud of it," Bryant told Bobby.

* * *

After his initial conversations with Bryant, Bobby set out to try and verify some of his claims. With the help of a police detective friend from White Plains, he was able to determine using high school yearbooks that Adolph was Adolph Hasbrouck and Burr was Burton Tinsley. He finally located a phone number for Hasbrouck in Bridgeport, Connecticut, and taped a conversation with him. He admitted knowing Bryant and being friends with Tinsley, with whom he continued to talk regularly. Then Bobby asked whether he'd been in Greenwich on the night of Martha Moxley's murder. Hasbrouck replied that "unfortunately" he wasn't. He claimed to have only recently learned about it. Adding that he hadn't been in touch with Tony Bryant since the 1970s, Hasbrouck asked for his phone number, which Bobby gave him.

Hasbrouck agreed to meet at his home with Michael's investigator, Vito Colucci. Starting to talk immediately, he changed his story three times over the course of some seventy minutes as to his whereabouts the day of the murder. Asked if he'd be willing to take a polygraph, Hasbrouck demurred, saying he gets really nervous and would "probably flunk it."

According to Hasbrouck, he had consulted with Tinsley for the purpose of reconciling their stories. Next Bobby reached out to Tinsley in Portland, Oregon, who said Hasbrouck had mentioned he might be calling. Had he gone to Greenwich on the night before Halloween in 1975? Bobby asked. "Halloween, it seems to me we were going up there. I have a hard time remembering." But yes, they'd attended a dance there during that period, confirming Bryant's recollection.

Bryant, Tinsley, and Hasbrouck were all available to testify at the hearing on Michael's petition for a new trial in 2006. They had each invoked their Fifth Amendment right. When the judge asked prosecutors why they hadn't offered Bryant immunity to compel his testimony, the state's attorney explained that it deemed Bryant's account "wholly incredible." The judge pointed out in a blistering dissent that Bryant's account had far more corroboration than the weak case against Michael Skakel.

After Michael's arrest, Bryant had authorized Mills to convey to the police—without identifying him by name - the information he'd passed along about Hasbrouk and Tinsley being the killers. Frank Garr, the state's lead investigator, testified that, yes, he'd received this in advance of Michael's criminal trial—but never followed up on it.

In its ruling dated October 25, 2007, the trial court concluded that the statements of Bryant, Hasbrouck, and Tinsley were admissible as statements against their penal interest. Yet the court concluded with a contradictory finding that the new evidence wasn't credible enough to have prompted the jury to change its verdict. Michael's attorney appealed to Connecticut's Supreme Court, arguing that this and other newly discovered evidence entitled him to a new trial. Now, almost twelve years since his false conviction, he'd gotten his wish.

On August 8, 2014, after Connecticut prosecutors filed an appeal of a judge's grant of habeas corpus, "that action caused me to begin work on this book," Bobby wrote. Despite a deadline looming for his family memoir, he decided to put most everything else aside for a year and assemble what in 2016 became *Framed: Why Michael Skakel Spent Over a Decade in Prison for a Murder He Didn't Commit*. The book was brought out by the maverick Skyhorse Publishing company. Although it received relatively few reviews, *Framed* made the *New York Times* Bestseller List.

Aside from the personal reasons—"Michael's freedom, reputation, and constitutional rights"—Bobby wrote in his Introduction, "There are broader issues, as well, that need airing, including the abuse of police and prosecutorial power and the role of the media in our democracy. Michael's ordeal is a parable about how mercilessly the flames of passion and prejudice consume even the most privileged individual when democracy's firewalls—police, prosecutors, the justice system, the press—give way to the clamoring of the mob. The inferno that devoured Michael is no anomaly. It feeds every day on the economically disadvantaged and minorities. Only visibility distinguished Michael. Mostly the casualties of their broken institutions are the invisible and discarded—people living in ghettos and fringe communities, from Ferguson to Baltimore." After Bobby and Cheryl Hines married in the summer of 2014 and moved to the West Coast, just as he decided to embark on the book himself, the subject matter proved a fitting one for her to assist with. According to Bobby, "Cheryl really helped me with that book." She had met Michael and reflected: "I made a few notes here and there, yes. It's nice because Bobby understands that I'm coming from a very different point of view than he is. So when I read something he's written that *seems* like everyone should know it—if I don't, then I think a lot of people won't."

The media had, of course, long trumpeted Michael as a "Kennedy cousin" and created backdrop images of all the Kennedys on the beach in Hyannis Port. As Bobby explained to TV host Jack Ford in 2016, "there are twenty-nine grandchildren of Joseph and Rose Kennedy who are generally thought of as Kennedy cousins, and none of those grandchildren had ever met any member of that generation of Skakels until 1983."

Why then, Ford asked, did Bobby make it a personal crusade to exonerate Michael? Bobby outlined his long friendship with Michael and their attending dozens of AA meetings together "where you share your most intimate thoughts and secrets." He knew the whole story of the murder, "all the things that later on the police and prosecutors and Mark Fuhrman said were recent fabrications Michael made up to cover the possibility that his DNA would be found on Martha Moxley's body. I knew that wasn't true, and I watched an innocent man get convicted.

"If you see a mugging on the street, do you put your head down and keep walking and say 'I don't want to get involved with this because there may be a cost?' Or do you turn around and get involved? I didn't feel like I had any choice. Believe me, it's not something I wanted to do. Writing the article was a couple of months' work and doing the book took a year. It takes me away from what I want to be doing, which is litigating against polluters."

Bobby described attending the arguments on Michael's appeal before a six-justice panel of the Connecticut Supreme Court on February 24, 2016. As the prosecutor "recited a cruel battery of calumnies and lies about Michael," Bobby scribbled down a passage from the Old Testament and handed it to his cousin:

> He was despised and rejected—a man of constant sorrow, acquainted with deepest grief. We turned our backs on him and looked the other way. He was despised, and we did not care.
>
> —Isaiah 53:3

Michael scrawled a note back to Bobby: "I'm familiar with the sorrow. Having someone to shoulder the burden with me has cut its weight in half. Thanks for our friendship—and for all the laughs."

At the time, Michael's fate remained uncertain. The Connecticut Supreme Court disagreed with the 2013 decision to release him pending a retrial. Michael's lawyers requested a review of the latest ruling, which, if it failed, would mean he'd go directly back to prison to complete his 20-years-to-life sentence. But by 2018, the makeup of the state's highest court had changed with the retirement of one of the justices who'd reinstated Michael's conviction two years earlier. In a 4-to-3 ruling, the court once again overturned the conviction citing shortcomings in his defense. Justice Richard N. Palmer wrote in the majority opinion that Michael's conviction was founded on a case "devoid of any forensic evidence or eyewitness testimony linking the petitioner to the crime."

And on October 29, 2020, nearly forty-five years to the day that Martha Moxley was killed, prosecutors announced that Michael, now sixty, would not have to face a second trial. Their reasoning, they said, was that proving his guilt would be impossible because many key witnesses in the case had died. Outside the courthouse, Michael did not comment. His defense lawyer, Stephan Seeger, called it "no stretch of the imagination that they can't prove the case within a reasonable doubt. It's not simply a technicality that ends the case."

From left to right: Kathleen, Mother, David, Bobby, Father, and Joe, at Hickory Hill in McLean, VA

Family photo (left to right): Kathleen, Bobby, Kerry, Mother, David, Father, Courtney, Michael, and Joe

A Kennedy Christmas card

Bobby with his cousins in Hyannis Port, MA

Back in Hyannis Port in 1960 with the
presidential nominee

Bobby with his Grandfather Joe Kennedy

A President gets his advice from many
sources - To Bobby - in the way head as others
John F. Kennedy

Returning from the 1960 convention with the President John F. Kennedy

Bobby gifting President John F. Kennedy a salamander in the Oval Office

Family photo in Hyannis Port. L to R: Bobby, Kathleen, Kerry, Dad, Mom, Chris, Joe, Courtney, David, and Michael

Golf cart ride with the president, Hyannis Port, MA

n the San Juan Peninsula in Washington state, on a
ing trip with Supreme Court Justice William O. Douglas

Twelve-year-old Bobby with Kerry and with a Red
Tail Hawk named Morgan at Hickory Hill

Millbrook School in NY, 1969

Pit Viper in Tanzania

Augur Buzzard in Kenya

Bobby (second from left) rafting on the Apurímac River in Peru

Bobby with Lem Billings

With Maasai in Kenya, 1973

Bobby in the Llanos of the Andes in Colombia with
Lem Billings

Bobby carrying his daughter Kick while on a falconry expedition

Bobby and his brothers Max and Michael, on the Bíobío River with the Pehuenches Chief

Bobby (right) with local fisherman at Laguna San Ignacio during fight against Mitsubishi in Mexico

Bobby holding Bobby III with his godmother, Jacqueline Kennedy Onassis, in Bedford, NY

Bobby and Emily in Charlottesville, VA

Uncle Ted, Bobby, and Aunt Eunice Kennedy Shriver in Hyannis Port, MA

John Kennedy Jr., Bobby, and Chris Lawford

Carolyn Bessette, Bobby, John Kennedy Jr., and Mary Richardson

Bobby with the family and second wife Mary (from left to right: Conor, Bobby, Finn, Kyra, Aidan, Mary) in Martha's Vineyard, MA

Conor, Bobby, and Bobby III at Standing Rock Reservation in North Dakota

Bobby, a lifelong falconer, training a bird of prey

Bobby with Aidan and Finn during annual Hudson River camping trip on Magdalen Island near Saugerties, NY

From left to right: Kerry Kennedy, Ana Bernard, Tim Haydock, Bobby, Katherine Smith, Michael Kennedy, and Lem Billings (front) in Hyannis Port, MA

In Montgomery, Alabama with Harvard friend, Peter Kaplan, Alabama Governor George Wallace, and Bobby, who was writing his first book on civil rights icon Judge Frank M. Johnson Jr.

With patrol boat for nonprofit Hudson Riverkeeper

With John Cronin, Hudson Riverkeeper, on the Hudson River

With Jerry Garcia of the Grateful Dead. Jerry was a passionate supporter of Riverkeeper.

WATERKEEPER ALLIANCE 2013

th many from our Waterkeeper organization at our annual conference near Atlanta, Georgia. There are more than three hundred Waterkeepers protecting waterways across the planet. For more info visit www.waterkeepers.org

NASDAQ Bell ringing for Waterkeeper Alliance

In North Carolina fighting Big Agriculture and factory farming

With Bill Clinton, Doug Spooner, and Senator Sheldon Whitehouse in New York

With Dewayne Johnson, who prevailed in the lawsuit against Monsanto after their products caused him to be diagnosed with non-Hodgkin's lymphoma

Cheryl Hines, Julia Louis-Dreyfus, Bobby, Larry David, and Brad Hall

Bobby and Cheryl on the water at Cape Cod

Bobby and Cheryl's wedding in Hyannis Port, MA, 2014

In Nashville for a fundraiser for nonprofit Tennessee Riverkeeper with Cheryl, Reba McEntire, and David Whiteside

In Boston after Bobby's presidential announcement on April 19, 2023

Bobby and friends seine netting on the Hudson River

Bobby with children at a rally for Children's Health Defense

Laura Bono, Bobby, and Mary Holland with Children's Health Defense

Bobby delivering an Earth Day themed speech in Centennial Olympic Park, Atlanta, 2018

Reached by a *New York Times* reporter, Bobby said, "we would have made a mockery" of prosecutors had they chosen to retry the case. The *Times* article, misleadingly headlined "Michael Skakel, Kennedy Cousin, Will Not Face 2nd Murder Trial," did not mention anything about the alleged perpetrators who remained free more than a decade after being identified by Tony Bryant.

In *Framed*, Bobby had written: "With his new lawyers and the exculpatory evidence now in hand, and the evidence of Adolph Hasbrouck's and Burton Tinsley's likely culpability already judged admissible, prosecutors would have almost no chance of convicting Michael in a new jury trial." With the evidence that he'd cited in the book, he believed that "prosecutors have sufficient cause" for the others' indictments. Bobby expressed hope that "perhaps the Connecticut State's Attorney will decide to prosecute the likely perpetrators of the Moxley murder and finally bring Martha's killers to justice."

But Bobby didn't think this would ever happen. "The State will never go after them," he said, "because it's an embarrassment. They got the wrong guy, and they know it. They're in a terrible position because the lawyer for the other side is going to say, 'Hey, you're now telling us it's these guys, but two years ago you said it was Michael Skakel.'"

And so, at long last, in large part due to Bobby's multiyear intervention, the defense rests.

CHAPTER TWENTY

LEGAL EAGLE: ALL IN A DAY'S WORK

In 2002, Riverkeeper had upped the ante against Hudson River polluters. The actor Harrison Ford, who maintained a residence close to the river and a private school that two of his kids attended in Manhattan, had been sending checks to the organization for some years when he first met Bobby at a fundraiser in the spring. "Anything I can do to help," Ford told him. Bobby said he might be looking for pilots to fly the Hudson waterway looking for signs of toxic dumping. Ford jumped at the chance, offering to fly his Bell 407 helicopter on aerial surveillance runs with a photographer friend. That November, Bobby and Ford would appear together on the cover of *Men's Journal* headlined as "The River Warriors."

I'd visited the Pace University Environmental Litigation Clinic in White Plains in September, attending a midafternoon seminar at Aloysius Hall, where Bobby sat behind a "judge's bench" with codirector Karl Coplan. Across the room, ten third-year law students took turns at a podium, practicing their oral arguments for a forthcoming case. On behalf of Riverkeeper and four fishermen's organizations, the clinic was suing New York City for discharging sediment-laden, highly turbid water into a renowned Catskill trout stream, Esopus Creek, without a required federal permit under the Clean Water Act.

Bobby's penetrating blue eyes focused on a student as she concluded her argument. "Your presentation was great," he told her. "Great eye contact, using your hands—you didn't seem to be reading your notes." He addressed the need to paint a picture to the judge of the poor design of the intake pipe.

With the next student, Bobby started playacting devil's advocate, firing questions and posing hypothetical situations to him. "Doesn't a government agency have the right to make the calculation that bringing water to children and the elderly is more important than a trout?" The student, John Paul, responded eloquently, discussing a minimal cost to ratepayers if the environmental regulations were followed. Instructor Coplan started to respond, "Now if you can do that again in front of a judge. —" when Bobby interjected: "Then we won't get sued for malpractice." The class burst out laughing.

Four Clinic students (including Paul) went on to conduct most of the arguments to a US District Court the following January. Judge Frederick Scullin would hand down the highest penalty ever awarded in a citizen's suit against a municipality ($5.7 million), as well as order New York City to finally obtain a discharge permit.

But the latest victory for the Pace team of student litigators was still a ways off, as the two-hour class ended and Bobby walked back to his office a couple of buildings away. The outer entryway is dominated by a fish-filled aquarium, where he pointed out a rare Hudson sturgeon. His rather spartan office is at the end of a long hallway past a long mural of the river's history that Bobby designed. Inside, a 1967 photograph of his father during a Scenic Hudson Preservation Tour is framed on the wall alongside some nineteenth-century nature prints.

I've come here to write a profile of Bobby for *E The Environmental Magazine*. It was about a 20-minute drive to his home in Mount Kisco. Out back of the house, oak, hemlock, and cedar trees crowned a path toward a thirty-acre lake stocked with largemouth bass, yellow perch, pumpkinseed sunfish, and more.

Bobby hoisted his one-year-old son Aidan into a backpack and went to transfer two pet hawks from their outdoor weathering perch to an indoor mews. At his father's prompting, the little boy began to mimic the bird's cries. "He doesn't talk yet, but he imitates all the animals," Bobby said proudly. Two peacocks wandering the yard paused to observe the ritual.

After dinner, having given Aidan a bath and with Mary putting the other children to bed, Bobby settled into his "junk room." Underneath the anaconda skin that stretched across most of the ceiling, we continued to talk. He reflected first on his father's legacy. "The principal issue that came to govern my father's life, I think, was civil rights. And I believe there's no more critical civil rights issue than environmental protection. If you look at environmental degradation, access to public lands, toxic waste—all of those burdens fall heaviest on the shoulders of the poor and minorities in this country. Four out of every five toxic waste dumps in America are in a black neighborhood and probably the biggest health care crisis we have is the 44 percent of African American youth who suffer the effects of lead poisoning. I recently visited a Navajo reservation in Arizona where, because of thousands of tons of toxic uranium tailings that have been dumped on their land, the youth have seventeen times the rate of sexual organ cancer as other Americans. And there are 150,000 Hispanic farm workers who are poisoned by pesticides every year. So all of the communities my father was deeply interested in are communities where the principal burden they are facing today is not violence in the ghetto. There are more people now dying of brain tumors in American ghettos than there are of bullets."

I asked about Vieques and his time in jail on the island, had this transformed him in any way? Bobby smiled and replied: "You know what? Doing time there was really like a vacation for me. I got to play basketball and share meals with some fascinating people. Of the 140 in my cellblock, 60 of them were political prisoners. I was only allowed ten minutes a day on the phone. So there was no outside intrusion, nobody asking me for a decision. I got to read biographies of Napoleon, Buddha, St. Augustine. If I could do this once a year for a month, I probably would—except my wife would kill me!"

The next night, Bobby's latest speech brought him to a fundraiser at a mansion overlooking the Hudson Valley. The host, Seema Boesky, is a prominent philanthropist. Several hundred guests were assembling to hear Bobby address concerns about the nearby Indian Point nuclear power plant. Actor Chevy Chase, on hand to introduce him, stood admiring some of Boesky's "Old Master" artwork. "You know, if you take a coin and scratch the paint, you can tell whether one of these is real or not," Chase dead-panned in my direction.

The discussion that followed felt all too real. "There is no power plant in the country that is more vulnerable, or a more attractive target for terrorists, than Indian Point," Bobby told the overflow crowd gathered under a large tent outside. "It's located twenty-four miles north of New York City. If a meltdown occurred, within the fifty-mile 'kill zone' are twenty-one million people. A week after September 11, the Nuclear Regulatory Commission admitted that nobody knows whether or not a large commercial aircraft could break through the containment dome."

To illustrate what he was talking about, Bobby told me how he'd recently enlisted his brother Douglas, now a reporter for Fox News, to take a film crew and rent a small Cessna at the Westchester County Airport. While they circled about a thousand feet above the nuclear plant for around twenty minutes, "nobody even came out and waved them away." Finally the pilot grew nervous and asked to see the brother's identification—realizing that Douglas could have easily overpowered him and plunged the aircraft directly into the fuel rod pools.

A number of other unprotected areas existed at Indian Point, including the cooling water intake on the Hudson River. Bobby's friend Ron Gatto had told him: "I lose sleep knowing how vulnerable this whole system is. It's absolutely insane—they're only open 'cause they've got pals in Washington."

"Incidentally, Indian Point has the worst safety record of any nuclear plant in the United States," Bobby continued. "I think it should be closed down immediately," as was the Shoreham facility on Long Island (before it could produce electricity). Within two years, according to Bobby, new high-efficiency natural gas turbine plants could replace all the power generated at Indian Point.

By the end of the evening, between $50,000 and $70,000 would be raised for an advertising campaign to alert the public about the potentially catastrophic situation. It took another fifteen years, but following recurring emergency closures, in January 2017 Riverkeeper entered into an agreement with the Entergy company and New York State for a planned shutdown of Indian Point. The first of two working reactors stopped producing energy in April 2020, followed by the second a year later.

Bobby felt compelled to take an opposing stand on the Cape Wind Project, a massive $700 million wind farm that a private company proposed to erect in Nantucket Sound. The 130 wind turbines spread across twenty-five square miles (and as tall as forty stories) would be the first such renewable energy development in US waters. Supporters, including many environmentalists, touted its role to provide electricity for much of Cape Cod while cutting pollution and combating climate change. Foes worried about potential harm to prime fishing grounds and sea birds, as well as to tourism and scenery.

Bobby, whose family's summer home on the Sound would be in the proximity, was being accused of NIMBYism (Not in My Backyard). Asked about this, he responded: "I'm a strong advocate of wind farms on the high seas. But there are appropriate places for everything. We wouldn't put one of these in Yosemite, and I think environmentalists are falling into a trap if they think the only wilderness areas worth preserving are in the West. The most important are the ones close to our cities, where the public has access to them. And Nantucket Sound is a wilderness, which people need to experience. I always get nervous when people talk about privatizing the commons. In this case, the benefits of the power extracted from Nantucket Sound are far outweighed by the other values our communities derive from it."

Things came to a head with other environmentalists in August 2002, when Bobby launched the Nantucket Soundkeeper in Hyannis Port aboard a 120-foot sailing sloop, hosting a press conference, cocktail party, and dinner speech. "Greenpeace protested me," Bobby wrote in his journal. "They sent a fleet of Zodiacs and a sailboat out with signs supporting Cape Wind and scolding me, 'Bobby you're on the wrong boat.' I invited them aboard my boat to hear my speech and went out to meet them afterward. I was grateful the Cape Cod commercial fishing fleet showed up in force to support me putting themselves between the Greenpeace fleet and our ship and blowing their horns and putting up tall wakes. It's odd that Greenpeace is shilling for a private developer that is trying to privatize twenty-four square miles of public trust waterways."

Such were the dichotomies of Bobby's complicated world, where his allies might not be the environmental groups so much as the blue-collar workers they claimed to protect.

* * *

Arriving back at his house around 9 p.m., and yet to have any dinner, Bobby pointed to a baby owl in its cage right outside the front door. "I have to feed this guy first," he said. The owl, found abandoned in the woods, is now a part of his licensed wildlife rehabilitation center. With his wife still attending a school open house for their daughter Kyra, Bobby wandered into the kitchen and located a salad in the refrigerator.

The phone rang. It was his brother Max, twelve years Bobby's junior. A professor at Boston College Law School, where he teaches a course called "Nature in American Culture," Max is also a codirector of the Urban Ecology Institute, which works with 17 Massachusetts schools, getting students involved in hands-on efforts to study local water quality and catalogue species diversity. Max was calling Bobby, as he often does, for some advice. "Would I be doing this if it weren't for Bobby? I'm almost certain I wouldn't be," Max would tell me later. "I don't know if I'd have gone to law school. For sure I wouldn't be a falconer, or have all the animals we keep at our house."

Now, on the kitchen TV, a commercial came on promoting the latest SUV. Bobby shook his head and said: "The checkbook diplomacy between Detroit and the White House, including about $30 million in campaign contributions from the big auto companies, has bought the auto industry immunity from making any sacrifices for our country. Franklin Roosevelt got on the radio and asked Americans to conserve gasoline, but now we have a president who goes on TV, asks Americans to go shopping, and gives them a $100,000 tax deduction for the worst gas-guzzling SUVs."

Privately, Bobby pulled no punches about what he saw happening under the current George W. Bush administration. He had declined an invitation to the White House to view Kevin Costner's movie on the Cuban Missile Crisis, *13 Days*, based on his father's same-titled book, as a quiet protest against the president's environmental policies. Then came the September 11 attacks. "Armageddon!" was the only word Bobby penned in his journal. The next day, he wrote privately that Bush was "an idiot and a puppet, and it's painful watching him on TV." On the 15th, he described Bush as a simpleton who is "promising revenge in a crusade. He will drop million-dollar bombs on ten-dollar houses in Kabul and bomb [them] back to the dirt age. It's agonizing having him as our leader, and I know the forces of darkness in his administration will turn this awful tragedy to their advantage."

Bobby had met Bush for the first time on November 20, at a ceremony where the Justice Department building was being renamed after his father. "While my brothers and sisters shook hands with the president in a quiet gathering, my sister Kerry was outside the building leading a protest against Bush administration torture policies." In his journal, Bobby added without judgment: "Bush was charming."

Things had continued to deteriorate since then. In our conversation ten months later, Bobby continued: "I lived through the Reagan years, through [James] Watt and [Newt] Gingrich, and nothing has ever been as bad as this. What's going on right now is the worst assault in history on our environment. These people have such narrow minds and are so filled with fear and really have no faith in our country. Industry gave them $300 million to win the last presidential election, and you and I and our children are going to pay them back a hundred times."

His voice quavering with rage, Bobby continued: "Using what happened on September 11th to push forward their agenda is the most cynical thing I've seen in American history. Every time there's any kind of crisis in this country, the Bush administration sees it as an opportunity to attack the environment."

Out the picture window in his "junk room," night had enveloped the trees and the lake stocked with fish. Bobby continued to talk fervently about Bush's White House Commission, "whose purpose is to figure out ways to dismantle the National Environmental Policy Act." And about the efforts to gut key provisions of the Clean Water Act and Clean Air Act, allowing more mountaintop mining and letting coal-burning power plants off the emissions hook.

"Our biggest problem is getting the message out," Bobby said. "If the American people, Republicans and Democrats, hear the message, we win on the merits. The trouble is, industry has all the money and an extremely sophisticated public relations machine. That includes spending millions to create these phony think tanks in Washington, DC. They fill them up with these 'biostitutes,' marginalized scientists who make pronouncements that global warming doesn't exist and the ozone hole is a myth and the oil industry is good for the fish."

Bobby's knowledge, charisma, and ability to articulate all of this raised the obvious question again: can he, should he, run for political office himself? John Adams, NRDC's president, had known Bobby for twenty years and told me: "I think he has the potential to be a great political leader. If he were, he'd also be a great leader when it comes to values. He's enormously attractive to a lot of people who care about the future of the planet, and people all across America talk to me about wanting to support him if he does run for office. But I think he has considerable reluctance. He's got a wonderful family. And it could be dangerous."

Author Jack Newfield, who accompanied Robert Kennedy on his last campaign and later came to know Bobby, believed more was involved than the dangers inherent in a Kennedy running for high office. "I've sensed a real ambivalence in him about politics, and what you have to endure to go through an election, in terms of negative campaigning and fundraising. But given his name, intelligence, and character, he could probably run for and win any office."

For his part, Bobby, who had already endorsed John Kerry for president in 2004, acknowledged having long considered entering the political arena. "But really I just live my life one day at a time, trying to be effective doing what I'm doing. I have benefits from this lifestyle, which allows me to hang out with my family and be a good father. I think that would be more of a challenge if I ran for office. But maybe at some point I'll get so angry about the way our politicians behave that I would make another choice."

As he spoke, bathed in the glow of the lamplight, the momentary resemblance to his father was positively eerie. He continued: "What I seek to impart to my students is the same thing I try to teach my own kids—to instill them with noble thoughts. Which is, I think, the principal objective of parenthood, to make them feel like they can be heroes. And that the object of life is to transcend narrow self-interest, and to spend your resources on behalf of the community. That's the key to personal happiness and fulfillment. "You work as hard as you can for the right thing, and then let God be in charge of the results. We still have time to preserve a planet for our children that provides at least some of the opportunities for dignity and enrichment as those our parents gave us. My job is to be able to look myself in the mirror and say that I spent my short time on this planet trying to make it a better place for my children. I have to look my children in the eye. And I will be able to do that. Because I know I will never sell out, and I'm going to spend my life fighting as hard as I can. I will die with my boots on. That's really all I care about."

He was starting to research a new book, chronicling the damage that the Bush administration was causing to the environment, fighting in every way he could—through teaching his law school students, giving interviews and speeches, writing op-eds, bringing lawsuits, running environmental nonprofits, devoting his life to service and to family.

CHAPTER TWENTY-ONE

THEMES OF THE FUTURE: ORIGINS

In the aftermath of the 2000 election debacle, where Bush finally prevailed when the Supreme Court ruled against Gore's challenge to the Florida vote count, Bobby was announced as the leader of a new environmental group enlisting trial lawyers to file huge lawsuits against polluters as an alternative to enforcement of regulations by the incoming Bush administration.

For the next two years, Bobby would travel the country on an informal speaking tour—not unlike the one President Kennedy undertook during the last months of his life warning about air and water pollution, pesticide poisoning, and species extinction. He addressed hundreds of audiences, including conservative women's groups; public school teachers; civic, religious, and business groups; trade associations; farm organizations and rural coalitions; and numerous colleges. He heard a recurring refrain: "Why haven't I heard any of this before? Why aren't the environmentalists getting the word out?"

It wouldn't be long before Bobby signed a book contract to expose what the Bush people were doing. In a prequel to *Crimes Against Nature*, he wrote a long piece for *Rolling Stone* with the same headline. Appearing in November 2003, it began: "George W. Bush will go down in history as America's worst environmental president. In a ferocious three-year attack, the Bush administration has initiated more than 200 major rollbacks of America's environmental laws."

The second paragraph took up a theme, children's health, that would become predominant for Bobby as the decade progressed. "I am angry as a citizen and a father. Three of my sons have asthma, and I watch them struggle to breathe on bad-air days. And they're comparatively lucky. One in four African American children in New York shares this affliction; their suffering is often unrelieved because they lack the insurance and high-quality health care that keep my sons alive."

The previous summer, Bobby's cousin Maria Shriver had approached him on Cape Cod, avowing that if elected, her husband, Arnold Schwarzenegger, was determined to be "the best environmental governor in California history." Although Schwarzenegger was a Republican, Bobby agreed to help him and worked with a sympathetic group from both parties to draft his environmental platform. In October 2003, Schwarzenegger won the election. Bobby went on to note in "Crimes Against Nature" that the fishermen he represented in legal actions are traditionally Republican, but "they see this [Bush] administration as the largest threat not just to their livelihoods, but their values and what it means to be American."

In the March 8, 2004 issue of *The Nation*, Bobby delved deep into a subject he had touched on for *Rolling Stone*. His piece, headlined "The Junk Science of George W. Bush," began on a personal note from his youth. "As Jesuit schoolboys studying world history, we learned that Copernicus and Galileo self-censored for many decades their proofs that the earth revolved around the sun and that a less restrained heliocentrist, Giordano Bruno, was burned alive in 1600 for the crime of sound science. With the encouragement of our professor, Father Joyce, we marveled at the capacity of human leaders to corrupt noble institutions. Lust for power had caused the Catholic hierarchy to subvert the church's most central purpose—the search for existential truths."

Bobby's article continued: "Today, flat-earthers within the Bush Administration—aided by right-wing allies who have produced assorted hired guns and conservative think tanks to further their goals—are engaged in a campaign to suppress science that is arguably unmatched in the Western world since the Inquisition. Sometimes, rather than suppress good science, they simply order up their own. Meanwhile, the Bush White House is purging, censoring and blacklisting scientists and engineers whose work threatens the profits of the Administration's corporate paymasters or challenges the ideological underpinnings of their radical anti-environmental agenda. Indeed, so extreme is this campaign that more than sixty scientists, including Nobel laureates and medical experts, released a statement on February 18 that accuses the Bush Administration of deliberately distorting scientific fact 'for partisan political ends.'"

In what can only be described as one of recent history's ironies, precisely sixteen years later (February 18, 2020), a group of twenty-seven scientists and medical experts signed a letter at the onset of the COVID-19 pandemic saying: "The rapid, open, and transparent sharing of data on this outbreak is now being threatened by rumors and misinformation around its origins. We stand together to strongly condemn conspiracy theories suggesting that COVID-19 does not have a natural origin."

This time, the deliberate distortion of scientific fact came from the scientists themselves. The letter was drafted by Peter Daszak, whose EcoHealth Alliance had been directly involved with China's Wuhan Institute of Virology in creating manmade lethal coronaviruses, under grants from Dr. Anthony Fauci's government agency. Professor Jeffrey Sachs, later appointed to chair a COVID-19 commission, believes it likely that the Wuhan lab is where the virus came from. As the respected publication *Current Affairs* put it in August 2022: "He believes that there is clear proof that the National Institutes of Health and many members of the scientific community have been impeding a serious investigation of the origins of COVID-19 and deflecting attention away from the hypothesis that risky US-supported research may have led to millions of deaths."

And this time, the suppression of good science and the "purging, censoring, and blacklisting" is coming not from the right-wing and conservative think tanks, but the liberal establishment of which Bobby has long been a part. But we are getting ahead of our story. Bobby's first "junk science" investigation described how he'd just opened an office at 115 Broadway, catercorner from the World Trade Center, when the September 11, 2001, attack occurred. Upon returning to the office in October, Bobby's law partner, Kevin Madonna, "suffered a burning throat, nausea, and a headache

that was still pounding twenty-four hours after he left the building."
Though the Environmental Protection Agency claimed that New York
City air quality was safe, Bobby noted that "Kevin refused to return, and
we closed the office." Of course, most workers did not have that option.
We now know that the government was lying to us. An Inspector General's
report released in August of 2003 revealed that the EPA's data did not
support their assurances and that "its press releases were being drafted or
doctored by White House officials intent on reopening Wall Street."

Bobby also recalled the case of Dr. James Zahn, a widely respected
microbiologist with the Agriculture Department's research services, who
had identified antibiotic-resistant bacteria that can make people ill in
the air around industrial hog farms. In April 2002, Zahn had accepted
Bobby's invitation to speak at a gathering of over a thousand family farm
advocates in Clear Lake, Iowa. The day of the conference, Zahn canceled
under orders from the USDA. Bobby's research uncovered a trail of faxes
showing the order being prompted by lobbyists from the National Pork
Producers Council, which also scuttled over a dozen other Zahn speeches
to similar groups. "Soon after my conference, Zahn resigned from the
government in disgust," Bobby wrote. He sought to fill the void in Iowa,
exhorting the crowd: "I am more frightened of these large multinationals
than I am of Osama bin Laden. I got a standing ovation from all the farmers in the room, but I got six months of abuse from the Farm Bureau."

These were only two of numerous examples cited in Bobby's *Nation*
article, highlighting the Bush Administration's "plans to systematically
turn government science over to private industry by contracting out thousands of science jobs to compliant consultants already in the habit of mas-
saging data to support corporate profits." Millions had been invested over
the previous two decades to corrupt science and "distort the truth about
tobacco, pesticides, ozone depletion, dioxin, acid rain, and global warming." Research budgets had been gutted and science politicized. The article
concluded presciently: "The very leaders who so often condemn the trend
toward moral relativism are fostering and encouraging the trend toward
scientific relativism."

Bobby's book, *Crimes Against Nature*, was a devastating indictment of the Bush administration's impacts on the environment, from

climate change to the corporate takeover of regulatory agencies. Parts
of the book were deeply personal. "There was no environmental issue
about which my father cared more passionately than strip-mining,"
Bobby wrote. "He visited Appalachia in 1968 and told me how the
coal companies were using this technique to put miners out of work.
In the process, they were also destroying our historic landscapes and
permanently impoverishing the region." Bobby described having
flown over the hills of West Virginia, Kentucky, and Tennessee in May
2002 and observed "a sight that would sicken most Americans. The
[coal] mining industry is dismantling the ancient mountains and pristine streams of Appalachia through a form of strip-mining known as
mountaintop removal."

                              * * *

Crimes Against Nature included a telling anecdote from Roger Ailes, the
former president of Fox News. As noted earlier, Roger was an old friend
of Bobby's with whom he spent a summer camping in Africa almost thirty
years ago. "He is jovial, animated, and genuinely funny, and we loathe
each other's politics," Bobby wrote. When Bobby asked Ailes why the
networks didn't cover environmental stories, the former network executive considered the question for a moment before emphatically responding: "It's because environmental stories are not fast-breaking!"

That October, I attended a fiery hour-long speech delivered by Bobby
in Pittsburgh to the annual conference of the Society for Environmental
Journalists. He began by calling the media complicit in not reporting
critically enough on the Bush Administration's environmental policies.
The president's decision to exempt coal-burning power plants from the
Clean Air Act meant that the nation's major source of mercury emissions
would continue unchecked to increase learning disabilities in children.
Journalists needed to "connect the dots," tracing corporate political contributions and industry insiders being appointed to regulatory positions
with increasingly lax environmental standards. Reporters often failed the
public by not trying to find the real science and scrutinizing origins of the
corporate science behind what the administration claimed. "The truth is
often far away from balance," Bobby said.

It was a subject that came up again in the question-and-answer session following a talk that same October at Rivier College in Nashua, New Hampshire, when someone asked why haven't we heard more from the Democratic nominee for president, John Kerry, about the environment? Bobby replied that Senator Kerry was talking about it, but the press wasn't covering it. "The White House press corps is just a bunch of stenographers," he said. "They think they've done their job if they print press releases from the Bush side and Kerry's. The Fourth Estate is asleep. The failure of democracy is the failure of the press."

Bobby traced the root of the problem to 1988, when Ronald Reagan abolished the Fairness Doctrine. This rule had been in place since 1924, at the advent of commercial radio. "It acknowledged that the airwaves belong to the public. They're part of the commons, a public trust asset. The Fairness Doctrine required that broadcasters air issues of public import. Last year, only 4 percent of the 15,000 minutes of network news were devoted to the environment. And those are mostly human-interest stories like whales caught in sea ice, not the ones about all the kids with asthma and the money that changed hands so the utilities didn't have to remove the ozone and particulates. Do you remember when [after 9/11] we had anthrax on TV every day for months and only six people died? Well, what if a terrorist poisoned all the lakes and streams in nineteen states so fish would no longer be safe to eat? Somebody did that, yet the press doesn't have the investigative reporters anymore who'll connect the dots for the American people."

The Fairness Doctrine also required media outlets to tell both sides of major issues they were covering. You could not have had a Fox News or a Rush Limbaugh when the Fairness Doctrine still existed. Nor could you have had a Rachel Maddow. Finally, the Doctrine required local control—crop reports in North Dakota, tornado warnings in Kansas, country music in the South - and diversity of ownership, "not homogenized broadcasting dictated by a couple of corporate epicenters." According to Bobby, Reagan had abolished the Fairness Doctrine "as a favor to some of the big studio heads who helped him get elected. As a result, Bobby told the environmental journalists, six companies owned virtually all the press outlets in America—all 6,000 TV stations, almost all the 15,000 radio stations and most of the 16,000 newspapers. "The news departments have become corporate profit centers. They no longer have an obligation to serve the public interest, so the only interest they're serving is their shareholders. How do they do that? By increasing viewership, that's the only thing that motivates them. You do it by appealing to the prurient interests all of us have in the reptilian core of our brain—sex and celebrity gossip. And they give us nothing about the kids with asthma or the one of out of six American women who can't safely bear children. To save money, they've fired their investigative reporters, their documentary producers who had the time to go out and tell those stories, and [closed] their foreign news bureaus." We are now in the absurd position where, in the nation that invented the free press, it's hard to get reliable news.

Bobby was asked; "Is the press keeping you quiet? I've never heard your message before." He replied: "Listen, do you think National Public Radio is kind of left-of-center? Well, NPR refused to advertise my book because they thought the title was too controversial. I went on Charles Grodin's show in 1996 and talked about some of the pollution issues caused by General Electric polluting the Hudson with PCBs. But GE owns the NBC network, and he was fired as a result of my interview. People like Aaron Brown will not let me on. He said, 'I don't want any Bush-bashing on my show,' and this is CNN, which is supposed to be a middle-of-the-road network."

Another member of the audience took the microphone. "Is it true that you won't run for public office because of your children?" she asked. "You seem so strong about all these issues. I would vote for you!" The rest of the packed house applauded. "I like what I'm doing now," Bobby answered. "If I can be effective doing what I'm doing, I would prefer to do it like this." But it was starting to become clear that it would be difficult for him to be as effective as he wanted to be without personal political power.

When Bush defeated Kerry by a narrow margin and gained a second term as president, Bobby began wondering if he could remain on the political sidelines much longer. Early in 2005, Bobby was considering running for attorney general of New York, possibly pitting him against his brother-in-law Andrew Cuomo, who was in the process of divorce from Bobby's sister Kerry. The New York Times reported that he'd held several meetings with politicians and statewide-elected officials, including Senator Hillary Clinton and Attorney General Eliot Spitzer, whom Bobby would succeed if he decided to run and then win. The attorney general position, as everyone knew, was the customary jumping-off point for becoming governor, or even president.

An interview appeared in the *Times* the next day. Bobby said he was receiving encouragement from a number of elected officials. But a host of factors, both personal and political including his ties to Andrew Cuomo, would influence a decision he planned to make within two weeks. "Basically I'm concerned about elevating discussion about the environment and the corrosive impact of corporate power on American democracy," he said, adding that Cuomo's rival candidacy "increases the risk that the race will become tabloid fodder rather than a serious debate on these critical issues."

On January 22, a piece appeared in *Common Dreams*. It noted that, while Bobby had endorsed John Kerry early in the campaign against Bush, he'd come away disappointed in his candidacy. Bobby was quoted saying: "The Republicans are 95 percent corrupt and the Democrats are 75 percent corrupt. They are accepting money from the same corporations. And of course that is going to corrupt you." George W. Bush, in his view, was "the most corrupt and immoral president that we have had in American history."

On the 24th, Bobby let the *New York Times* break the story. After considering it for several months, Bobby had decided against running. He told the interviewer: "I talked about it a lot with my wife, with my kids, my uncle and my brothers and sisters. It's funny. All the family members who had not been in political office urged me to run and the ones who had served in office urged me not to." While he'd received hundreds of calls and emails from politicians and ordinary New Yorkers urging him to do otherwise, Bobby reflected: "the kinds of hours I would put into this job would mean that I would never see my kids. I feel certain that I will run for office one day. But I think I need to wait a few years until my younger children get a little older."

For the moment, though, another river adventure in Latin America loomed in March. This time, he would bring Mary, their nine-year-old daughter Kyra, and thirty-three friends including tennis star John McEnroe and actors Dan Aykroyd and Julia Louis-Dreyfus on a ten-day whitewater rafting trip down Chile's Futaleufú waterway that was threatened by the building of three new hydropower dams. It was an expedition sure to make the pages of *Town & Country Travel*, for which Bobby himself wrote a long article that the magazine headlined "Bobby's Big Adventure" (Winter 2005). It would take some years, but eventually the Endesa company would withdraw its hydropower plans, citing "international opposition." The celebrity adventures were essential to getting the word out.

CHAPTER TWENTY-TWO

MERCURY IN VACCINES: OPENING PANDORA'S BOX

After he bowed out as a potential political candidate early in 2005, Bobby took up another subject that would increasingly concern him as time went by. He'd first raised the specter of fascism steadily creeping into America in his "Crimes Against Nature" piece for *Rolling Stone* in 2003. "Growing up, I was taught that communism leads to dictatorship and capitalism to democracy," he began. "But as we've seen from the Bush administration, the latter proposition does not always hold. While free markets tend to democratize a society, unfettered capitalism leads invariably to corporate control of government."

Bobby then launched into a history lesson, writing: "America's most visionary leaders have long warned against allowing corporate power to dominate the political landscape. In 1863, in the depths of the Civil War, Abraham Lincoln lamented, 'I have the Confederacy before me and the bankers behind me, and I fear the bankers most.' Franklin Roosevelt echoed that sentiment when he warned that 'the liberty of a democracy is not safe if the people tolerate the growth of private power to a point where it becomes stronger than their democratic state itself. That, in its essence, is fascism—ownership of government by an individual, by a group or by any controlling power.'"

Then this insightful warning from Bobby: "The rise of fascism across Europe in the 1930s offers many informative lessons on how corporate power can undermine a democracy. In Spain, Germany, and Italy, industrialists allied themselves with right-wing leaders who used the provocation of terrorist attacks, continual wars, and invocations of patriotism and homeland security to tame the press, muzzle criticism by opponents, and turn government over to corporate control. Those governments tapped industrial executives to run ministries and poured government money into corporate coffers with lucrative contracts to prosecute wars and build infrastructure. They encouraged friendly corporations to swallow media outlets, and they enriched the wealthiest classes, privatized the commons and pared down constitutional rights creating short-term prosperity through pollution-based profits and constant wars. Benito Mussolini's inside view of the process led him to complain that 'fascism' should really be called 'corporatism.'"

Bobby continued hammering home that theme in dozens of speeches across the country in 2005, often referencing his *American Heritage Dictionary*'s definition of fascism: "The domination of government by large corporations driven by right-wing ideology and bellicose nationalism." The Healthy Forests initiative promoted destructive logging of old-growth trees. The Clear Skies program suggested repealing key provisions of the Clean Air Act. "Streamlining" or "reforming" regulations actually meant weakening them. The language of the Bush White House could have been lifted right from George Orwell's *1984*: "war is peace, freedom is slavery, ignorance is strength."

The Superfund program to clean up toxic waste sites had gone bankrupt in 2003 after Bush failed to renew a tax on oil and chemical companies. The EPA announced that, for the first time since passage of the Clean Water Act more than three decades ago, American waterways were getting dirtier.

Bobby was particularly enraged about the ever-increasing levels of mercury in the environment. Bemoaning the sad fact that "few children today can enjoy that quintessential American experience, going fishing with Dad and eating their catch," he wrote in *Crimes Against Nature*, "most bodies of water in New York—and all freshwater bodies in 17 other states—are so tainted with mercury that one cannot eat the fish with any regularity. Forty-five states advise the public against regular consumption of at least some local fish due to mercury contamination."

The reason was simple: "Mercury is a potent brain poison," he asserted. "Even minuscule amounts can cause permanent IQ loss, along with blindness and possible autism in children who are exposed while in the womb." Bobby had recently had his own blood tested for mercury, and the analysis showed that while his levels were just below the cause for concern, a pregnant woman with the same result was very likely to have a cognitively impaired child.

The Clinton Administration had begun regulating mercury emissions from coal-burning power plants as a hazardous pollutant, but the Bush administration scrapped that and proposed new rules with allowable releases nearly seven times as high. According to the EPA, those power plants accounted for 40 percent of the airborne mercury in the US and almost 70 percent east of the Mississippi.

Bobby was litigating at that time on behalf of various waterkeepers in the United States and Canada against thirty-eight coal-burning power plants and cement kilns, mainly for discharging mercury. Going around the country talking about those lawsuits, he began to see a curious phenomenon wherever he went. Well-dressed women of similar age sat in the front rows of the venues, listened attentively, and afterward approached him for a brief but impassioned conversation. Most of them were professionals - pharmacists and doctors and lawyers. "They were a very unusual group," Bobby recalls, "very talented, eloquent, articulate, and generally respectful. They were also mildly scolding." One of those women, Rita Shreffler, remembers sprinting down the aisle to corner Bobby after his talk at a college in Columbia, Missouri, in March 2005. "I said something along the lines of, 'It's nice to hear you bringing attention to mercury and the harms associated with it, but you're missing something really important—the vaccine part.'"

* * *

Dr. Robert Mendelsohn, a famed American pediatrician who wrote five bestselling books in the 1970s and 1980s, was known as the "People's Doctor" who pulled back the curtain on certain realities of his profession—including that modern medical methods were often more dangerous than the diseases that they were designed to diagnose and treat. Mendelsohn wrote: "Immunizations have been so artfully and aggressively marketed that most parents believe them to be the 'miracle' that has eliminated many once feared diseases. . . . For a pediatrician to attack what has become the 'bread and butter' of pediatric practice is equivalent to a priest denying the infallibility of the Pope." Dr. Mendelsohn died in 1988, a year before the US embarked on a massive campaign to increase the number of vaccines given to children.

Autism isn't usually diagnosed for a few years, and a majority of the mothers seeking Bobby's ear had children born after 1989, which a study conducted by Health & Human Services tagged as the red-line year that signaled an increasing uptick in autism cases. As the number of vaccines increased after the manufacturers were given liability protection in 1986, so did the rates of autism and other neurodevelopmental disorders among children.

Rita Shreffler had a boy and a girl born seventeen months apart in the early 1990s, who after normal happy babyhoods suddenly underwent frightening behavioral changes—crying, screaming, biting, beating their heads against the wall. The symptoms, she would come to learn, were remarkably similar to those found in mercury poisoning. And her children, like thousands of others it turned out, had received pediatric vaccinations containing a mercury-based preservative called thimerosal that provided longer shelf life, reducing production costs. Thimerosal's mercury content was much higher than what you might inhale living several miles downwind of a coal-burning power plant.

There was precedent in the medical literature for such seemingly unrelated factors or environmental impacts. In 1953, Dr. Morton S. Biskind had suggested in the *American Journal of Digestive Diseases* that the polio virus might have morphed into paralytic poliomyelitis following the sudden widespread prevalence of DDT in the environment, known to cause spinal lesions in both animals and humans. Dr. Biskind thought DDT's capacity to degenerate the spinal cord might also be a potent escort for childhood polio. While mass immunization with the Salk vaccine is lauded as the miracle cure, the virus completely disappeared after the EPA banned

DDT - although stubbornly persisting in areas like India and Congo, where the pesticide remained in use.

Shreffler was the executive director of a new nonprofit, the National Autism Association, and had been on her way to Washington to lobby Congress with a group of other parents when she learned about Bobby's scheduled appearance in Missouri. He expressed sincere sympathy for Rita, as he had with other polite but desperate mothers who approached him. "Privately, I was skeptical," he wrote in 2005. "I doubted that autism could be blamed on a single source, and I certainly understood the government's need to reassure parents that vaccinations are safe; the eradication of deadly childhood disease depends on it. I tended to agree with skeptics like Rep. Henry Waxman, a Democrat from California, who criticized his colleagues on the House Government Reform Committee for leaping to conclusions about autism and vaccinations. 'Why should we scare people about immunization,' Waxman pointed out at one hearing, 'until we know the facts?'"

As Lyn Redwood, cofounder of another autism nonprofit called SafeMinds and chief organizer of the women's "march on Bobby," recollects: "We'd thought this man could potentially be an advocate for us. However, he did *not* want to get involved, which isn't surprising given his background where the Vaccines for Children program had been a strongly Democratic initiative." (Established by Congress in 1993, the program provided no-cost immunizations to families that couldn't afford them.) In addition to running SafeMinds, Redwood was a public health official in Georgia, married to a prominent emergency room physician, and a nurse practitioner who had administered dozens of vaccines. Then her child developed autism and, after she brought a lock of hair in for analysis, discovered these were "off the charts for mercury."

SafeMinds became the first organization to blow the whistle on thimerosal as a probable cause after the organization published a landmark paper linking mercury to the development of autism. She testified before Congress in 2000 at a hearing on "Mercury in Medicine: Are we taking unnecessary risks?" SafeMinds asked the FDA to immediately recall any vaccines containing thimerosal and ban its further use the same year, but the request fell on deaf ears.

Redwood remembers: "We knew the Centers for Disease Control (CDC) was looking at the issue, and when we took our concerns to them, they gave us a study saying there were no problems associated with thimerosal whatsoever. But the data looked very suspicious, and so we filed a Freedom-of-Information-Act request."

Redwood fortuitously lived in Atlanta, where the CDC is headquartered. In June of 2001, on her way to Florida for her mother's eightieth birthday, she stopped off and paid the agency more than $1,500 to release several large boxes of documents. That weekend, Lyn decided to open what would become a Pandora's box. A particular document "looked like the easiest to digest because it was all minutes to one meeting versus thousands of emails. So about ten o'clock I started reading. I was still awake at four a.m. I couldn't put it down. It was riveting in a terrible way, and I was in tears reading it." The smoking gun was that 260-page transcript chronicling a two-day meeting of leading government scientists and health officials. It had been convened by the CDC's National Immunization Program in June 2000, at the Simpsonwood Conference Center deep in the Georgia pine forest. Returning home, Redwood made copies of the report and dispatched these to four allies in SafeMinds. They were similarly stunned and angered when they read it. She also provided a copy later to David Kirby, who was working closely with Redwood on a book called *Evidence of Harm*.

Redwood was still mulling over what more to do with the trove when her movement received a harsh blow in May 2004. The respected Institute of Medicine (IOM) issued a report stating that the bulk of evidence "favors rejection of a causal relationship" between thimerosal and autism. At this point, perhaps in response to the gathering storm of angry mothers, in 2001 thimerosal was removed from most routine vaccinations given to American children. But it remained in flu shots that hadn't been part of the original schedule. The CDC now recommended that babies and pregnant women should be receiving the new flu vaccine.

This was when one of the mothers on Redwood's burgeoning Listserv reached out. Her name was Sarah Bridges. A psychologist and writer from Minneapolis, she had a son named Porter, diagnosed with autism, who had received a thimerosal-containing vaccine a decade earlier. At first it was thought the little boy's seizures were due to epilepsy. But the hospital record clearly stated: "Brain damage from pertussis vaccine." The mother was told that a report had gone to the CDC about Porter's adverse reaction. A sympathetic physician alerted her to the existence of the National Childhood Vaccine Injury Act.

Passed by Congress in 1986, the legislation wasn't as altruistic as it sounded. It came about under pressure from pharmaceutical companies who, after years of being sued and ordered to pay damages for injuries and deaths caused by their products, threatened to stop making vaccines altogether if the government didn't intervene. Fearing a vaccine shortage that might result in widespread disease, Congress obliged. Released from future liability, the drug companies rushed to add new (and lucrative) vaccines to the children's schedule. In exchange, the legislation established a "vaccine court" that could award taxpayer-funded damages to families who could prove that their child's disability came from vaccines. And Sarah Bridges had developed plenty of evidence.

The family set about filing a claim. Summarizing all of Porter's medical records, it was obvious that his seizures were intractable and his other disabilities unchanged as he grew older. Special education testing revealed what they'd suspected: their son's brain had been severely impaired, and he was presumed to have developed autism due to the DPT shot he'd been given. "The government received our petition and immediately conceded," Sarah wrote in a five-thousand-word article for the *Washington Post* in 2003. "As part of the settlement the government set up a trust to provide for all of Porter's medical needs."

Some years earlier, attending Wellesley College, Sarah had become close friends with Max Kennedy's future wife Vicki, whose efforts with the Vaccine Court Sarah followed closely. "You really need to connect with Bobby," Vicki told her friend, and they cooked up a plan. Sarah would pay a visit to the Hyannis Port compound when Bobby was around, and Vicki would make sure to introduce them.

Lyn Redwood recalls Sarah informing her: "Hey, I have a connection to Bobby Kennedy, what do you want me to take to him?" Redwood remembers "going through all my stuff, saying okay, he's got to see this, he's got to see that. We sent a big box to Sarah, and she was able to bring it to the Cape." It was the summer of 2004. Bobby remembers his sister-in-law coming to him: "Vicki said there's a friend, she wants to talk to you, she's a psychologist, she has a kid who was vaccine-injured, and she won a damage settlement from the Vaccine Court."

Sarah remembers: "That piqued his interest. What inspires the court to give out these settlements acknowledging that kids got autism? So he agreed to get together, but very begrudgingly. He wanted me to know he was very content trying to clean up rivers. One afternoon after sailing, we went to sit down and I brought out the box. I told him these were studies that my organization had collected, peer-reviewed information about what was going on with thimerosal. He said, 'Sorry, but I don't want to get involved with something that's vaccines, our family have always been huge advocates for public health.' We went back and forth for a while. Finally in exasperation I told him, 'I'm not going to leave unless you at least read through some of this.'"

* * *

There are different ways of interpreting what happened to Bobby when he sat down that night with a pile of papers in front of him, poring through an eighteen-inch-thick stack of documents that had only surfaced due to a group of accomplished women with mentally disabled children, women "who had enough resilience in their life, they were not going to be defeated by this," as Bobby put it in 2022.

After becoming president, JFK founded a National Institute of Child Health and Human Development to support continued research. He also created a Presidential Panel on Mental Retardation and signed the first comprehensive child health legislation. In 1965, Senator Robert Kennedy made an unannounced visit to the Willowbrook State Development Center on Staten Island—and, refused entry, kicked down the door. This was a "snake pit," he declared, where thousands of residents were "living in filth and dirt, their clothing in rags, in rooms less comfortable and cheerful than the cages in which we put animals in a zoo." Inside, pharmaceutical companies conducted cruel and often-lethal experiments on incarcerated children. Suddenly in the national spotlight, Willowbrook became a symbol "in the evolution of the legal rights of people with disabilities to live in dignity."

As a boy, Bobby had worked as a counselor and coach of special needs children at Camp Shriver. His aunt Eunice had founded the Special Olympics. And yet, as Bobby would reflect, "I never saw anybody with autism. We prided ourselves at the Special Olympics on being able to take care of every child, no matter how serious their disabilities. Even kids who were functionally vegetative, we would figure out games for them, sit them on a little stool or table and have them push a bean bag onto the ground and then everybody would applaud. So there was no kid we ever ran into that we couldn't accommodate in some way. A kid with full-blown autism, there was nothing we could have done—because they're sometimes violent, some bite and scream, they have terrible tactile light and audio sensitivities. And that's why the parents become the sole care-takers because nobody can handle them. We could not have handled these kids, but we never saw them."

There was another personal reason for Bobby to overcome his resis-tance. "My son Conor had severe food allergies and twenty-nine emer-gency room visits by the time he was two years old," he explained. "There were fewer foods he could eat than he was allergic to; just to keep him alive was a daily terror. But my wife had the ability to know immediately when Conor was going to get sick before he even did. She could put her hands on him and know which allergen was hurting him. It was completely invis-ible to me, but she knew. So I knew that I should listen to these mothers."

Bobby's efforts to curtail pollution of the Hudson River played a role, too, because there were often significant discrepancies between what scien-tists at the regulatory agency claimed and what the fishermen observed. In one example, the Department of Environmental Conservation (DEC) maintained that goldfish—which began disappearing when a new sewage treatment plant changed its chlorine protocols—didn't exist in the river. But Bobby had witnessed a once-flourishing goldfish population that fish-ermen had made a living off of selling to the pet trade. Another time, the DEC commissioner maintained that it hadn't included eel on a list of fish whose catch was being banned due to high levels of PCB contamination "because they hadn't yet determined whether an eel was a snake or a fish," Bobby recalls with a laugh. "Well, that had already been decided. So, I was skeptical of regulatory agencies and their 'experts.'"

Bobby's book *Crimes Against Nature* had included a couple of disturbing revelations about the Centers for Disease Control. A month after Bush took office, the health agency selected five individuals with ties to the lead industry to serve on its Advisory Committee on Childhood Lead Poisoning Prevention. More recently, the CDC replaced an environmental health advisory panel with industry representatives, a move denounced by ten leading scientists in the journal *Science*. They wrote that "regulatory paralysis appears to be the goal here, rather than the application of honest, balanced science."

Now his eyes were being opened to documents that the CDC had reluctantly released. He told me: "I knew how to read science and was reading mainly the abstracts—because they were stunning, like a kick in the gut. The science was so clear, just study after study after study saying the same thing about the impacts of vaccine mercury."

Then he came to the CDC's Simpsonwood transcript. "Until I read that, I wasn't planning to do an article. Then I said to myself, 'I've got to let people know what happened.'"

Four years had gone by since the CDC issued private invitations to 52 attendees from its own agency and the Food and Drug Administration (FDA), as well as the leading vaccine specialist from the World Health Organization and representatives of every major vaccine manufacturer. Everything at Simpsonwood was to be strictly "embargoed," the CDC announced—no photocopies, no papers leaving the room with the participants.

At Simpsonwood, controversial discussion centered around an internal CDC study by Tom Verstraeten, an epidemiologist whose team had scrutinized the Vaccine Safety Data Link, the nation's largest data base containing some 100,000 children's medical records. Verstraeten compared outcomes between infants who received the thimerosal-containing hepatitis-B vaccine in their first thirty days to those who got the shot later or not at all and found a substantially elevated risk for autism (along with attention-deficit disorders, speech delays, and hyperactivity) in the first group. Verstraeten told his peers he was stunned by what the analysis showed. Perhaps adding insult to injury, in 1991 the CDC and FDA recommended three additional mercury-containing vaccines be given to very young infants.

The Simpsonwood meeting minutes recorded one participant responding to Verstraeten's findings: "What if the lawyers get ahold of this? There's
not a scientist in the world who can refute these findings." A number of
other experts also raised alarms. Dr. William Weil, a prominent Michigan
pediatrician, told his colleagues: "There's a host of neurodevelopmental
data that would suggest we've got a serious problem." Dr. Bob Chen, chief
of the CDC's vaccine safety division, expressed relief that "given the sensitivity of the information, we have been able to keep it out of the hands
of, let's say, less responsible hands."

Bobby told me in 2022: "They spent the first day talking about how
the science is clear and there's no way around it. Something really horrible
has happened, potentially poisoning a generation of children. And the
second day, they talk about how to hide it from the American people. It
was shocking to see these regulators, most of whom presumably got into
public health because they were concerned about children, now seeing
their own skins on the line and deciding to cross the medical border into
a vortex that has now consumed public health in this country."

The box held damning emails, as well. "I'm not sure there will be an easy
way out of the potential perception that the FDA, CDC, and immunization-
policy bodies may have been asleep at the switch re: thimerosal until now,"
wrote Peter Patriarca, director of viral products for the FDA. He added that
the coziness between regulatory officials and the pharmaceutical companies
"will also raise questions about various advisory bodies regarding aggressive
recommendations for use" of thimerosal in child vaccines.

Bobby would write about the follow-up to Simpsonwood: "The CDC
paid the Institute of Medicine to conduct a new study to whitewash
the risks of thimerosal, ordering researchers to 'rule out' the chemical's
link to autism. It withheld Verstraeten's findings, even though they had
been slated for immediate publication, and told other scientists that his
original data had been 'lost' and could not be replicated. And to thwart
the Freedom of Information Act, it handed the Vaccine Safety Datalink
program, a database that contained the medical records of over 400,000
children, over to a private company, declaring it off-limits to researchers.
By the time Verstraeten finally published his study in 2003, he had gone
to work for [pharmaceutical company] GlaxoSmithKline and his initial

findings had been reworked four times between 1999 and 2003 to bury the link between thimerosal and tics."

That paragraph appeared in Bobby's article headlined "Deadly Immunity," which came out simultaneously in *Rolling Stone* and on Salon. com, on June 16, 2005. He recalled years later: "That took me down a wormhole on this issue, ended up changing my career in many ways and taking me away from what I'd really like to be doing - which is to be suing water polluters and working on clean energy and environmental issues."

* * *

After his initial deep dive into the documents at the family compound, he spent months researching the piece, writing in the conclusion that he "devoted time to study this issue because I believe that this is a moral crisis that must be addressed." The big scientific question Bobby felt he needed to answer was this: What happens to the mercury from thimerosal once it enters a child's body? Since mercury is the most toxic non-radioactive element known to science, a thousand times more neurotoxic than lead, couldn't it be causing IQ loss or long-term brain injury?

Bobby was used to consulting experts, and he began by calling Francis Collins at the National Institutes of Health (NIH) and Katherine Stratton with the IOM. They both passed the buck, suggesting he contact Dr. Paul Offit, who had emerged as the public face explaining why vaccines did not cause autism. Bobby knew that Offit held a $1.5 million research chair at Children's Hospital funded by Merck, as well as holding a patent on an antidiarrheal vaccine he'd developed with the pharmaceutical giant.

Bobby had also seen a leaked memo written in 1991 by Dr. Maurice Hilleman, then the president of Merck's vaccine division. Hilleman warned the company that administering its shots to six-month-old babies could expose them to dangerous levels of mercury. Thimerosal should be discontinued, "especially when used on infants and children," and "the best way to go is to switch to dispensing the actual vaccines without adding preservatives." Merck, however, had kept on supplying thimerosal-containing vaccines until the fall of 2001.

So Bobby thought that reaching out to Dr. Offit as the last word on mercury "would be like me contacting EPA to ask about a coal regulation and they send me to Don Blankenship, who runs Massey Coal." But he made the call and related to me: "Offit was very complimentary. He said he got into public service because he was inspired by my father. I'm susceptible to that kind of flattery, so I was inclined to like him and listen. Then I asked, how can you regulate mercury in fish by telling pregnant women not to eat it, at the same time you're recommending they get flu vaccines that are loaded with mercury? And he replied in a very patronizing way, 'Well, Bobby, there are two kinds of mercury, a good mercury and a bad mercury.' At that point I knew that his argument was not with me, it was with the periodic tables. No toxicologist or expert in metals will tell you there's such a thing as a 'good mercury.' I asked, 'Why do you say that?' Offit said, 'Because ethylmercury is expelled from the body so fast, it doesn't have time to do injury.'" While "bad" methylmercury lodges in fish tissue for some time, "good" ethylmercury used in vaccines quickly disappears from the blood.

Bobby knew this was what Eli Lilly, the corporation that invented thimerosal in the 1920s, had been claiming ever since. The first clinical study of thimerosal in vaccines was published in November 2001 by a lead author, Michael Pichichero, who had an acknowledged financial tie to Eli Lilly. It appeared in the prestigious medical journal *Lancet* and certainly seemed to offer validation for what Dr. Offit and others were saying. Infants who received vaccines with thimerosal possessed minuscule mercury levels in their blood, well below EPA safety limits.

In retrospect, the timing is interesting. Behind the scenes hovered Eli Lilly and its ties to the Bush family. Before becoming vice president under Ronald Reagan, George H. W. Bush had been a director on the company's board. His son George W. Bush's budget director, Mitch Daniels, arrived fresh from being an Eli Lilly vice president. And Eli Lilly's chairman and CEO Sidney Taurel had been appointed to a presidential council advising Bush on domestic security in June 2001, a few months prior to the 9/11 attacks.

In 2002, at the last minute, three anonymous riders were attached to the massive bill that created the Homeland Security department. These riders, it came out, emanated from the White House by way of a "cutout," soon-to-retire Republican Majority Leader Richard Armey, and were directly aimed at dismissing any thimerosal lawsuits filed in civil court. The bill sailed through the House and, despite SafeMind's sounding the alarm about the riders, also passed the Senate. That same day, a Justice Department order sought to permanently seal all thimerosal discovery materials presented in Vaccine Court.

In January 2003, Lyn Redwood and other outraged moms generated enough blowback in the media to get the "Lilly rider" overturned and force the Bush administration to withdraw the sealing of thimerosal records. Edward Kennedy, longtime chairman of the Senate Health Committee, helped upend the skulduggery and keep open the door for families to be compensated by the government's Vaccine Court if thimerosal causation of their injuries could be proven.

The Pichichero study hadn't really explained how the mercury left the body, since his team didn't find it in the blood, the sweat, the hair, the urine, the fingernails, or the fecal material. To try to answer this nagging question, the National Institutes of Health (NIH) commissioned a series of studies, foremost by a developmental toxicologist at the University of Washington named Thomas Burbacher. His mercury experiment utilized macaque monkeys, verifying that indeed ethylmercury was gone from the blood within a week. Bobby recalls: "But when Burbacher sacrificed the macaques and autopsied them, he found that the ethyl mercury had *not* left the body. It had crossed the blood-brain barrier and lodged in the brain. It turned out to immediately transform into organic mercury, which is the most toxic form."

After Bobby came across the Burbacher findings, he called Paul Offit and asked again how the doctor knew that the mercury leaves the blood. Offit immediately cited the Pichichero study. "I said, 'But Dr. Offit, the Burbacher study showed that Pichichero got it wrong.' There was dead silence on the phone because he was shocked that I knew the science and was conversant in it. We both knew in that moment that I'd caught him lying. Finally he said, 'Well you're right, Bobby, it's not a single study, it's a whole mosaic of studies.' I said, 'Can you provide me any citations on the other ones?' He said, 'I'll send them to you.' That's the last I heard from him. And that's when I realized, my God, the emperor really has no clothes."

In his article, Bobby would describe internal corporate documents showing that Eli Lilly was long aware of what thimerosal could do to humans and animals: "In 1930, the company tested thimerosal by administering it to twenty-two patients with terminal meningitis, all of whom died within weeks of being injected—a fact Lilly didn't bother to report in its study declaring thimerosal safe. . . . During the Second World War, when the Department of Defense used the preservative in vaccines on soldiers, it required Lilly to label it 'poison.'"

Bobby had offered a preview of his piece in an op-ed for Knight Ridder in mid-April 2005, touching on the Simpsonwood saga and introducing what had happened to Sarah Bridges's son: "After a seven-year legal fight, the US government acknowledged that Porter was damaged by his vaccines. There are now 520,000 autistics in the United States with 40,000 new cases each year."

Both Salon and *Rolling Stone* scrupulously fact-checked Bobby's article. But there was a firestorm following publication, with the pharmaceutical industry and government health agencies bombarding the two outlets with furious letters. "Salon and *Rolling Stone* corrected six minor errors," Bobby later wrote. "None of those errors were even remotely material to the article's central propositions. Four were minor clarifications or corrections or inadvertent editing or punctuation errors and one was a wrong name applied to a congressional staffer. The only error to rise above the level of trivial nitpicking was the assertion that a six-month-old infant could receive a level of mercury from vaccines that was '187 times greater than the EPA's limit for daily exposure to methylmercury, a related neurotoxin.' As the email record exchanges with editors from the time period confirm, that error, along with most of the others, was made by *Rolling Stone* and Salon editors as they cut my 16,000-word submission to 4,700 words. In the days after the publication of the correction in both *Rolling Stone* and Salon, editor Joan Walsh apologized to me, in writing, for introducing those errors themselves."

* * *

The backlash continued at a rapid clip. Ten days after "Deadly Immunity" appeared, the *New York Times* ran a front-page article headlined "On Autism's

Cause, It's Parents vs. Research" (June 25, 2005). "They have failed," the reporters said of the parents, "because the CDC, FDA, WHO, Institute of Medicine [IOM] and the American Academy of Pediatrics have all largely dismissed the notion that thimerosal causes or contributes to autism. Five major studies have found no link. . . . The issue has become one of the most fractious and divisive in pediatric medicine." The lengthy *Times* piece didn't mention any of Bobby's evidence, only a one-liner about his "arguing that most studies of the issue are flawed and that public health officials are conspiring with drugmakers to cover up the damage caused by thimerosal." The chairwoman of the IOM's thimerosal panel, Dr. Marie McCormick, was quoted: "It's really terrifying, the scientific illiteracy that supports these suspicions." The *Times* briefly touched on the fear of public health officials that the anti-thimerosal campaign "is causing some parents to stay away from vaccines, placing their children at risk for illnesses like measles and polio."

The *Columbia Journalism Review* would conclude in an article by Daniel Shulman (November/December 2005) that the "one-sided" *Times* story "casts the thimerosal connection as a fringe theory, without scientific merit, held aloft by angry, desperate parents," while portraying legitimate scientists with a contrarian viewpoint as eccentric or even crazy. Shulman wrote: "Several reporters I spoke with who have covered the thimerosal controversy described the *Times* story as a smear. One called it a 'hit piece.'"

In his article "The RFK Jr. Tapes," David Samuels wrote that "the case [Bobby] made . . . was no more or less plausible and empirically grounded than the cases that he and dozens of other environmental advocates had been making for decades against large chemical companies for spewing toxins into America's air, water, and soil." Regarding the backlash Bobby faced, Samuels argues that "it doesn't take an alarmist to recognize how fast and far the term 'conspiracy theory' has morphed from the way it was generally used even a decade ago . . . [it] has become a flashing red light that is used to identify and suppress truths that powerful people find inconvenient." He continued, "Whereas yesterday's conspiracy theories involved feverish ruminations on secret cells of Freemasons, Catholics or Jews . . . today's conspiracy theories include whatever evidence-based realities threaten America's flourishing networks of administrative state bureaucrats, credentialed propagandists, oligarch, and spies."

Bobby had come to expect this from other media. "In November 2003," he recounted in *Crimes Against Nature*, "when environmentalists around the country were engaged in fighting the Cheney energy bill, the NRDC was anxiously trying to get me airtime because no one was talking about the bill on TV. Fox TV host Bill O'Reilly agreed to schedule me, but only with the explicit proviso that I wouldn't say critical things about George W. Bush. I would first have to do a pre-interview to make sure I was capable of talking about the environment without bad-mouthing the president. Later, Fox decided that even this was too chancy; they would just tape the show, rather than risk me going off the reservation on live TV. The same week, Tom Brokaw, a committed environmentalist and fly fisherman, scheduled me for a segment on *NBC Nightly News*—but the producers bumped me for yet another Michael Jackson story."

But Bobby had believed the *New York Times* was different. Until now, he had been a kind of darling with the paper's editorial staff. Among outside contributors, only he and Henry Kissinger were invited to publish an op-ed every six months, the maximum number that the *Times*'s rules allowed. Bobby initially responded with a letter to the editor that the paper published (July 2, 2005), where he pointed out that the four European studies relied upon by the CDC and IOM to defend thimerosal "were written principally by vaccine industry consultants and employees without revealing the bias of their authors. They are all flawed. . . . The institute selectively ignored the hundreds of biological, toxicological and epidemiological studies linking thimerosal to the wide range of neurological disorders, including autism."

Next Bobby asked for and received a meeting with the paper's executives and science writers. He brought along Sarah Bridges as well as Dr. Boyd Haley, chairman of the chemistry department at the University of Kentucky, who had conducted numerous lab experiments revealing the toxicity of thimerosal but had been dismissed as a quack in the *Times*'s article. The trio was ushered into a small room at the headquarters in Times Square, where Bobby expected a private meeting with the editor. Instead, "they had assembled a group of science editors that were so hostile and antagonistic, it was like talking to a brick wall," he remembered. "They were absolutely determined that there would be no public discussions in their paper about mercury and neurological disorders."

Bridges recollected that "Bobby's sentences were cut short by rapid retorts, as if the room was laced with invisible mines." Dr. Haley had a phonebook-sized stack of articles perched on his lap ready to share. Bobby offered DNA, animal, genetic, epidemiological, and biological studies, only to be repeatedly met with the statement "The CDC says the vaccines are safe." Surprised and even somewhat mystified at the response, he didn't take it lying down.

Bobby pointed out that one of the *Times*'s own reporters, Judith Miller, had promoted going to war with Iraq for a solid year, highlighting the lie that Saddam Hussein possessed weapons of mass destruction. For this, the paper eventually issued a public apology. "And I said, 'you're going to have to issue another apology someday - because that war is going to cost us $3 trillion, but the treatment of this generation of children is going to far exceed those costs.' This was when my relationship with the *Times* changed. I think I may have published one or two more pieces with them after my [thimerosal] article came out, but after that it was just hostility."

The apparent hostility surfaced publicly on Bobby's next birthday in January 2006, when *Times*'s columnist John Tierney published an op-ed titled "Not in Their Backyard." He cited Bobby's earlier op-ed that a wind farm planned for Nantucket Sound would "damage the views from sixteen historic sites," one of which happens to be the Kennedy compound at Hyannis Port. He didn't specify the damage, but this is what it would amount to: If you stood in Hyannis Port, six miles from the wind farm, the turbines on the horizon would appear to be a half inch high, about the size of a fingernail. As we have seen, that was not Bobby's point. His complaint was that the windmills' proximity to the shore would endanger the livelihoods of fishermen. Shortly after his ill-fated meeting in Times Square, Bobby joined a rally organized by SafeMinds on Washington's Capitol Steps. He told a gathering estimated in the thousands: "I've worked environmental issues for twenty-five years. And I know what captive agency phenomena is. . . . I've made a profession of reading phony science, because I see it every day. I've sued over 400 polluters, and this is how they defend themselves." The dozens of mothers he'd encountered "were not hysterical people. They were people who had their feet on the ground. They had 'calmly and deliberately gone through the scientific literature' and concluded that 'we're making the sickest generation of children in the history of our country.'"

It was an impassioned speech, quoting from the Simpsonwood transcripts that so shocked him. And Bobby pulled no punches about who he believed was to blame. The CDC and IOM, having disappeared the original data in the Verstraeten study, proceeded to work with the pharmaceutical companies to "gin up four phony European studies," published in the American Academy of Pediatrics magazine, "which receives 80 percent of its revenue from the vaccine industry." The mainstream press contented itself with reading only the CDC's descriptions of the studies, not the studies themselves. None of the scientists involved disclosed any of their myriad conflicts, and Bobby singled out Paul Offit as "the poster child for the term 'biostitute.'"

Was Bobby overreacting, taken in by well-intentioned but distraught mothers looking for something to blame for their child's affliction? Was he jumping to conclusions that polluters and vaccine makers were all part of the same mix? He'd done wonders for rivers with his tactical legal brilliance and made such a name for himself that assuming high political office did not seem out of the question. Might he now be risking everything he'd achieved for a quixotic cause?

But risk taking was something that by then Bobby could have patented. In a way, ever since his father's assassination, he felt he had nothing to lose. Yet his very survival gave him a sense of destiny and duty, a belief that you had to "die with your boots on" for a more just and safer world. The rampant corruption and lack of accountability that he saw wherever he looked pained him deeply. And as the Bush administration began its second term, Bobby became convinced that the president's reelection was not legitimate.

CHAPTER TWENTY-THREE

INVESTIGATING THE 2004 ELECTION

Snowed under by promoting his new book and studying the box of revelations pointing to a cover-up of mercury's impact on many vaccinated children, Bobby admittedly hadn't paid much attention to the integrity of the 2004 presidential election. Democratic candidate and Massachusetts Senator John Kerry had spent fishing time with Bobby over the years on the Cape, and Kerry had windsurfed in 2002 for forty-five miles over open ocean from Woods Hole to Nantucket with Bobby's brother. So they knew each other well. Now the country's vote had been extremely close, especially in Ohio, where the results swung the election in George W. Bush's favor.

"I kept the same kind of deliberate blinders on that much of the media did," Bobby later said. Then, over the Christmas holidays a year after the election, he was staying in Sun Valley, Idaho, at the home of Larry and Laurie David, when the "Seinfeld" creator urged him to read a book called *Fooled Again* written by NYU professor and media critic Mark Crispin Miller. It was subtitled *"How the Right Stole the 2004 Election and Why They'll Steal the Next One Too (Unless We Stop Them)."*

Bobby found Miller's findings impressive enough to bring up the subject a few days later when he went skiing with Jann Wenner, the *Rolling Stone* founder/publisher and a longtime friend. "He encouraged me to do a piece and said, 'We'll print one if you write it,'" Bobby recalled. After initial hesitation, Bobby agreed.

This led him to the official record of a congressional inquiry into the election, presided over by Rep. John Conyers of Michigan. In testimony released by the House Judiciary Committee in January 2005, witnesses from both parties—elected officials, poll observers, voting machine company employees, and numerous voters—spoke about the harassment and actual vote suppression they'd endured in Ohio the previous November. Bobby started calling them up. He would spend the next six months researching what happened.

Like many of what became at least temporary crusades, this one had roots in his family history. "I'd spent a lot of time thinking and writing about populism and racial justice in the South, because of my family's involvement," he told me in 2022. "My father and uncle and Martin Luther King had all come to the same conclusion, that the key to civil rights in our country was voting rights. So the initial hook that got me into this was realizing how the blacks were being systematically disenfranchised by right-wing groups.

"First there were the so-called felon purges, with a wide net, because they would find a felon named Washington or Jefferson or Lincoln and purge everybody with that name from the voting rolls. They were making it harder for the military to vote, because so many were black. Then I saw how some of the counties would send a single voting machine to a black neighborhood where there'd be eleven-hour lines, and ten machines to a Christian college, which made voting very easy. These manipulations were basically overruling all the things my father had done to make it possible for blacks to vote—abolish poll taxes and literacy tests and all the tricky mechanisms used during the Jim Crow era."

As Bobby told an interviewer upon publication of his latest article, "Was the 2004 Election Stolen?" in *Rolling Stone* in June 2006: "The Republican Party has been using old-fashioned, Jim Crow, apartheid-type maneuvers to steal the last two national elections. If you look at who's being denied the right to vote, on absentee ballots, on provisional ballots, it's Hispanics, it's blacks and it's Native Americans, and the Democratic Party ought to be touting this as the biggest civil rights issue of our time. But they are ignoring it, and that really is shocking."

Bobby said his investigation had uncovered more than 350,000 principally Democratic voters in Ohio either being prevented from casting ballots or not having their votes counted. Those votes would have been more than enough to put John Kerry in the White House, given that President Bush won the state by only 118,000 votes.

Bobby cited three factors in maintaining that "our democracy is broken": a campaign finance system "of legalized bribery allowing corporations and the very wealthy to control the electoral results," "the failure of the American press" to expose a "culture of corruption," and the suppression and fraud surrounding the voting process. As Russ Baker writes on the WhoWhatWhy website (October 10, 2022): "Among 'reasonable' people, it's taken as axiomatic that our elections are secure. But that's not true. It's a crisis for our country, beyond the hundreds of election deniers running for key positions overseeing future elections, and the situation sets us up for future disasters."

Baker is talking about precisely what Bobby was—"the vulnerabilities of electronic voting systems . . . that can be manipulated in various ways that are difficult to detect." In June 2006, Bobby had filed what's called a *qui tam* lawsuit (meaning an action against a person or company on the government's behalf) against some of the voting machine companies. He pointed out that the Help America Vote Act passed in 2002 was Orwellian sleight-of-hand. Orchestrated by Rep. Bob Ney of Ohio, a key provision discouraged vote verification by paper ballots while requiring states to buy voting machines from the Diebold corporation.

Bush's predecessor, Bill Clinton, read Bobby's article and urged others to, believing it "made a very persuasive case." "What was clear," Clinton wrote, "is that the Secretary of State [Ohio's Kenneth Blackwell] was a world-class expert in voter suppression." The former president wasn't alone. Senator Howard Dean added that he was "not confident that the election in Ohio was fairly decided." Bobby quoted Lou Harris, the father of modern-day political polling, stating that "Ohio was as dirty an election as America has ever seen."

Interestingly, it was Salon—copublisher of Bobby's piece on mercury and vaccines a year earlier—that now led the backlash against his voting rights effort. Farhad Manjoo, in a critique as long as the article itself, suggested that Kennedy's arguments fell short and, in any case, wouldn't fix the problem.

A few days later, the two squared off on Salon in a pair of articles, with Bobby expressing how it was good to see Manjoo weigh in. Bobby wrote: "Unlike reporters in the mainstream media, Manjoo has displayed a willingness to actually read the published reports that document the electoral travesty that occurred in Ohio. It is a shame, however, that in his attempt to debunk my article, he commits precisely the sins of omission and distortion that he accuses me of having perpetrated."

Without citing evidence on either side of the argument, the *New York Times* went after Bobby in a Sunday Styles piece headlined "Another Kennedy Living Dangerously" (June 25, 2006). Reporter Mark Leibovich claimed that "Mr. Kennedy [was] hitching his name to a cause that has largely been consigned so far to liberal bloggers and which nearly all Democratic leaders and major news media outlets have ignored and which, unsurprisingly, Bush supporters have ridiculed." After quoting Manjoo on the "numerous errors" and "deliberate omissions," the reporter continued: "It is impossible to read the *Rolling Stone* article without wondering how Mr. Kennedy's audacious accusations might relate to his philosophical evolution or even affect his political viability." Bobby, without a doubt, wasn't thinking about "political viability." He was just telling the truth about what he believed his research showed.

Bobby addressed the Ohio findings on his "Ring of Fire" weekend radio show, part of the Air America network. He agreed to an interview with Tucker Carlson on Fox, who at the end called this "not a crackpot theory, but a serious issue" and urged viewers to read Bobby's ten-page-long piece. On the other end of the political spectrum, he went on Thom Hartmann's radio show, and Thom reminded him how the *New York Times* back in November 2001 headlined: "Study of Disputed Florida Ballots Finds [Supreme Court] Justices Did Not Cast the Deciding Vote." The paper's implication was that, by the thinnest of margins, Bush had assuredly defeated Gore and democracy remained intact. However, Hartmann noted that buried in paragraph seventeen was a sentence reading: "An approach Mr. Gore and his lawyers rejected as impractical—a statewide recount—could have produced enough votes to tilt the election his way, no matter what standard was chosen to judge voter intent."

Bobby concurred. "They deep-sixed that, deep, deep in the story. This was the product of a one-year investigation by a whole raft of newspaper syndicates that actually went out and counted all of Florida. And the product of that investigation wasn't released until a month after 9/11, and the editorial staff of the *New York Times* decided to bury it because they thought it was a time of national emergency. But it was a momentous story that should have been on the front page - that the wrong person was in the White House." No one should doubt that it can happen in America.

Early in August, Bobby started work on his second piece for *Rolling Stone*. With the midterm elections looming and 80 percent of the ballots to be tallied by four touch-screen electronic voting machine companies (three of which had strong ties to the Republican Party), it seemed not only important, but urgent.

A key source was a consultant for Diebold Election Systems in Georgia named Chris Hood, who had helped the company promote its new touch-screen electronic voting machines. Hood was an African American whose parents fought for voting rights in the South during the 1960s. Hood painfully explained how he'd become part of a scheme to illegally install uncertified software in the machines capable of altering the ballots—prior to surprise victories by Republican candidates in the senate and the governorship. "What I saw," Hood said, "was basically a corporate takeover of our voting system."

A substantial percentage of Ohio's 2004 votes were also counted by Diebold software and Opti-scan machines, which malfunctioned particularly in the Democratic stronghold of Toledo. Not only was the man in charge of oversight, Ohio Secretary of State Kenneth Blackwell, Bush's campaign manager for the state, but he also turned out to own Diebold stock.

Bobby's article ran in the October 5, 2006, issue, headlined "Will the Next Election Be Hacked?" Who knows whether it opened some eyes, but in the election a month later, the Democrats won control of both houses of Congress. In late 2006, Diebold removed its name from the front of the voting machines in what a spokesperson called "a strategic decision on the part of the company."

Diebold sold its voting machine operation to ES&S in 2009. In October 2013, Federal prosecutors would file charges against the Ohio-based security and manufacturing firm, alleging that Diebold had bribed government officials and falsified documents to obtain business in China, Russia, and Indonesia. Diebold agreed to pay $50 million to settle the two criminal counts. Senator Kerry later said: "I know what happened. I know we won the election. We had all these lawyers on the ground, but they couldn't come up with a smoking gun. Not until months later, and then it was too late. Did you see that Ohio threw out all the analysis?"

Bobby's research and articles and his willingness to stand up and speak out raised important questions. What was the greater threat to democracy? Was it coming out and telling the truth about what you knew to the American people, taking the risk that many might wonder if their vote still meant something? Or was it going along with the myth that the powers that be in this country would never stoop so low as to allow an illegitimate president to be in office for two full terms?

CHAPTER TWENTY-FOUR

TAKING ON INDUSTRIAL AGRICULTURE

Bobby's longest-running fight during the first decade of the twenty-first century was against the factory farming industry, a campaign that came to involve rural citizens from thirty-four states with the Waterkeeper Alliance. It brought out such enmity that Bobby found himself shadowed and harassed by a corporate hireling who trailed him from speech to speech. And it eventually saw Bobby traveling all the way to Poland to address that nation's parliament about the dangers of the practices being exported by a hog farm monopoly in North Carolina.

When Bobby was growing up, the area around Hickory Hill remained very rural. The Kennedy home had cornfields on three sides and a raspberry patch on a fourth. Hickory Hill itself wasn't a production farm but had barns and pastures with farm animals as well as a garden. Bobby took care of the chickens and had a cow that he raised from a calf and showed off at the 4-H fair.

In the mid-1990s, as we have seen, as Riverkeeper's lead attorney, Bobby forged an unlikely alliance with upstate farmers in helping craft the landmark New York watershed agreement. So it was not a stretch for him to relate to Rick Dove shortly after that. Dove had been raised along the rural western shore of Chesapeake Bay, whose waters he'd seen contaminated by pollution from the Bethlehem Steel plant. He'd gone on to earn a law degree before joining the Marine Corps, where he served twenty-five years, including two tours of duty in Vietnam, and had been promoted to colonel and military judge by the time he retired in 1987.

"I wanted to follow a childhood dream to be a commercial fisherman," Dove recalls. "So I went out with my son on the Neuse River in North Carolina, where I'd moved. Those first three years were probably the happiest time of my life; we had three boats and a seafood store. Then all of a sudden the fish began to die. There were huge sores all over their bodies, and I got the sores on me, and my son got sick too. And we had to give it up."

This was in 1991, when the Neuse River suffered one of the largest fish kills ever recorded. More than a billion fish perished in a matter of days that September. Oxygen levels were normal, and nothing previously known could explain why this happened. Some fishermen became afflicted with respiratory problems and even memory loss. For a time, the stench of rotting fish was so bad that nearby residents wouldn't venture outside.

The culprit turned out to be a recently identified single-celled creature so tiny that 100,000 could fit on the head of a pin, known as *Pfiesteria piscicida*. It produced a neurotoxin in the water that paralyzed fish and was also getting into the air. And its arrival was related to runoff that had increased dramatically due to all the industrial swine and poultry farms that had replaced small family-run operations since the mid-1980s.

Dove was practicing civilian law when he happened upon an advertisement in the local paper about an organization called Riverkeeper that was looking to hire someone to patrol the Neuse River. Dove successfully applied in 1993. He first met Bobby at a meeting in Maine of fourteen then-existing Riverkeepers from around the US. The crew-cut Dove remembers walking into a room wearing his Marine uniform, right down to combat boots and a cap. "Everybody must have said to themselves, what the hell is this? But very soon we formed a bond," Dove remembers. Some of the other Keepers turned out to also be former Marines. By the close of the gathering, they began talking about forming an organization. According to Dove, "Bobby said, let me do that, and the idea for Waterkeeper Alliance was born."

It would take a while to formalize, but in 1999, the Alliance soon incorporated in New York and set up headquarters in the basement cellar of Pace University's Environmental Law Clinic, where Bobby was teaching. Dove became part of its first board of directors and set about educating Bobby, who recalls: "Rick was on fire. He put together Keepers on fourteen rivers across North Carolina and had them all in uniforms and highly organized. Then he assembled an air force, with more than twenty small planes flying over the big hog farms every day and catching them breaking the laws on dumping their waste into the waterways."

During the 1970s, three agribusiness barons in nearby states—Frank Perdue, John Tyson, and Bo Pilgrim—had devised a new way to raise poultry in small cages in a factory-like atmosphere and put a million independent farmers out of business in the process. A North Carolina state legislator, Wendell Murphy, figured why not translate such a methodology into his pork business? Traditional family farms had used time-honored techniques of good animal husbandry, recycling manure to fertilize the soil to grow feed crops and raising their animals in a humane manner. Murphy's "innovations" forced hogs into football field-sized warehouses where their concentrated waste started being dumped into open-pit "lagoons" about which regulators paid little attention. Their feed was saturated with toxic chemicals and antibiotics that made dumping even more toxic and, of course, dangerous.

In 1997, the North Carolina Legislature placed a moratorium on new swine farms. As the book *Wastelands* recounts, by the end of that year "Murphy was the unqualified champion of pork production, the 'Ray Kroc of Pigsties,' according to *Forbes*. He had 275,000 birthing sows, double the capacity of the runner-up, and 6 million hogs across his supply chain, from farrow (birth) to finish (slaughter). *Forbes* calculated Murphy's stake at a hefty $1 billion."

Murphy stepped down as a state senator and went into business with Smithfield Foods, having convinced the meat-packing owner Joe Luter to build the world's largest slaughterhouse with a killing capacity of 32,000 hogs a day. "They dropped the price of pork from 65 cents to 5 cents a pound, and it cost the farmer about 32 cents a pound to raise that pig to kill weight . . . so literally no hog farmer could stay in business unless he signed a contract with Murphy," Bobby pointed out. "They basically became indentured servants on their own land—2,200 factories in the

eastern part of the state with ten million hogs. It was the end of family farming for hogs in the state, and the beginning of a huge pollution source."

The statistics were staggering. A hog produces ten times the amount
of fecal waste by weight as a human being. So a facility with ten thousand
hogs is creating as much sewage as a city of a hundred thousand people. In
1999, devastating floods from Hurricane Floyd sent enormous amounts
of hog farm pollution into communities and waterways. Dove took Bobby
up in his Cessna to see the devastation for himself. He pointed out the
"witch's brew of nearly four hundred toxic poisons," as Bobby described
it in his book *Crimes Against Nature*, being loaded into the waste pits
and sprayed onto fields, saturating them with toxins in concentrations far
greater than growing crops could assimilate. Outraged, Bobby felt it was
time for another war against what Teddy Roosevelt once called the "male-
factors of great wealth."

* * *

In the summer of 2000, Bobby brought onboard a woman named
Nicolette Hahn. She was a lawyer from Michigan, inspired after hearing
Bobby speak there, to go to work for the National Wildlife Federation.
When they met after another of his talks two years later, Bobby encour-
aged her to apply to Waterkeeper, where she could "come in on the ground
floor." Hahn was soon promoted to senior attorney.

"Bobby talked to me about wanting to start a national campaign
against industrial agriculture," Hahn recalls. "At the very beginning,
I didn't understand why he was so interested in this. But to see the raw
unvarnished power being exercised by those corporations was shocking
and against Democratic values. Clearly, that's a lot of what was motivating him. I remember discussing with Bobby that the major environmental organizations were primarily urban-based and not looking at issues in
rural areas. Smithfield Foods had deliberately chosen areas of operation
that didn't just have low land value but where the people had little political influence. Especially communities of color."

In December 2000, Bobby called a press conference in Raleigh, North
Carolina, to announce his Alliance's first lawsuits against Smithfield Foods
for disposing hog waste in violation of several federal laws. "I was standing

at a podium in a large conference room overflowing with local press, when people I hadn't expected to see started filing in," he later recalled. "They were black, white, men, women—rough and weather-beaten farmers, many in overalls, who had driven from all over the state to thank my Waterkeeper colleagues and me for standing up to an industry that had bullied them for years. Many came from families who had occupied the same piece of land for generations." He described the scene as one of the most moving moments in his twenty years of environmental advocacy: "As I started to speak about the rural communities and once pristine waterways of eastern North Carolina, and how they have been ruined by industrial pollution, I was astonished to see tears flowing down so many wrinkled faces."

Hahn had spent weeks with Rick Dove organizing a Summit for Sustainable Hog Farming, to take place in midwinter, when farmers were best able to attend. Almost a thousand people gathered early in 2001 in Rick Dove's hometown of New Bern. They came from twenty states, including faith leaders, scientists, sociologists, and citizen activists. Bobby arrived at midday to interact and give the keynote speech. Attendees were stunned to learn that a factory with fifty thousand hogs produces the same amount of waste as a city of half-a-million people. The audience heard from Elsie Herring, the granddaughter of a slave whose family's land deeds had been illegally changed for the benefit of the pork industry. The African American minister of Herring's church brought a busload of people, a déjà vu of the civil rights movement.

According to Hahn, "we wanted to make it clear we weren't against meat and the rearing of livestock, but needed well-raised animals to make the system ecologically viable." So there were spokespeople on hand from Northern California's Niman ranch, a model of humanely raised, antibiotic, and hormone-free beef cattle and hogs. "For many years afterwards," Hahn says, "I was told what an impact this conference had on what farmers were doing and made them feel much less isolated."

The following September, a Reagan-appointed federal judge, Malcolm Howard, shocked the agribusiness moguls by ruling that they were in violation of the Resource Conservation and Recovery Act for illegally operating without permits under the Clean Water Act. This meant that almost none of Smithfield's 1,500 corporate-owned or controlled facilities were in compliance with the law.

A few weeks later, however, the Bush administration met with lobbyists from the affected industries, and the Mercatus Center think tank presented a plan to alter the Clean Air and Clean Water Acts. The EPA then proceeded to develop a new set of regulations, written in concert with the agribusiness giants. No sooner had the weakened rules been promulgated than Smithfield finally obtained permits for all its facilities. Their tons of toxic waste oozing from the massive lagoons in North Carolina and elsewhere would no longer be illegal.

In April 2002, Waterkeeper held another well-attended conclave in Clear Lake, Iowa, where Bobby received a standing ovation. Shortly before this, the Pork Producers Council had created a front group calling itself the Truthkeepers. Its president was a former pork factory manager from South Dakota named Trent Loos. When Loos arrived early at the Summit, making sure to get a front-row seat in the packed auditorium, Don Webb walked over to say hello.

Webb had sold his hog business near North Carolina's Outer Banks after he sought state officials' advice on alleviating their nasty odor but whose solution (stirring yeast into the waste "lagoon") only made it far worse. Webb went on to purchase a large piece of lakeside property, only to find Wendell Murphy building a new factory farm right next to him. That's when Webb joined Bobby's campaign. According to Dove, "He was like a preacher when he talked, a . . . very smart, educated man. Whenever he appeared, it was always in his farming clothes; Don said he liked people underestimating him."

As Bobby described Loos's antics in *Crimes Against Nature* in 2004: "Loos follows me around the country, shadowing me at public appearances in New York, California, Illinois, and Kansas. Sporting a handlebar mustache, cowboy boots, a dandy scarf, and, appropriately, a black Stetson, he sits in the front row at my speeches and conferences, holding a microphone and a tape recorder. In Minnesota he followed me into a restroom and lurked behind me at the urinal. In Iowa he waited for me in a dark parking lot and made menacing comments. At an agricultural conference in Gettysburg, Pennsylvania, he trapped me in a hotel corridor and I had to threaten fisticuffs to access my room. I was recently relieved of Trent's company when he was charged with cattle theft in Nebraska. He pleaded guilty, and his probation prohibited him from leaving the state."

According to Dove, "He tried to get the best of Bobby, but he never did." As Nicolette Hahn puts it, "Don Webb had a huge influence on Bobby." And on Rick Dove, who, when he observed a Smithfield lawyer showing up at every presentation he gave, would call him out: "Hey, look everybody, he's trying to find something he could sue me for. But you know what? I'm not worried a bit, because everything I'm saying is true." As Dove says, "That was Bobby's philosophy as well."

At the Iowa event, almost nine months after 9/11, Bobby called the factory meat integrators more of a threat to America than Osama bin Laden. The *Des Moines Register* then published his op-ed under the headline "I'm serious: Hog lots threaten democracy," where he addressed how the same consolidation had happened in Iowa, which lost 41,000 independent hog farmers since 1984 with almost half of the remaining 10,500 controlled by the big integrators. Smithfield Foods had expanded into Missouri, Iowa, and other states, while its PAC plowed money into state and federal elections.

Bobby and Dove commenced a tour of numerous states in the Midwest and East, giving talks in a variety of venues. "Bobby put the fear of God into Smithfield Foods and Wendell Murphy," Dove enthused in 2022. "They were beside themselves because he was such a good spokesperson. He just nailed the issues."

Bobby wrote of driving through the industrial farm country with Dove and observing "the desperate, ruined communities the meat moguls have left in their wake: deserted main streets; bankrupted hardware and feed stores; banks, churches, and schools shut down. The heartland and its historic landscapes are being emptied of rural Americans and occupied by corporate criminals. These industries are murdering Thomas Jefferson's vision of an American democracy rooted in tens of thousands of independent freeholds owned by family farmers, each with a strong personal stake in our system."

Dove's favorite stop was an event invited by the Hog Industry Growers Association in Northfield, Minnesota, which had invited Bobby figuring he'd be heckled out of the room. "I'll never forget that meeting," Dove recalls. "It was full of agitated and concerned hog growers. Bobby laid it on the line. He said here's who you are and what you're doing, and here are the negative consequences of your actions. They didn't like it one bit, but what always impressed me most about Bobby was his fearlessness. I think he would've made a great Marine."

After all the media coverage generated and the billboards that Dove and friends plastered across the state, scientists at Eastern Carolina University surveyed residents and found that their number one concern wasn't crime or taxes. Ninety percent wanted to get rid of the hog lagoons and pollution and restore their damaged rivers.

By the time of Waterkeeper's third Sustainable Summit in June 2003 at Gettysburg, a consortium of environmental groups joined with other sponsors. With a crowd of about 250 people from across the US, the conference included workshops on farming practices that included allowing animals to roam pastures rather than be confined. Bobby coauthored a *New York Times* op-ed with Eric Schaeffer, former director of the EPA's office of regulatory enforcement, headlined "An Ill Wind From Factory Farms" (September 20, 2003).

It remained an uphill fight, with President Bush receiving the 2004 Friends of the Pork Producer Award from its national Council for his "tireless efforts to use reason and science in shaping environmental policies impacting agriculture." Then, early in 2005, the Second Circuit Court of Appeals ruled that the Bush administration's farm pollution rule violated the Clean Water Act by allowing large-scale livestock farmers to apply manure to land without public input and federal or state oversight. This came in response to a March 2003 lawsuit (*Waterkeeper Alliance v. EPA*) filed along with the NRDC and the Sierra Club. While expressing his gratitude to the court, Bobby customarily pulled no punches: "These regulations were the product of a conspiracy between a lawless industry and compliant public officials in cahoots to steal the public trust."

On October 19, 2007, Waterkeeper Alliance and Smithfield Foods announced agreement on new measures to enhance environmental protections at approximately 275 hog production facilities in North Carolina. These measures resulted from settlement of two lawsuits filed by Waterkeeper six years before. Murphy-Brown, a Smithfield subsidiary, said it would fund major programs to identify and eliminate potential lagoon risks to groundwater; monitor surface water runoff from land; and increase stream buffers, wetlands, and other methods for protecting the public's waterways. The company would also implement enhanced manure management at all its facilities. Bobby saw these as "positive steps" to improve the current system.

As the end of Bush's presidency approached in November 2008, the EPA finally curbed factory farm pollution with new requirements calling for a "zero discharge standard" on large factory feedlots. Of course, those didn't extend beyond American borders. And Smithfield's brand of conspiracy possessed octopus-like tentacles.

* * *

On May 5, 2009, the *New York Times* ran a major exposé of Smithfield that began: "Almost unnoticed by the rest of the Continent, the agribusiness giant has moved into Eastern Europe with the force of a factory engine, assembling networks of farms, breeding pigs on the fast track, and slaughtering them for every bit of meat and muscle that can be squeezed into a sausage. . . . In less than five years, Smithfield enlisted politicians in Poland and Romania, tapped into hefty European Union farm subsidies and fended off local opposition groups to create a conglomerate of feed mills, slaughterhouse and climate-controlled barns housing thousands of hogs. It moved with such speed that sometimes it failed to secure environmental permits."

Starting around the time of Waterkeeper's court victory forcing tighter EPA regulations, the overseas expansion by Joe Luter's $12 billion (annual revenue) outfit had a familiar refrain: "In Poland, there were 1.1 million hog farmers in 1996. That number fell 56 percent by 2008, as the advent of modern farming methods transformed agriculture." The *Times* came right out and said it: "Facing more restrictions in the United States . . . the company moved nimbly through weak economies and political and regulatory systems."

Synchronously, Bobby happened to be in Poland when the article appeared—and was about to address the country's parliament on this very subject. He'd been invited by Andrzej Lepper, a populist politician who had served as the country's deputy prime minister and minister of agriculture and rural development for a brief period in 2006. A farmer himself, Lepper knew about Bobby's efforts in North Carolina and Iowa.

"The head of Smithfield Poland had offered him a million dollar bribe," according to Bobby, "in order to pass legislation that, through a series of subterfuges, would have effectively shut down all of the independent [slaughterhouses] and put the industry at the mercy of Smithfield's one slaughterhouse, just like they'd done in North Carolina. That's how they controlled the price of pork. And Smithfield wanted to make Poland a platform for taking over all hog production in Europe." Lepper turned down the bribe.

Bobby's trip to Poland was marked by gourmet feasts. "It was actually the Communists who invented factory farming, because they were trying to industrialize and put it under centralized control," Bobby explained to me. "But because Poland was so poor, the local farmers were a barrier to that. They couldn't afford pesticides or fertilizers, so their meats and foods were the purest. In fact, there was a great food tourism industry, where people from all over the world stayed on a different farm every night and ate local homegrown products." That's just what Bobby did under Anderzei Lepper's tutelage, fondly remembering the extraordinary meals of kielbasa, sausages, pork chops, and a variety of cheeses.

In 1964, at the age of ten, Bobby had traveled to then-Communist Poland with his father, stunned by the warm reception of the people despite official hostility and a media blackout. In Warsaw, waiting for an audience with Stefan Cardinal Wyszynski (then under house arrest, now a Catholic saint), Bobby later recalled: "a handsome young priest sat with us in the kitchen and made us all chicken sandwiches and talked about how much he loved skiing. Fourteen years later, he would be Pope John Paul II."

After four days traveling through Poland's agricultural areas, Bobby came once again to Warsaw, where forty-five years earlier people cheering his family had mobbed every street into the central square. There is a YouTube video of Bobby passionately debating the head of Smithfield Poland before a stone-faced group of parliamentarians, informing them how once-independent farmers become serfs on their own land, first by

Smithfield's driving the price of pork so low that the only ones able to stay in business were those signing contracts with the company. Suddenly Smithfield was instructing them on how to raise their uniform GMO hogs in their little cages. Bobby told the Polish politician: "The farmer owns the house and owes money to the bank. Smithfield says, we own the feed and the pig, but you own the manure and we're not going to pay you enough to legally dispose of it. We're going to sign a one-year contract, even though it will take you twenty years to pay off your debts, and we can change the terms anytime we want." Even that wasn't enough. Smithfield wanted to insure a monopoly for its massive slaughterhouse. "There is no stronger advocate for free market capitalism than myself," Bobby continued, "because I believe it's the most efficient and democratic way to distribute the goods of the land. But Smithfield doesn't want a free market. Poland recently overthrew the shackles of communism, and it cannot now accept the shackles of corporate monopoly."

Bobby didn't know it yet, but Smithfield had him over a barrel, as well. While the Communist government of 1964 had grudgingly accepted a Kennedy visit, now Bobby basically had to flee the country. On his way out, he learned that Smithfield Poland had filed suit against him. Under the rather archaic laws of the "new" Poland, truth is not a defense against libel, as it is in the US. And a public prosecutor isn't needed to file a criminal action. Private corporations can do so themselves against their critics, who will then be held in prison until the issue is resolved.

"I could have lost the suit because I had insulted them, even though it was true," Bobby says. "So I had to get out of Poland. And I couldn't go back again for ten years." Was it worth it? "They succeeded in a limited way because they still opened up those big factory operations. But we were able to stop them from closing the abattoirs."

* * *

By 2013, Smithfield had moved its headquarters to Virginia and been bought out by the Chinese conglomerate WH Group. Back on their home turf, Bobby and Waterkeeper Alliance lent legal assistance to the North Carolina firm Wallace and Graham, which had decided to represent clients in North Carolina "nuisance cases" over the hog industry's toxic odors. Multiple juries awarded substantial damages to more than five hundred suing citizens. Smithfield's effort to overturn these through the conservative Fourth Circuit Court of Appeals failed. While the settlement agreements with the plaintiffs were confidential, Rick Dove estimates that Smithfield had to pay out close to a billion dollars.

Since 2000, the meat industry had contributed more than $5.6 million to politicians in the state. But, according to the nonprofit Clean Water for North Carolina, "a growing body of public health research shows that people who live closer to industrial animal operations get sick more often, stay sick longer, and die more often than people who live further away." As the online magazine of the Institute for Southern Studies reported in the summer of 2022, some of the notoriously smelly waste pits had closed via a Swine Floodplain Buyout program, after a lagoon failure on a Smithfield contract farm in December 2020 spilled a million gallons of waste water into a tributary of North Carolina's Trent River.

Among Waterkeeper's early recruits to fight the industry, Nicolette Hahn eventually left to get married and moved out to the family's ranch in Northern California, where she authored two books (*Righteous Porkchop* and *Defending Beef*) and has written articles on sustainable agriculture for numerous publications.

At eighty-three, Rick Dove says "[I] still got my boots on," as a senior adviser for Waterkeeper's Pure Farms Pure Waters campaign. He works part-time, flying and taking pictures, tracking the industry, "and trying to put a thorn in their side every chance I get." And he hastens to add: "From that little existence we had in 1993, we've now grown to covering 2.6 million square miles of waterways around the world, more than three hundred waterkeepers and groups, and over 1.1 million volunteers in forty-five different countries." Waterkeepers didn't always win the fights, but it was clear that they would never quit. Neither would Bobby.

Dove closed our interview with a fond memory. "I remember Bobby saying way back when: 'You know, someday I'm gonna realize my dream to have waterkeepers all over the world.' I looked at him and said: 'Bobby, we're never gonna be able to do that. We don't have any money, there's only fourteen of us, how we ever gonna grow an organization like that?'

"And Bobby just stared back at me and said, 'It can be done.'"

CHAPTER TWENTY-FIVE

SHIFTING TIDES

In the meantime, Bobby had started writing children's books. His first, published in 2006, was a handsomely illustrated volume about Saint Francis. The new book told the story of Joshua Chamberlain, a citizen-soldier hero of the American Civil War whom Conor had studied in school while serving as his dad's research assistant. Bobby described Chamberlain's exploits as "mystical, in many ways." He closed his introduction to the book with this paragraph: "It is the fantastic bravery of a long line of stalwarts like Joshua Chamberlain, and their love of principle, their commitment to ideals, and their willingness to sacrifice, which has defined our people and guided our nation's destiny. It's worth considering today how grievously we would dishonor the memory of these gallant heroes if we should ever let America become a nation governed by fear, or if we willingly compromised the rights they gave so much to guarantee."

As the 2008 election approached, the inevitable question arose, once again, as to whether Bobby might finally run for political office. It first surfaced in one of the most revealing interviews Bobby had ever given, to Oprah Winfrey for the February 2007 issue of her new magazine *O*. The grande dame of TV talk shows visited his home, where she was photographed landing and then holding a falcon on her gloved hand as Bobby looked on approvingly.

Asked if he ever wondered what the world would have been like had his father lived, Bobby replied: "You know, I think about that a lot as it applies to the kinds of decisions being made in our country today. To the fact that America is now involved in torturing people, that habeas corpus, which is a fundamental civil right guaranteed since the Magna Carta, has been abandoned, that we're imprisoning people without proper trials. My father thought of America as the last best hope for humanity. He believed we had a historical mission to be a paragon to the rest of the world, to be about what human beings can accomplish if they work together and maintain their focus. He was never afraid of debate; he was willing to debate with Communists because he believed this country's ideas were so good that we shouldn't be scared of meeting with anybody."

The most bitter pill to swallow was what had happened to the public love for America that he'd witnessed in his youth: "In six short years, through monumental incompetence and arrogance, the White House has absolutely drained that reservoir dry." Worst of all, using fear as a governing tool and throwing away our commitment to human rights as a reaction to terrorism.

So why wasn't he running for office? Oprah asked. Bobby candidly replied: "At this point I would run if there were an office open, because I'm so distressed about the kind of country my children will inherit. I've tried to cling to the idea that I could be of public service without compromising my family life. But at this point, I would run." Hillary Clinton hadn't yet announced whether she would run for president; if so, Bobby said he'd consider seeking her senate seat, or perhaps pursue the New York governorship, but he didn't want to challenge Eliot Spitzer there. And he didn't really want to be the state's attorney general, a post then held by Andrew Cuomo. Bobby loved taking his kids on trips and involving them in his work, "but now America has changed so dramatically that I'm asking myself: *What's going to be left of this country?*"

He realized that the spasmodic dysphonia plaguing his voice would be an obstacle. "It began as a mild tremble for a couple of years," he explained to Oprah. "After people would hear me speak, I'd get all these letters, almost always from women. 'I saw you on TV and you were crying—it was so good seeing a man share his feelings.' I'd think, *Oh God*. I knew for every woman who wrote, there were men saying, 'Look at this friggin' crybaby!'" Oprah laughed. Oprah asked if it was degenerative, would it get worse? "I'm told that it's not supposed to, but I think it has." But on the contrary, it has in fact improved.

Besides Oprah, Bobby appeared on the cover of *Vanity Fair*, alongside Al Gore, George Clooney, and Julia Roberts in a special "green issue" (May 2007), which headlined, "A threat graver than terrorism: Global Warming . . . the call for a new American revolution." As a contributing editor for *Rolling Stone*'s special report on "The Climate Crisis," Bobby penned a lengthy article, "What Must Be Done" (June 28, 2007).

For its May 2008 issue, *Vanity Fair* invited Bobby to write "The Next President's First Task: A Manifesto," in which he laid out the need to end subsidies to coal and oil producers and revamp the high-voltage power transmission system to deliver solar, wind, geothermal, and other renewables around the country. "Washington is a decade behind its obligation, first set by Ronald Reagan, to set cost-minimizing efficiency standards for all major appliances," Bobby wrote. "For roughly a third of the projected cost of the Iraq War, we could wean the country from carbon." And he continued to barnstorm universities and other venues, packing a Virginia Tech auditorium with almost three thousand people to hear his speech "Our Environmental Destiny."

* * *

The subject of his political future didn't languish as Barack Obama and Hillary Clinton ran neck and neck in the primaries. Responding to rumors in March 2008 that a victorious Obama might ask him to administer the EPA or Department of Interior, Bobby had said: "If asked, I will serve." But then the *New York Daily News* reported he was taking himself out of the running for the EPA job. "I have six mouths to feed," he was quoted.

At the time, Bobby was in a quandary. "My dilemma with Obama was particularly acute and painful, and really heartbreaking," he said in 2023. "My father had made this weirdly prescient prediction in 1968 that within forty years, there would be a black president of the United States. And that's exactly where we were when Obama appeared on the scene. He embodied everything that my family had been working for since the 1960s. He also was the embodiment of the American experience and American idealism. His mother was from Kansas, the reddest state, and his father was once a goatherd from Kenya. He'd lived partially in Hawaii, partially in Indonesia. So many aspects of his character and experience spoke directly to the melting pot that is America. He could talk as well to an audience on the south side of Chicago as he could to Nebraska farmers and truck drivers from Alabama—an extraordinary ability to speak to a wide range of Americans the same way my father could. He was the first person since my father who could inspire white working-class Americans and minorities to follow him. From his shared experience and unmatched eloquence, to communicate with a vision of a future rooted in the finest features of America as an exemplary nation."

Bobby described that he felt a strong connection to Obama because of his friendship with Tom Mboya, the man responsible for Obama's coming to America. But he felt torn. "The rest of my family had a very strong relationship with the Clintons," he said. "They were so kind to my mother. Bill Clinton wrote her letters constantly. He invited her to the White House, visited our home, really treated her almost like she was his own mother. She loved him. And that kindness really impressed me."

"Teddy and Caroline and Max, the majority of my family, endorsed Barack and threw themselves into his race because he was historically so important and such an impressive person. But I just couldn't bring myself to abandon the Clintons. I was loyal to a fault. In retrospect I wish I had, because the moment I became more outspoken on vaccines, instead of calling and asking me about it, Hillary turned her back on me."

When Obama won the presidential election and appointed Hillary his secretary of state late in 2008, New York Governor David Patterson called Bobby that same night. The governor had to name her successor to the senate and asked if Bobby would take the position. "Of course I was very tempted, because how often do you get a US senate seat without having to run a campaign? Bobby thought about it for a couple days, but, in the end, the same considerations persuaded him that he just couldn't do it. "My kids were all in vulnerable positions," he later told me, "so I had to walk away from it."

Bobby said that, once elected, Obama "showed a kindness to my mother that was even greater [than the Clintons]. He came to Hickory Hill repeatedly, and she went to the White House for conversations. They shared

Valentine's cards. He would pick up the phone when she called, and he called her often. It was really an extraordinary relationship, and every time she saw Barack, she got a thrill almost as strong as when Jack was in the White House. She was so proud and so pleased with him, and he was so indulgent with her, that it was impossible for me not to love him."

What Bobby had sensed during Obama's candidacy only accelerated during his eight years in office. "He talked about race in a way I have not seen any other liberal or African American politician do so skillfully. He always downplayed the racial aspect of controversy and made it a human issue, as he'd done after hurricane Katrina in 2005. When he'd toured the stadium in New Orleans filled with ten thousand displaced African American refugees, he talked about the corruption and the failure of government. He was always conscious of his role to bridge divides and bring America together, and he possessed an uncanny wisdom and self-discipline to do that."

* * *

The Clinton Administration, recognizing the gravity of the mercury problem, had reclassified it as a hazardous pollutant under the Clean Air Act. But the industry, having donated $156 million to President Bush and his party since the 2000 election cycle, succeeded with eviscerating that rule and replacing it with one written by utility company lobbyists.

The issue of mercury in vaccines that had sparked an uproar when Bobby published "Deadly Immunity" in *Rolling Stone* and Salon back in 2005 seemed to have taken a backseat. Bobby chose not to address a *New York Times* announcement about a new study in the *New England Journal of Medicine* determining that the thimerosal preservative appeared to be harmless (September 27, 2007), although the study "did not address thimerosal's association with autism directly." At this point, "nearly five thousand families had filed claims with the federal government contending that vaccines caused their children to become autistic," the *Times* noted in a paragraph toward the end of its story.

Scrutinous Bobby watchers, however, would surely take notice when he appeared on the cover of *Spectrum* (April/May 2008), a magazine "For Families of Individuals with Autism & Developmental Disabilities," under the headline "RFK Jr. Won't Stop." The article was written by Sarah Bridges, the mother who had turned Bobby's head around in presenting the box full of documents to him at the Kennedy compound four years earlier.

After publication of his "Deadly Immunity" article, Bobby said he had received thousands of letters and emails from all over the world. "The astounding thing was how alike all of them were," Bobby said, "and that people from Mississippi to New Delhi shared such identical experiences. Here is the typical scenario I heard: A mother took her toddler to the doctor, where he received a spate of vaccines, became ill that night, often with a fever, sometimes with seizures, then lost the language he had, developed stereotyped behavior, and regressed into a looking-glass world of debilitated relationships and social isolation. Essentially, their lives were plunged into unimaginable agony."

He felt it imperative to keep getting the story out to prevent the catastrophe from damaging other children. Bridges wrote: "However, nothing prepared him for the resistance and anger he faced when discussing autism with politicians and the media. The unbelievable thing is how these children's stories are suppressed by the medical community, Big Pharma and the American press. There is a total refusal to have the discussion and derision toward anyone who tries."

The autism rate in children had almost tripled since 2000 (one in 88 children in 2008), and Bobby found the claim by government officials that the increase was an artifact of new measurement methodology to be "patently ridiculous. As Dr. [Boyd] Haley says, if new diagnostic techniques are causing more autism, then where are all the thirty-year-old people with diagnosis? Why haven't the new techniques helped us find all those people with the disorder born before 1989?"

The CDC, supposed to be a watchdog, had been captured by corporate interests, with studies of poisons in food safety being done by Monsanto and of drug safety being conducted by Pfizer. The flu vaccine still contained 50 percent of the mercury being given to kids before thimerosal was removed from other childhood vaccines and was being recommended even for pregnant women and infants.

"It would be bad enough," Bobby added, "if thimerosal were our children's only toxic exposure. However, we know that there are ten thousand pesticides for every man, woman, and kid in the country and many of them untested—yet we don't know the synergistic effect of these on developing brains. Obviously, thimerosal is a big piece of what has happened to the kids, but it is interacting with and compounded by the continued environmental assaults the children are receiving."

When parents asked about the best way to raise awareness about autism, Bobby suggested using every available means: "As Martin Luther King said, 'you need agitation, legislation and cogitation.'" Bobby thought much could be gleaned by studying the fights against tobacco and climate change, where the cozy relationships between big companies and government needed to be rooted out.

In a summation that could as easily have been made twelve years later with the coming of COVID-19, Bobby said: "I'm continuing to push for studies, particularly of unvaccinated cohorts from America, and staying in touch with various advocacy groups on their progress. I am used to the ridicule that inevitably comes with speaking out on unpopular issues; in fact, it typically gets loudest when we are closest to the truth. With each new study showing thimerosal's devastating effects, I am more resolute than ever that these children get the justice they deserve. I won't stop until they do."

* * *

The same month that Bobby's interview in *Spectrum* appeared, the CDC publicly conceded that thimerosal-containing vaccines may have injured a nine-year-old girl from Athens, Georgia, in early childhood and agreed to pay her family for her care. A settlement had been reached the previous fall, but only disclosed in recent days, the *New York Times* reported (March 8, 2008). However, the paper quoted CDC Director Dr. Julie L. Gerberding: "Let me be very clear that the government has made absolutely no statement indicating that vaccines are a cause of autism." The little girl had an underlying mitochondrial disorder that vaccines aggravated, the government believed.

Nine months later, in the wake of more rallies by vaccine critics and their appearances on talk shows like *Oprah* and *Imus in the Morning*, the *Times* ran a front-page story headlined: "Book Is Rallying Resistance to the

Anti-vaccine Crusade" (January 13, 2009). The author was Dr. Paul Offit, the pediatric spokesperson whom Bobby had been urged to contact when researching his "Deadly Immunity" article. Titled *Autism's False Prophets*, there would be no book tour for the author because "he has had too many death threats." While dismissed by many parents as "'Dr. Proffit' because he received millions in royalties for his RotaTeq vaccine," Offit countered: "Opponents of vaccines have taken the autism story hostage. They don't speak for all parents of autistic kids, they use fringe scientists and celebrities, they've set up cottage industries of false hope, and they're hurting kids."

A month after that (February 13, 2009), the *Times* reported: "In a blow to the movement arguing that vaccines lead to autism, a special court ruled on Thursday against three families seeking compensation from the federal vaccine-injury fund. Both sides in the debate have been awaiting decisions in these test cases since hearings began in 2007; more than five thousand similar claims have been filed."

On January 6, 2011, the *Times* would feature an analysis by British writer Brian Deer of the already "widely discredited research" of Dr. Andrew Wakefield, which in 1998 had suggested a link between autism rates and the measles, mumps, and rubella (MMR) vaccine. Deer concluded that "Dr. Wakefield and his colleagues had altered facts about patients in their study."

Bobby's investigations into thimerosal had nothing to do with Wakefield's work, but a book published that same month led to a sudden retraction by Salon and removal from its web archive of Bobby's 2005 article. *The Panic Virus, The True Story Behind the Vaccine Autism Controversy*, written by Seth Mnookin, contained a chapter on Bobby titled "A Conspiracy of Dunces." Mainly, the author criticized Bobby for using quotes from the Simpsonwood transcript "ridiculously out of context. . . . The piece took quotes that were dozens of pages apart and connected them with an ellipsis." But this is hardly unusual for a journalist when excerpting quotes from the same individual in a 289-page document, and the quotes Mnookin cites didn't support his contention to any significant degree. Factual errors noted by Mnookin were largely made during the extensive edits by *Rolling Stone* and Salon editors, for which Salon editor-at-large Joan Walsh apologized to Bobby at the time.

"Salon precipitously took down the piece on Sunday, January 16, 2011, during the Martin Luther King Jr. holiday weekend," Bobby later wrote. "My only prior warning was an email from Salon's new editor, Kerry Lauerman, whom I had never met, sent the night before. In that email, Lauerman explained that Salon was timing its retraction of my article to accompany the launch of *Panic Virus*, a new pro-thimerosal book by Seth Mnookin [who had begun his journalistic career with Salon]. Lauerman's letter informed me that the retraction of my piece would accompany Salon's publication of an interview with Mnookin at noon the following day. In his introduction to that interview, Lauerman described Mnookin as a personal friend 'and a friend of Salon's.' Lauerman refused to take my calls protesting the retraction."

Bobby continued: "In a published note explaining the retraction . . . Joan Walsh never mentioned any new errors that Salon may have found in my piece. Nor does she cite any scientific studies. Instead, she parrots CDC's talking points that thimerosal is safe, citing 'continued revelations of the flaws and even fraud tainting the science' linking thimerosal to neurological disorders. That vague proclamation serves as Salon's official rationale for defenestrating my article. Walsh duly promised, in the future, to exercise 'critical pursuit of others who continue to propagate the debunked, and dangerous, autism-vaccine link.'"

Bobby said that "after reviewing Mnookin's book and Salon's strange explanations, mystified *Rolling Stone* editors elected to not remove my piece." Indeed, even before Salon made its move, it was Mnookin who admittedly started a rumor that *Rolling Stone* had already chosen to pull it. This brought a retraction about the retraction *from* the magazine saying: "Editor's Note: The link to this much-debated story by Robert F. Kennedy Jr. was inadvertently broken during our redesign in the spring of 2010. (We did not remove the story from the site, as some have incorrectly alleged, nor ever contemplated doing so.) The link to the original story is now restored, including the corrections we posted at the time and the subsequent editorial we published about the ensuing controversy."

Another who weighed in to defend Bobby's integrity was David Talbot, Salon's founder in 1995 and former editor in chief. Talbot addressed a letter to Bobby on April 6, 2015, that stands as an epitaph to this saga:

"I was dismayed when I first heard that Salon had removed your article about the hazards of thimerosal from its web archives. As you know, I was no longer the editor of Salon when your article was published. And I am not an expert on the subject. But without taking a position on mercury preservatives in vaccines, I know enough about the debate—and about the pharmaceutical industry's general track record on putting profits before people, as well as the compromised nature of regulatory oversight in this country when it comes to powerful industries—to know that 'disappearing' your article was not the proper decision."

Talbot continued: "I founded Salon to be a fearless and independent publication—one that was open to a wide range of views, particularly those that were controversial or contested within the mainstream media. Removing your article from the Salon archives was a violation of that spirit and smacks of editorial cowardice. If I had been editor at the time, I would not have done so—and I would have offered you the opportunity to debate your critics in Salon's pages.

"In my day, Salon did not cave to pressure—and we risked corporate scorn, advertising boycotts, threats of FBI investigations by powerful members of Congress, and even bomb scares because of our rigorous independence. Throwing a writer to the wolves when the heat got too hot was never the Salon way. It pains me, now that I'm on the sidelines, to ever see Salon wilt in the face of such pressure."

But newspaper editors and TV producers would repeatedly cite Salon's action as justification for their decision not to run Bobby's editorials, articles, and letters to the editor, or allow him to talk about vaccine safety on the air.

Of what happened, Bobby told me in 2022: "It was clearly political and had nothing to do with inaccuracies in my article." After it was published, Bobby felt he'd done what he needed to do and didn't want to spend the rest of his life on the issue. "I want to go back to the environment," Bobby explained: "So I left it for five years and then Salon sucked me back in. When they pulled the piece down, I thought, okay, now I need to defend this again and show people what's happening."

And so he would, no matter the personal cost. For some years after the Clintons left the White House, they had cohosted with Bobby the St.

Patrick's Day parade in Mount Kisco. That ended in 2014, when Bobby published a book, *Thimerosal: Let the Science Speak*. According to Bobby, "the Clinton Global Initiative had partnered with Bill Gates and the Rockefeller Foundation to vaccinate the world, and all of a sudden I was persona non grata. I couldn't get the Clintons on the phone. I was radio-active." And it wasn't just the Clintons. It was literally everyone—friends, family, the media, his business partners.

CHAPTER TWENTY-SIX

TRAGEDY STRIKES

When his son Aidan was born in 2001, Bobby was still in jail on Vieques for his act of civil disobedience. He wrote in his journal: "I'm so proud of my Mary. She has become the woman I fell in love with—through hard work. She has overcome her fears, enshrined her faith . . . immersed herself in gratitude and God gave her a baby . . . a beautiful and serene and happy soul . . . I couldn't be happier or more grateful for the life and the wife God has given me." It was clear then that Mary too loved Bobby, though she wished he didn't travel so much, and that their children were happy and well cared for.

A few years after that, Mary began suffering from a mental illness that her friend Kerry Kennedy said "made her so unlike herself—kind, generous, a wonderful mother, perfect friend, force for good on our Earth - that it was as if she'd been invaded by a foreign body."

Bobby and Mary separated in May of 2010. Two years later, at fifty-two years old, Mary died by suicide. "Estranged Wife of Robert F. Kennedy Jr. Is Found Dead at Home in Westchester," the *New York Times* would headline. Bobby reflected in 2015: "Mary was an extraordinary woman. She was brilliant and always picked the right words. She had a gift for caricature and for remembering events and describing them accurately and with hilarity. We were always laughing during those days. She was incredibly strong and deeply spiritual."

In 2022, Bobby marveled at other qualities of hers. "Mary had a contagious passion for beauty," he said. "At her suggestion we went to a lot of museums and art galleries together. . . . Her encyclopedic knowledge of art history imbued all these works with life and background and comprehension that made me fall in love not only with them, but with her. . . . She had more friends than anybody I've ever met and a photographic memory for names and faces. And she was a wonderful raconteur, with a story for every occasion. Mary just had a social genius."

St. Patrick's Roman Catholic Church on the village green in Bedford was not far from the Mount Kisco home. Bobby and the children followed the hearse in a black town car, arriving shortly after 10 a.m. Ethel Kennedy, then eighty-four, rode in a black SUV behind them. Bobby walked ahead of the oak casket, which the three oldest children helped carry with ten other pallbearers into the church. Some four hundred mourners included Bobby's older brother Joe, Caroline Kennedy, Maria Shriver, and friends Susan Sarandon, Chevy Chase, Dan Aykroyd, John McEnroe, Julie Louis-Dreyfus, Larry David, and Edward James Olmos.

In the same church where her children received their first communion, the cover of the funeral program depicted Mary smiling and wearing her familiar dark sunglasses. Color photos on the walls showed Mary fly-fishing, riding a camel through the desert, and, above all, frolicking with her children. Actress Glenn Close opened the service by leading everyone in singing "Amazing Grace." Her son Conor read from the Book of Job: "The souls of the just are in the hand of God and no torment shall touch them . . . they are in peace." Her daughter Kyra recited Psalm 42, and Aidan gave a reading from Letter of Paul to Romans. Family members offered the Prayer of St. Francis. "She lived her life to the fullest," Kerry Kennedy said. "She did everything she could for her friends and she loved her children more than anything else." Kerry described the depression "that so many, many, many Americans suffer from" and its demons that her best friend had battled all her life. Bobby delivered the final eulogy and revealed parts of their last conversation. "You know me better than anyone in the world," he recalled Mary saying. "I was such a good girl." Bobby replied, "I know you are—and you still are." After a half-hour greeting and thanking guests, the Kennedy family boarded a coach taking them to Massachusetts, where Mary was buried in a Cape Cod cemetery a few miles from the seaside Kennedy compound. Bobby and the children knelt beside her coffin at the gravesite, saying a final good-bye.

CHAPTER TWENTY-SEVEN

A NEW LEASE ON LIFE

In the aftermath of his wife's death, Bobby entered a period of seclusion. He left New York, and took his children with him to begin the process of healing at his brother's home in Florida. They spent time together in nature, which had always brought solace to Bobby.

After a considerable time of mourning and trying to comfort his children, Bobby found the energy to restore himself through his relationship with Cheryl Hines. "She was really like an angel sent by God coming into my life," he says. "She gave me hope and something to fight for and live for. She took care of my kids, who were at risk because they'd been through a very tough couple of years culminating with their mother's suicide. At one point Conor said to me that he felt uncomfortable that I'd moved so quickly into the relationship with Cheryl. I said, 'If you tell me that you don't want me to do this, I won't.' That was a huge risk for me. But he came back two days later and said, 'Dad, I know she's making you happy and we want you to be with her.' Conor sat Cheryl down and told her how important she was to them, too. And he gave the most extraordinary speech about Cheryl at our wedding."

Born in central Florida in 1965, more than eleven years after Bobby, Cheryl couldn't have grown up in a more different milieu. Her "redneck dad" (as she described him) drove a silver El Camino around Tallahassee. Cheryl told an interviewer years later: "He thought that was a little classier than a pickup truck."

She had two brothers and a sister, receiving not enough support from her father's irregular self-employed construction, so their mother always worked at least one job. "We didn't have much money," Cheryl told me. "When I was in elementary school, I was on free lunch. But it was actually a joyful childhood. My grandparents lived in a trailer on their orange grove, and on Saturdays the fun was riding around in the back of grandpa's pickup, which he'd stop and pretend had stalled when the sprinklers came on. We were raised Catholic, and the priest would come over after mass to watch football and drink beer with my dad."

Her parents divorced when Cheryl was fourteen, and she shared an apartment bedroom with her mom through the first year of high school. She already aspired to be an actress. After a stint at beauty school, she cut hair and waitressed for a year, saving enough money to put herself through college and eventually graduating with a degree in TV and film production from the University of Central Florida. Because she had to work nights, Cheryl couldn't be cast in any of the school's theater productions. That broke her heart, but her passion propelled her to Hollywood.

After working awhile at Universal Studios in Florida, in 1994 Cheryl drove her Toyota Tercel with 100,000 miles on it out to Los Angeles and started bartending at a hotel downtown (the same one where the jurors in the O. J. Simpson trial were sequestered, she recalls). There she met the sister of *Saturday Night Live* cast member Phil Hartman, who told her about an improvisational theater company called The Groundlings that had launched the careers of such comics as Will Ferrell and Kathy Griffin. Studying under Lisa Kudrow, star of the sitcom *Friends*, Cheryl performed and learned to write comedy sketches. She went on to work for several years as a personal assistant to Rob Reiner and appeared in a few TV shows.

Then, in 1999, Cheryl got her big break. Auditioning for the role of Larry David's wife in the TV sitcom *Curb Your Enthusiasm* that he created, Cheryl didn't have much hope for landing the role. But Larry David wanted to cast an unknown, and Cheryl was not only beautiful but carried the mix that he was looking for—assertive though not overly pushy. "We got along so well and had so much fun, and they called me the very next day," Cheryl remembers. She would go on to twice receive Emmy nominations for Outstanding Supporting Actress in a Comedy Series during the twelve seasons over twenty years of the wildly popular HBO series.

Cheryl's character in *Curb Your Enthusiasm* was also a committed environmentalist, about which Larry David and his then-wife Laurie proceeded to educate her. In 2002, they brought Cheryl along to a fundraiser weekend for Bobby's Waterkeeper Alliance at a ski resort in Banff Springs, British Columbia, also attended by Pierce Brosnan, Glenn Close, and other celebrities including moonwalker Buzz Aldrin.

By the time the 2010 Waterkeeper fundraiser in Deer Valley, Utah, came around, Bobby had separated from Mary. "At the last minute on these events, we were always scrambling to get more celebrities there," Bobby said. His administrative assistant at Pace, Mary Beth Postman, suggested that Cheryl be invited. When Mary Beth called her, she said she didn't know if she could make it because there was a party she'd already committed to. Then Mary Beth said to me, 'I think if you push her, she'll come.' So I called her and, to my great surprise, Cheryl said okay. I sat next to her at dinner the first night at the lodge, and I immediately had chemistry with her. She made me laugh. A *lot!*" Cheryl had a similar reaction. "People in my circle tend to look at people in politics as being boring," she told the *New York Times*, and she was surprised at what a simpatico sense of humor Bobby had.

"I knew I wanted to date her, but I wanted to ask Larry," Bobby said, "even though I know it's his TV wife, I needed to get square with him whether I might be crossing a boundary." So they met at the Carlisle Hotel in New York, and Bobby said, "Larry, what do you think?" Bobby would never forget Larry David's response: "Larry said, 'Cheryl is the best person I've ever met, the only person in the industry who is universally loved, and doesn't have a single enemy. She has a level of professionalism that's so solid. She's never late for an appointment, always knows her lines.' And I've come to agree—Cheryl is the finest human being I know."

Cheryl knew very little about the Kennedys, "other than what was in the history books." She felt: "Maybe it was better that way, because I had no expectations about who Bobby was when I met him. At that time, he was living in New York, and I was living in Los Angeles. I had my career and my daughter there. So we really had separate lives. After Mary died, I knew that everyone was in mourning and processing. So it was difficult because I didn't hear from Bobby that much for some months, and at the same time I understood it because I knew he had a lot to care about; he really had to be there for the kids. But I worried about who was helping *him*."

Bobby reflected in 2022: "Cheryl doesn't have any guile. She has a lot of compassion and always knows exactly the right thing to say. If you watch her on TV, no matter how difficult the question is, she always comes out with a graceful, elegant, funny answer, and it's just because she has such an amazing relationship with the truth. I've never heard Cheryl say one word that was even slightly dishonest or pretentious. Because of that, she has a real clarity of thought and of judgment. And that was very helpful to me at that time, and since."

As Bobby told talk-show host Megyn Kelly: "Cheryl has an acute sense of what's right and wrong in any situation. She shared that with my children during that terribly difficult time and has been a loving friend to all of them. My kids are now flourishing, all doing well in careers, school athletics and so on—and a lot of that is due to the strength and stability that Cheryl brought them."

She and Bobby announced their engagement in April 2014 and married on August 2nd of that year at the Kennedy compound in Hyannis Port. They soon bought a house at Point Dume in Malibu, close to a surfing beach, and Bobby moved to join her in Southern California. When Cheryl received her "star" that year on the Hollywood Boulevard Walk of Fame, Bobby, her former husband, and Larry David all sat together in the front row. Inclusivity was what Cheryl Hines was all about.

CHAPTER TWENTY-EIGHT

THE ASSASSINATIONS: QUESTIONING THE OFFICIAL NARRATIVE

It is more comfortable to sit content in the easy approval of friends and of neighbors than to risk the friction and the controversy that come with public affairs. It is easier to fall in step with the slogans of others than to stand on judgments of your own.

—Robert Kennedy

Perhaps it was coincidental timing, coming so soon after Mary's passing, that Bobby began investigating the official narratives about the assassinations of his uncle and his father. While Bobby was no stranger to death, Mary's suicide and its subsequent turmoil marked the first family tragedy since those of his youth to be so deeply personal. It seemed to throw him back upon questions that he and his siblings had long felt too painful to raise. Were Lee Harvey Oswald and Sirhan Bishara Sirhan really isolated deranged killers driven by ideology and possessed by private demons? Or could there have been conspiratorial forces at work?

Even as a child, Bobby later said, he was skeptical that Oswald acted alone. In December 1992, he told an interviewer from *Rolling Stone*: "From what I know about President Kennedy's assassination, it makes sense that there was more than one person involved. My father never discussed it, because that whole event was so painful to him. You could see it every time you looked into his eyes."

As Bobby would write in an unpublished manuscript in 2017, "Abolishing all reminders of Dallas was part of my mother's strategy to pull my father from his grief. We learned from her signals. We switched the TV off quickly at the mention of the assassination. The subject was taboo. We never spoke of the assassination among ourselves. That was our family's ritualistic orthodoxy. Two generations and half a century later, I think we all still have a kind of PTSD. A year before his death, I revived the subject with Teddy [in 2009]. He tried, but his eyes filled with tears and he shook his head, unable."

In adulthood, Bobby would recoil whenever the "familiar frames of the ubiquitous Zapruder [film] clips ambush [him] from a TV screen." He described it this way: "A paralysis overtakes me as I wait for the slow motion spectacle to end. It's one of the few circumstances in life that can disable me."

In 1996, Bobby made a trip to Cuba on an environmental mission and met with Premier Fidel Castro, who told him: "I don't think they [the Kennedy brothers] had any choice [about the Bay of Pigs invasion] because all the plans had already been made and were in place when they took power. His death was hard for us to take." A strong indication that Cuba had not had any connection to the assassination.

He was approaching his midfifties when, in 2008, while preparing to give an environmental talk at the Franciscan Monastery in Niagara, New York, Bobby found a copy of a just-published book "on my greenroom table, left as an anonymous gift for me." It was titled *JFK and the Unspeakable: Why He Died and Why It Matters* by Catholic theologian James W. Douglass. Bobby found the book "a fascinating and meticulous dissection of the circumstances surrounding the assassination." Bobby spent a lot of time examining Douglass's thorough footnotes. He noted "the extraordinary analysis implicated rogue CIA operatives connected to the Cuban project and its Mob cronies."

Bobby was impressed enough to send the book to President Kennedy's speechwriter Ted Sorenson, who wrote him back in 2010: "It sat on a table for two weeks and then I picked it up. And once I started I couldn't put it down. And you know for so many years none of us who were close to Jack could handle ever looking at this stuff and all of the conspiracy books. Well, it seemed that nothing they had would stand up in court. All of us were, you know, 'it won't bring Jack back.' But I read this and it opened my eyes and it opened my mind and now I'm going to do something about it." Sorenson said he'd spoken to the author and planned to write a foreword for the paperback edition. "Thanks for getting the ball rolling," he wrote Bobby. However, Sorenson later told Douglass that his wife and daughter had persuaded him that his association with Jack had always been about the president's life and he should leave it at that. Sorenson died soon after that.

Bobby himself "embarked on the painful project of reading the wider literature on the subject." He also began contacting other surviving members of the Kennedy administration about his father's attitude toward his brother's murder. Adam Walinsky, a legislative assistant and speechwriter for Robert Kennedy, told Bobby: "Your dad was smart and he always knew what the stakes were. He knew there was more to it than a single insane gunman. Even though, at the time, he didn't know anything about all the monkey business with the autopsy and the forensics. But he knew he had to be circumspect. He couldn't go around Washington asking questions. He couldn't push LBJ."

By 1967, now in the senate, Kennedy had regained his strength emotionally. His press secretary Frank Mankiewicz recounted to Bobby that RFK dismissed the Warren Report's Oswald-acted-alone conclusion as "garbage." Back when his father was involved with the anti-Castro Cuban exiles, soon after the assassination, he had come to believe it was elements of that group and the CIA that had perpetrated JFK's murder. Frank Mankewiecz discovered Jack Ruby's phone records contained the numbers of many of the Mobsters that he'd gone after as attorney general—and that Oswald's killer two days after the assassination had been calling those organized crime figures in the weeks before November 22, 1963. After New Orleans DA Jim Garrison's new probe was revealed early in 1967, RFK had asked his former aide to quietly conduct a private investigation into who killed his brother. "If it gets to the point where

I can do something about this, you can tell me what I need to know," Mankiewicz remembered him saying. RFK told another close associate: "We have to wait till we have the White House."

Bobby also reached out to Robert Tannenbaum, who had served as deputy chief counsel to the House Select Committee on Assassinations when its reinvestigation in the late 1970s concluded there had been a "probable conspiracy" to kill JFK. Tannenbaum, later elected mayor of Beverly Hills, told Bobby: "Oswald was working for the CIA since at least 1957. We know it from manifold sources."

The year 2013 marked the 50th anniversary of the assassination and, in endorsing the paperback of Douglass's work, Bobby said it had moved him to visit the site of his uncle's assassination in Dealey Plaza for the first time. On January 11, Charlie Rose interviewed Bobby and his sister Rory at Dallas's Winspear Opera House, to kick off the city mayor's year-long program commemorating the Kennedy presidency. There, Bobby went on public record saying: "The evidence at this point I think is very, very convincing that it was not a lone gunman." He believed the same thing his father had concluded, that the Warren Commission's 1964 report constituted a "shoddy piece of craftsmanship."

Bobby later recalled "the thread that connects this whole story, the tension that existed between my family and the CIA." He spoke of how his grandfather had sat on a commission in the 1950s that recommended abolishing the clandestine services of the CIA, and Joseph Kennedy's strong belief "that imperialism abroad is inconsistent with democracy." His father, shortly before his death, had likewise told aides that he was going to stop "the mischief they [CIA] were creating around the world."

Bobby never mentions Sirhan's name in *American Values*. Bobby spoke of how he "never believed the orthodox version of my uncle's assassination and I don't think most people do." What would his father think of what's going on today in terms of civil rights? "I think my father would be happy we had a first black president in this country." More applause. "But I think he would be disappointed in the direction this country is going today."

Bobby recalled his forebears' belief that America was an exemplary nation that modeled democracy by not forcing it upon other countries. In 1780, we had been the world's first and only democracy; by 1865, there were six; by the time of JFK's presidency, as colonial empires broke apart, there was a cascade and many more democracies sprung up. "My uncle did not believe Communism was monolithic, he thought Castro had an absolute right to experiment with Marxism in Cuba if that's what the people wanted." American involvement would only fortify tyrannies by radicalizing them against us as an outside enemy. Why not let them fall apart on their own?

* * *

In terms of his father's assassination, Bobby still accepted that Sirhan was responsible. After all, a dozen witnesses in the Ambassador Hotel pantry at the moment of the shooting in Los Angeles saw Sirhan fire his pistol. Six men forcibly removed the smoking gun from his hand. Sirhan had pled guilty and admitted the crime, saying he had no memory of the night and so had to take the word of people who saw him. The issue seemed so settled that the House Select Committee on Assassinations that looked again into the murders of JFK and Martin Luther King ignored Robert Kennedy's.

Bobby vaguely remembered hearing the story of how Sandra Serrano, an RFK campaign volunteer, had seen Sirhan with a woman in a polka-dotted dress and a male friend escorting Sirhan toward the hotel kitchen, then fleeing after the shooting with the woman exclaiming: "We shot him, we shot the senator."

In 1996, Bobby heard about a fifteen-year-old high school student, Scott Enyart, who took the only known photos of his father at the moment he was shot. He claimed to have been shadowed and then chased from the pantry by two men in suits and then tackled by LAPD officers who confiscated his film and destroyed the negatives and contact sheets. Almost thirty years later, Enyart won a $457,000 judgment against the LAPD.

But Bobby had still dismissed these claims as the ephemera of conspiracy theorists. Then, in 2015, Paul Schrade called to ask him for help. Schrade had been one of his father's closest friends. A prominent Labor Union organizer, he had successfully convinced the powers-that-be to support a charismatic young Chicano activist, Cesar Chavez, in organizing the United Farm Workers. Schrade had introduced RFK to Chavez and worked on the 1968 presidential campaign. The first bullet fired in the pantry struck Schrade in the forehead. Robert Kennedy's final words were: "Is everyone okay? Is Paul okay?"

Surviving the horror, Schrade had devoted much of his life since to keeping RFK's memory alive—and to determining what really happened that night. He insisted that Sirhan couldn't have killed Bobby's father and, in recent years, fought a seemingly quixotic campaign to secure the convicted assassin's release from prison. He wanted Bobby to sign a letter to the LA District Attorney requesting a reopening of his father's murder case.

"At first I was appalled," Bobby recalled. "I didn't want to get involved. I was particularly worried that embroiling myself would discredit my work on other issues and give ammunition to my many detractors in the pharmaceutical and carbon industries who would love to marginalize me as a conspiracy theorist. 'There are only a limited number of controversial issues that I can take on, Paul,' I told him, 'before people start fashioning a tin foil hat for me, and I am already at my quota.'" But in the end, Bobby couldn't turn his back on Paul Schrade. "I felt I had to at least go through the motions of hearing him out. I visited Paul in his Los Angeles home and looked at his evidence. Despite his age, Paul was clear-eyed and clear-headed and had near-total recall. What he showed me that day was persuasive, and I agreed to sign the letter."

Schrade had begun his presentation to Bobby with the autopsy report from LA Coroner Thomas Noguchi, which described four shots being fired at RFK from at most three inches away. Three of the shots appeared to be in contact with his back and shoulder, with one passing through the back of his head. Sirhan had fired his five bullets all from the front. "Paul showed me indisputable proof of police and prosecutorial misconduct in the original investigation, including the use of counterfeit bullets to paper over the fact that bullets fired by Sirhan's gun did not match the bullets taken from my father's body—and that *those* bullets did not match the ones recovered from the five other victims shot that night."

This was the first of many meetings with Schrade and the beginning of a saga where Bobby became convinced of Sirhan's innocence. He dived in as if he were a DA trying to prepare a case. As he started making phone calls, even his casual investigation surprisingly could find no one who had a direct perspective of both Sirhan's gun and his father amid the confusion in the crowded pantry.

In what many considered a highly suspect "defense," Sirhan's lawyer Grant Cooper had in 1969 decided not to contest the charge and instead set out to persuade the jury that Sirhan had "diminished capacity" and didn't know what he was doing, thus preventing him from receiving the death penalty. Because Sirhan pled guilty on his lawyer's recommendation, the prosecution never had to prove its case. But every one of the dozen closest witnesses testified that Sirhan was in front of Robert Kennedy at all times and never came closer than three to six feet. Indeed, all four shots that struck RFK came at point blank range *from behind*. With six wounds to other victims, that added up to a total of ten bullets. Counting the holes found in the pantry walls and ceiling, a minimum of twelve bullets were fired. Sirhan's gun carried only eight and he never had a chance to reload.

So there had to have been a second gunman. Suspicion fell on Thane Eugene Cesar, a uniformed security guard moonlighting that night for Ace Security, which the hotel had retained for the event. Cesar was stationed in the pantry before the crowd entered and moved in behind RFK as he came through the double doors, taking hold of his right elbow. "Cesar described himself standing immediately behind my father to the right, precisely the point from which the shots that killed him must have been fired, according to the autopsy," Bobby pointed out. "Cesar admitted pulling his gun but denied firing it," Bobby wrote. "However, initial reports indicated he had indeed fired." The LAPD never asked to examine his weapon. Its two officers overseeing the investigation, Manuel Pena and Hank Hernandez, were later determined to have previous ties to the CIA. As did Cesar.

Both defense and prosecutors concurred at his trial that Sirhan could not remember the events of that night. Sirhan's last memory before amnesia set in was "drinking coffee with a pretty girl." The LAPD lost his blood tests, but the arresting officer noted that his pupils did not react to light and were "real wide." Dr. Edmund Simson-Kallas, the chief of San Quentin prison's psychological testing program, examined Sirhan and concluded that he was under post-hypnotic suggestion—"perhaps hypno-programmed to shoot—as a kind of Manchurian candidate."

In one of history's bitter ironies, the day he died Robert Kennedy was staying at the Malibu home of John Frankenheimer, who directed the 1962 film *Manchurian Candidate* about an assassin operating under post-hypnotic command. Not until 1975 would the public learn that the CIA had, beginning in the early 1950s, invested massively in weaponizing hypnotism and developing protocols for preparing such "candidates" and then erasing memory in top-secret Cold War programs known as Project MK-Ultra and Project Artichoke. All three doctors who examined Sirhan agreed that he had likely been hypnotized. A few years later, in 1973, the CIA destroyed nearly all of its MK-Ultra files.

Bobby came across a lengthy affidavit filed in 2011 by Dan Brown, a professor at Harvard Medical School with vast experience in forensic psychiatry and hypnosis. Commissioned by Sirhan's appeal lawyer, Brown examined Sirhan for more than sixty hours inside San Quentin prison. "I have written four textbooks on hypnosis, and have hypnotized over six thousand individuals over a forty-year professional career," Brown wrote. "Mr. Sirhan is one of the most hypnotizable individuals I have ever met, and the magnitude of his amnesia for actions under hypnosis is extreme. . . . Mr. Sirhan did not act under his own volition and knowledge at the time of the assassination and is not responsible for actions coerced or carried out by others."

Colorado Senator Gary Hart, who investigated the JFK assassination as part of the Church Committee in the midseventies, told Bobby: "Nobody had any inkling how out of control the Agency [CIA] was with all these crazy, murderous people. And I think your dad and your uncle were kind of scrambling to get on top of these powerful and murderous forces and they couldn't keep up."

CHAPTER TWENTY-NINE

ENCOUNTERING SIRHAN

The Richard J. Donovan Correctional Facility, part of California's state prison complex, is situated on a mesa in the desert foothills outside San Diego, overlooking the Mexican border about a mile-and-a-half away. The prison houses almost four thousand inmates. Probably its most notorious is Sirhan Sirhan, who coincidentally (or not) was transferred to a private cell there on November 22, 2013—the fifty-year anniversary marking the assassination of President John F. Kennedy.

Just over four years later, during the Christmas holidays in 2017, Bobby and his wife Cheryl Hines pulled into the prison parking area after a three-hour drive from their home in Los Angeles. "I went with Bobby so I could support him emotionally," Cheryl told me in 2022. "I wasn't sure how it was going to unfold, but it seemed like it couldn't go well. I couldn't even imagine sitting across from someone who people think assassinated my father and shot other people. I wanted to accompany Bobby so we could talk and process it all."

While Cheryl waited in the car, Bobby sat alone with the seventy-four-year-old Sirhan in a private room for close to three hours. "I got to a place where I had to see Sirhan," he would tell a *Washington Post* reporter six months later. "I went there because I was curious and disturbed by what I had seen in the evidence. I was disturbed that the wrong person might have been convicted of killing my father. My father was the chief law enforcement officer in this country. I think it would have disturbed him if somebody was put in jail for a crime they didn't commit."

At the time, Bobby would not publicly discuss specifics of their conversation. Privately, he reported that Sirhan told him: "I didn't kill your father, and you are as much of a victim as any of us." Then, in 2022, he elaborated: "I was looking at a tiny, frail, elderly man who was crying, clenching my hands and asking for forgiveness from me, from my siblings and from my mother for his part in the tragedy," Bobby said. "He told me that every time he saw my mother on TV and thought about her and all of her children, it made him weep. I talked to him about the assassination and he was very calm, not excitable or trying to protest his innocence, but very firm that he does not believe he did it. He does not understand why he can't remember anything, but he's been consistent for fifty years that he doesn't."

The visit was very emotional for Bobby, but he has always had an incredible amount of will power. "Once I make up my mind to do something," he said, "if I decide it's the right thing to do, I kind of detach from fear or emotion. So I was more curious about *him* than I was emotional. I was looking for any insights into his character and his honesty, and his attitude and whether he was being authentic with me. And whether there was any evidence of guile." Bobby believed that there wasn't.

Bobby believed, "from all the external evidence and from my discussion with Sirhan and reading everything about him that he was absolutely sincere and telling me the truth." When asked whether Sirhan believed that he had been hypno-programmed, Bobby replied in the affirmative. "After he fell off a horse at a ranch where he worked as an exercise boy and hurt his head, he told me about going to see this weird doctor and being in a hospital. He thought he was in the Pasadena hospital. His lawyer went there to get his records and found that he was discharged with four stitches above his eye after an hour. But he was then taken someplace else that there is no record of. And other men from the racetrack were bedridden there with bandages on their heads. Doctors came in and gave him drugs every day. Sirhan thinks he was there for several months."

I asked Cheryl if the outcome of Bobby's meeting with Sirhan surprised her. "Not really," she said, "because Bobby has a very generous spirit. He sees kindness and has grace for others that most people don't. He just has a wider lens, seeing things that most others don't. I think what he went through as a boy is unimaginable, watching his father get shot on international television, and became part of the makeup of his personality and how he may understand people in a different way than other people."

* * *

At the end of May and beginning of June 2018, Tom Jackman of the *Washington Post* did a three-part series marking the fiftieth anniversary of Robert Kennedy's assassination, commencing with the headline: "Who Killed Bobby Kennedy? His son RFK Jr. Doesn't Believe It Was Sirhan Sirhan." A previous judge had rejected Sirhan's latest appeal, ruling that the second gunman theory "lacks any evidentiary support." But for the first time, a mainstream paper delved into the strong likelihood of a second gunman and the fact Sirhan may have been a victim of a CIA mind-control experiment. Lawyers for Sirhan had launched a long-shot campaign to have the Inter-American Court of Human Rights hold an evidentiary hearing.

Mike Rothmiller was a retired detective for the LAPD whose first book (*LA Secret Police*) had exposed the department's hush-hush Organized Crime Intelligence Division (OCID), which he'd joined in 1977 as the youngest member. Rothmiller told me: "From information I received, and from documentation I read that has never come out, and people I knew who were there in the law enforcement arena the night of the assassination—I believe there were two shooters. The report in the California archives now is not the complete deal. OCID did their own investigation in celebrity cases, and the guys in homicide never knew. OCID would employ tactics intimidating different informants, wiretaps, black bag jobs, things like that."

Rothmiller went on: "But I'll tell you right now, you will not get anything out of the old OCID files. Around 1980, the ACLU had a lawsuit regarding the other intelligence division, PDID, that was spying on everybody. So there was a panic within LAPD because nobody on the outside really knew about OCID. If that came out and subpoenas were served for OCID files, there would be a huge meltdown within the department about all the dossiers on the rich and famous that had nothing to do with criminal activity. So Chief [Darryl] Gates was panicked, and we started a purge. Roughly sixty guys got assigned to the files. We burned through two industrial-strength shredders. A lot more went to the city incinerator. And a lot of the real interesting stuff—dossiers, recordings—went to selected individuals within OCID to maintain at home. Gates selected people he believed would never turn anything over or admit they had it." Whatever was left, Rothmiller said, is still considered under "active investigation" and therefore can't be released.

When I asked again about the determination of more than one shooter, Rothmiller added: "There's no question, intelligence-wise, about that being the conclusion that was reached. There was always going to be a security detail around RFK. That was handled by OCID. Word came back that RFK did not want uniformed cops there. Okay, fine. But there were still OCID guys in the Ambassador [Hotel] the whole time, including one in the pantry [where RFK was shot]. They were, for lack of a better term, lingering around him. It was to see who was there and what was being said, what they could pick up on intelligence-wise."

Which, though Rothmiller wouldn't say so directly, apparently included their awareness of a second gunman. "Sirhan shot everybody else, no doubt in my mind. Should he have spent a long time in prison for that? Yeah. But did he pull the final trigger? No."

I asked Rothmiller if he recalled the name of Thane Eugene Cesar in those OCID files. "The security guard standing behind Bobby," he replied. "Well, there was a lot of background information in there, questions about him. From what I recall, it was all pretty much speculation."

Bobby later said this about Cesar: "After first agreeing to meet with me, he gradually escalated his price to $25,000 for the privilege of interviewing him," and then, "bedeviled by questions about his involvement in my father's murder, Cesar fled to the Philippines. I was in negotiations with him in 2018, before reports of his death in September 2019."

* * *

In 2018, a group of researchers into all four assassinations of political leaders in the 1960s (JFK, Malcolm X, Martin Luther King, and RFK) had formed a Truth & Reconciliation Commission modeled after South Africa's. The approach is one of "restorative justice," which differs from the customary adversarial and retributive forms of justice. According to its mission statement, the "process seeks to heal relations between opposing sides by uncovering all pertinent facts, distinguishing truth from lies, and allowing for acknowledgment, appropriate public mourning, public forgiveness and healing." After his time with Sirhan, Bobby became part of the exchanges among members. 'I'm thinking we should launch . . . a one-year online petition and continue to ask leaders and average Americans to sign on after MLK day," he wrote the group on January 15, 2019. "We did this on the global warming virtual march and got a couple million signatures along with an amazing list of business, cultural, sports and political leaders. . . . It would finally give the millions of Americans who hate the lies an outlet."

Sirhan had originally been sentenced to die in the gas chamber. But after California abolished the death penalty in 1972, this was commuted to life. He'd become eligible for parole in 1986 but been rejected fifteen times since. At his last parole hearing in 2016, the commissioners went through listening to more than three hours of intense testimony, before determining that Sirhan didn't show adequate remorse or understand the enormity of his crime and again denied him parole. On an afternoon toward the end of August 2019, Sirhan was stabbed by a fellow inmate and hospitalized in stable condition.

When Sirhan came up for parole again in 2021, Bobby wrote the commissioners a letter: "While nobody can speak definitively on behalf of my father, I firmly believe that based on his own consuming commitment to fairness and justice, he would strongly encourage this board to release Mr. Sirhan because of Sirhan's impressive record of rehabilitation." Sirhan was known to participate in more than twenty prison programs, including anger management, Tai Chi, and Alcoholics Anonymous.

Bobby was joined in his effort by his younger brother Douglas Kennedy, who told a two-person board panel that he'd been moved to tears by Sirhan's remorse at the latest hearing and asked that the seventy-seven-year-old be released if he no longer poses a threat to anyone. "I'm overwhelmed just by being able to view Mr. Sirhan face to face," Douglas said. "I've lived my life both in fear of him and his name in one way or another. And I am grateful today to see him as a human being worthy of compassion and love."

Prosecutors declined to support or oppose Sirhan's release, under a new LA County policy that the prosecutorial role should end at sentencing. So it came to pass that, following the support of two RFK sons, the California parole board voted on August 27, 2021, to free Sirhan after more than fifty-three years of imprisonment, concluding he was no longer a danger to society. Other Kennedy family members including Ethel and most of Bobby's siblings remained opposed to Sirhan's release. The ultimate decision, however, rested with California Governor Gavin Newsom.

Bobby received a now-rare opportunity to publish an op-ed in a major newspaper. He wrote in the *San Francisco Chronicle* (December 8, 2021): "The pain that we all feel from my father's death should not prevent us from pursuit of the truth. I firmly believe the idea that Sirhan murdered my dad is a fiction that is impeding justice. If Newsom overrules Sirhan's parole, he will become just one more California official who claims to love my father but persists in denying him justice."

On January 13, 2022, Governor Newsom denied Sirhan's parole.

CHAPTER THIRTY

ONCE MORE INTO THE FRAY

One afternoon late in 2022, Bobby sat on the couch at his home offering a comparative reflection between what he'd been through in helping craft the New York watershed agreement and his entering the highly controversial realm of vaccine safety. One of Bobby's big battles throughout his career has been with the concept that there's an engineering solution to every problem. "My enemy was often the Army Corps of Engineers," Bobby wrote, because they "wanted to channelize rivers and move water around." Building levees, he pointed out, didn't protect New Orleans. It actually destroyed the wetlands that could have prevented Hurricane Katrina from wiping out New Orleans.

These kinds of engineering solutions have been a catastrophe. "There are so many unanticipated results that go unaccounted for," Bobby pointed out. "Years down the road you see you made a big mistake thinking we could engineer our way around nature. I think that years of fighting the Army Corps, realizing how much tunnel vision they had and how blinded they were by the love of their own 'solutions,' fed my skepticism toward vaccines and the whole idea that these medical interventions are going to help."

Bobby paused for a moment before elaborating: "We should be looking at, how do you support people's natural immune systems, as we should support nature to keep the waterways clean in a way that creates dignified enriching communities. Do you build good public health by thinking it comes in a syringe? It doesn't. Four million years of evolution have perfected the human immune system. Viruses are part of who we are. It's almost impossible to kill a healthy human being with a virus. So how do you keep people healthy? What are the factors? Is it vitamin D, is it zinc, is it exercise, is it sunlight, is it fighting obesity? You need to weaken an immune system if a virus is going to hurt you."

"Then you start looking at the science." Bobby continued, "which is what I love to do. Reading science has gotten me in trouble. There was a famous study from 2000 done by the CDC and Johns Hopkins that looked at the decline in mortalities from all the infectious diseases—smallpox, polio, diphtheria, pertussis, tetanus, measles, mumps, rubella, and all the other ones—showing an 80 percent drop in the twentieth century. The question they asked was, is that because of medical intervention from antibiotics and vaccines? And when they looked at it, the CDC said that had nothing to do with it. It was all because of better nutrition and sanitation."

* * *

Like Lieutenant Ronald Gatto in the early days of Riverkeeper, Louis Conte was another law enforcement guy in Westchester County who ended up in a serendipitous and path-breaking relationship with Bobby. Conte was the father of triplet boys, two of whom had developed autism in 2002 at the age of eleven months. Conte was employed in the Westchester County Probation Office.

After 9/11, Conte attended a meeting where a federal official discussed the possibility of a "mercury bomb" causing symptoms of neurological impairment. Officially that never happened, although Conte got the sense that rumors of an attempted dirty bomb detonation in the New York City area were not far-fetched. When he asked the official where anyone could possibly get enough mercury for something like a bomb, he was told "they get it from medical products and vaccines."

Driving to work one morning, he heard author David Kirby on Don Imus's radio show talking about his book *Evidence of Harm* and mercury in vaccines. All of a sudden, Conte came to a sickening realization that this could be what caused autism in two of his triplets. Then in 2008, the parents of Hannah Poling, a nine-year-old girl from Georgia diagnosed with autism just after turning two, publicly announced that a federal vaccine injury court had awarded the family a settlement. Conte happened to catch the CDC director at a press conference saying, "We shouldn't look into this as being proof of anything."

By this time, Conte had advanced to the position of supervisor in the Probation Office, a seasoned investigator. He used those skills "burrowing through vaccine court decisions in the middle of the night" and came upon case after case describing settlements for kids with autism-like symptoms. Connecting with Mary Holland, a professor at the NYU School of Law and one of the activist "Autism Moms," "We put together a team and launched an 18-month investigation," Conte recalls. "The government was secretly compensating hundreds of families most likely. We were able to communicate with 194 of them, but by 2011 we'd identified 83 cases where vaccines seemed to have caused autism. One of the parents was Sarah Bridges, who knew Bobby, and that's how the connection grew."

Conte pulled together a paper he titled "Unanswered Questions from the Vaccine Injury Compensation Program." Mary Holland suggested that Bobby, then teaching at Pace, might be able to interest the school's *Environmental Law Review* in publishing it. The editor agreed, and, according to Conte, "the plan was for Bobby to do a press conference with us, but at the last minute, the journal panicked; one of the peer reviewers was afraid that it wouldn't be good for the vaccine program." One of the other reviewers had one question, and, when Mary answered it, Bobby was able to convince the editor to move ahead. The article passed peer review, and Bobby would not let fear prevent its publication. The press conference took place shortly thereafter.

Holland was simultaneously coauthoring a book titled *Vaccine Epidemic*, scheduled to come out in 2012 with Skyhorse Publishing. Tony Lyons, who'd founded the independent company in New York six years before, had already built a reputation for doing books that mainstream houses found too controversial. In his personal life, Lyons encountered a number of misdiagnoses and had little faith in mainstream medicine. His older brother Paul, for example, died of melanoma after being misdiagnosed. So Mary Holland introduced Lyons to Conte, who ended up going to work for Skyhorse part-time as an editorial consultant. Skyhorse later published his novel, *The Autism War*, "the story of a good suburban cop and father of a child with autism, [who] finds himself drawn into the controversy over the apparent but rarely acknowledged connection between childhood vaccines and autism." It was based on his personal experience.

By this time, Bobby and Conte had become friends. When Bobby told him that he'd written a "white paper" about thimerosal that he wanted to get distributed, Conte suggested he meet with Tony Lyons. As often happened, the two already shared much in common. Tony's publisher father, Nick Lyons, had been a close personal friend of Hudson Riverkeeper founder Bob Boyle. Tony had considered pursuing a summer internship with Riverkeeper while attending Albany Law School. Instead he spent the summer researching Hudson contamination issues for the Government Law Center. He'd heard Bobby speak on Long Island about mercury contamination in fish and vaccines. Before deciding to follow in his father's footsteps, Tony thought he might practice environmental law.

So it was an easy segue for Skyhorse to sign up Bobby in 2013 as the editor for *Thimerosal: Let the Science Speak*, a book that laid out all the evidence he'd been gathering in support of eliminating the mercury-containing preservative from the world's vaccine supplies. It was a scientific primer, citing hundreds of peer-reviewed papers, and included a preface by world-famous practitioner of functional medicine Dr. Mark Hyman; an introduction by Dr. Martha Herbert, a pediatric neuroscientist at Massachusetts General Hospital; and a foreword by Congressman Bill Posey. As Bobby wrote in his Introduction: "People who advocate for safer vaccines should not be marginalized or denounced as anti-vaccine. . . . I want our vaccines to be as safe as possible."

After the book came out early in 2014, Bobby traveled to Washington with Dr. Hyman and Lyn Redwood, the former public health official and founder of SafeMinds. They were scheduled to meet with Kathleen Sebelius, President Obama's Health and Human Services secretary and a strong vaccine advocate. She'd gone so far as to tell the media not to cover the dissenters. "I was still naive enough to believe that if people knew the truth, they would have the same reaction I did," Bobby told me. "I completely underestimated the corrupting influence of institutional and career self-interest."

When Sebelius stepped down a week before the April meeting amid controversy over Obamacare, Bobby and his two companions went ahead and met with representatives from the FDA and CDC. "We were trying to get them to finally remove thimerosal completely from vaccines," Redwood recalls. "They had left it in the flu vaccine, even those in single-dose vials." This only made sense to Redwood if the thimerosal was being used not as a preservative but an adjuvant (a toxin added to the vaccine in order to induce a stronger immune response to whatever is in the vaccine). "Thimerosal was not licensed by the FDA as an adjuvant, but they basically rubber-stamped it," because it made the immune response to the antigens in the vaccines stronger. Redwood's organization bought enough vaccine to have toxicology testing done for metal content, anticipating that mercury might be found at levels higher than the FDA allowed. It wasn't, but in one of the vaccines, levels of aluminum, another common adjuvant, were doubled what the manufacturer had published in the product information sheet.

Redwood told me: "Bobby and I both knew that mercury was just one of multiple harmful chemicals impacting our children's health, and aluminum in vaccines is also incredibly devastating. This came up at the FDA meeting, but they still thumbed their noses at us, so they didn't have to do anything."

It became clear that neither the FDA nor the CDC had any interest in looking at the science. Nor did the press. Thimerosal was universally paused, and the first negative reviews, which came out before publication, were based solely on the title. The publisher confirmed that no pre-publication copies had been sent out.

Bobby hadn't given up on exhorting the media to take a legitimate look at the situation. In the draft of "an open letter to American science journalists about mercury in vaccines," he outlined how the thimerosal debate "has precipitated a journalistic as well as a public health crisis." He described his shock that "science writers simply adopt the familiar mantras of badly compromised public health regulators and vaccine industry spokespeople. . . . In a weirdly impenetrable and Kafkaesque denial, many of them became irritable, curt, scornful . . . when I asked them to venture beyond the industry talking points."

When it came to scorn, this would prove to be a mild beginning. Questioning vaccine safety, or anything at all about vaccines, is obviously a bad career decision.

* * *

Around the same time in 2014, Bobby received a call from Eric Gladen. A Southern California-based civil engineer, Gladen had fallen ill after receiving a tetanus shot at the age of twenty-nine that he believed resulted from the malignant effects of thimerosal. He'd formed a nonprofit called Go Mercury Free and wondered if Bobby might consider joining the board. He was finishing a documentary called *Trace Amounts*, which included footage of Bobby's impassioned speech at a Washington rally in 2005. Gladen sent the film to Bobby, who was amazed at its quality and damning evidence about mercury in vaccines. The timing was "a fortunate coincidence," Bobby reflected, "or unfortunate for me"—because what he saw wouldn't allow him to, as Dylan Thomas wrote, "go gently into that good night."

This changed everything for Gladen. He later told the Associated Press that Bobby was "a machine," doing research, writing op-eds, delivering speeches, and connecting with well-placed people. There was "almost no limit" to whom they could reach. Bobby agreed to travel with Gladen for showings of the documentary in a number of cities. The premiere on February 4, 2015, presided over by Bobby, drew a packed house of celebrities to watch the film and hear a panel of experts. "The anti-vaccine crowd gets its Hollywood moment," the *Los Angeles Times* headlined, describing "a rare moment in the spotlight for a group that has been increasingly shunned and chastised." It came amid an "outbreak" of childhood measles that had led California legislators to advocate restriction of parents' ability to obtain a "personal belief" exemption from state-required vaccinations.

Bobby then accompanied Gladen to Oregon, where he met a prominent businessman and advocate named J. B. Handley. A Stanford-educated private equity entrepreneur whose fund had earned him a small fortune, Handley had a severely disabled son diagnosed with autism. He had created Generation Rescue with his wife, a nonprofit that proved extremely effective in organizing parents concerned about an autism-vaccine link and aimed at helping children recover.

Handley recalls that he'd "all but lost interest in spending time in the public realm, because [he] viewed it as a rigged game that the Democrats were largely responsible for." But Bobby struck him as "extremely inquisitive," saying, "I don't know if it's just his intellect or the fact that he's a former prosecutor, but he stayed on topic when he asked the next question and the next. So to have Bobby there asking real questions, looking you in the eye, made a very strong impression on me."

Bobby felt an instant rapport, as well. "J. B. was a major influence on me because he was so solid. He was super smart and had done a deep dive on the science, which takes years to master. I needed to make sure my ground was safe, that I wasn't saying things about a study that could be debunked. J. B. gave me a lot more confidence that this is something I was right about. Then it becomes totally compelling, because am I going to ignore what's being done to our children or does this knowledge compel me to go into the battlefield on an issue that I really don't want to have anything to do with? But now I know too much. Every time I see an injured kid, I know what's happened to them."

In January 2015, the powerful Oregon State Senator Elizabeth Steiner Hayward had gone on *ABC Nightly News* to announce the imminent passage of a law, Senate Bill 442, that would eliminate religious and philosophical exemptions from school shots in Oregon; parents would be able to opt out only if they could prove a medical reason. The Democratic legislative majority and the governor were on board, and, according to Handley, "it looked like a fait accompli." He spearheaded getting angered parents out to a public hearing and set up Oregonians for Medical Freedom to hire a lobbyist.

Then Handley floated an idea to Tim Knopp, the Oregon Senate's Republican Minority Leader. What if they brought in Robert Kennedy Jr. to address the Statehouse? Early in March, with the mandate bill about to be introduced on the floor, Handley picked up Bobby at the Salem airport. The scene remained vivid to Handley eight years later: the irony of the Republican leader waiting on the steps to greet Bobby, followed by his immersion in a sea of Democratic legislators all wanting to meet the son of one of their heroes. "We can't solve a credibility problem by forcing people to undergo a medical procedure without informed consent," Bobby said before the event.

"Bobby did a great job of threading the needle," Handley remembers. "If you walk in and say vaccines are causing autism, most people run in the opposite direction. So he wove a much more eloquent tale that relied on his prosecutorial argument that vaccine makers already have no liability and that we can't lose more checks and balances. The only thing standing between pharma and babies at this point are the moms, and we shouldn't take that away."

After his talk, Bobby spent about four hours meeting with legislators from both sides of the aisle. That evening, legislator Knopp hosted a screening of *Trace Amounts*. On the ride back to the airport, Handley recalls Bobby telling him, "This thing is dead." Incredulous, Handley replied: "Are you kidding me?" Bobby nodded and added: "Don't worry, the bill is never going to pass."

A few days later, Senator Hayward announced a new policy. While it would encourage more schoolchildren to get vaccinated, nonmedical exemptions could continue and alternative paths be provided for parents to comply with the law. "It's not in any way an exaggeration," says Handley, "to say Bobby literally saved the state from mandatory vaccination"— which remains true to this day. California was not so lucky, as then-Governor Jerry Brown signed one of the nation's strictest immunization bills into law in June 2015.

The two men remained good friends, as Handley went on to write two books (including the story of discovering a nonverbal method of communication that changed the life of his son). Handley noted that "Bobby's father is sort of a mythological figure to the rest of us, but to him it's his dad," and he "has completely absorbed the code that his dad lived by."

* * *

In 2015, Bobby wrote of his increasing marginalization by old friends and colleagues: "My grandfather taught me to fight as hard as I could and the result was not in my hands. It was the same kind of creepy feeling that I would get on the thimerosal or global warming issues, like a kind of mass hypnosis where people considered the unacceptable, acceptable. I felt very isolated." It wasn't the first time. He remembered running his uncle Ted's failed campaign in Alabama "and recalled the months of despair as [he] watched his polls sink, and people who at one time would walk over hot coals for our family turned their backs in disgust and walked away."

In January 2016, Bobby took time out from the vaccine front to publish in Politico's magazine a lengthy piece on a very different subject, "Why the Arabs Don't Want Us in Syria." His article dived into territory foreign to the standard American attitude about the perilous Middle Eastern landscape, pointing out that the hidden agenda revolved around the disputed Qatari pipeline. "Once we strip this conflict of its humanitarian patina and recognize the Syrian conflict as an oil war, our foreign policy strategy becomes clear," Bobby wrote. "We need to dramatically reduce our military profile in the Middle East and let the Arabs run Arabia. . . . We need to begin this process, not by invading Syria, but by ending the ruinous addiction to oil that has warped US foreign policy for half a century." His assessment became one of the most heavily read articles in Politico in 2016.

Another pipeline, this one domestic, brought Bobby's return to an earlier passion fighting for indigenous rights, a cause that had mushroomed into a crusade drawing hundreds of people to set up protest camps against the 1,200-mile-long Dakota Access Pipeline on the Standing Rock Sioux reservation in North Dakota. The pipeline, if allowed to be completed, would amount to an "unlawful taking" of tribal lands and would generate more carbon pollution than 27 coal-burning powerplants and disrupt twenty-nine waterways including the tribe's only water source at Lake Oahe. Not to mention the possibility of a devastating oil spill. Police and military power had been deployed, but protesters from around the country refused to leave.

In mid-November Bobby toured the protest camp with Sierra Club Executive Director Michael Bruin and held a press conference, calling on President Obama to halt construction immediately. He compared the Army Corps of Engineers' insistence that the tribe get out of the way to the 1890 slaughter at Wounded Knee after the discovery of gold. He referenced his father's holding senate hearings to investigate the violent attacks by growers against pickers in the produce fields around Delano, California. And he issued a formal statement saying: "Energy Transfer Partners' true interests reside solely with profits, not the public. They have desecrated sacred land and put millions at risk with their dirty business. Indigenous peoples from over 300 tribes, peaceful protestors, and allies took a brave stand against this outlaw corporation and prevailed. The standoff at

Standing Rock will go down in history as an historic and landmark event in the pursuit of justice and defines a new era in environmental advocacy."

In its final days, the Obama administration denied permits for the pipeline to cross the Missouri River and ordered a full Environmental Impact Statement to explore alternative routes and the impact on the tribe's treaty rights. Donald Trump, during his first week in office, then signed an executive order to let construction proceed. The pipeline would be completed in June 2017, but the tribe's challenge of the permits proved successful in 2020, and the company's appeal of the mandated environmental analysis was rejected by the Supreme Court in February 2022. Although the pipeline continued to operate with the review pending, Energy Transfer allowed in court papers that it was "vulnerable to a shutdown." For the North Dakota tribes, this had been a major victory.

The day after visiting the protest camps, Bobby had returned to officially announce the launching of the World Mercury Project (WMP). Filmmaker Eric Gladen's physical condition had worsened by the summer of 2016, when Bobby agreed to take over chairmanship of the group that Gladen had founded. Lyn Redwood, who'd already been doing an analysis of its strengths and weaknesses, then became the executive director of the organization. She quickly began to assembly a small team of other "mercury moms," including, Laura Bono, who cofounded the National Autism Association with Redwood.

Bobby's announcement on November 16, 2016, also marked the thirtieth anniversary of the 1986 National Childhood Vaccine Injury Act, "which launched the $32.2 billion yearly vaccine market and some argue indirectly spawned the epidemic of childhood neurological disorders including autism." This was what had granted vaccine makers their liability protection. Bobby went on to say: "That act and the amounts of pharma cash going to politicians, regulators, and the press have helped obliterate all the checks and balances that normally stand between a rapacious industry and vulnerable children in a free and democratic society."

* * *

Over the Christmas holidays in 2016, a newly elected president, Donald Trump, called Bobby. "He asked me to come to New York and meet with him at Trump Tower," Bobby remembers. "I had known Trump for many years. I had successfully sued to block him from building two golf courses in the New York upstate reservoir watershed. I knew he was no friend of the environment, but neither did he appear to be ideologically hidebound to a pro-pollution worldview. Indeed, he seemed less shackled to dogma, or obligated by encumbrances, than any other Republican presidential candidate."

So when Trump said he wanted to talk with him concerning vaccine safety, Bobby was willing to give it a shot. There were rumors that Trump's ten-year-old son Barron might have symptoms of the autism spectrum, and perhaps this was something closer to the man's heart than his pocketbook.

On January 10 of the new year, Bobby left his home in Malibu and flew to Manhattan. Others he remembered being at the meeting included Vice President Mike Pence, Steve Bannon, Reince Preibus and his sons, and Kellyanne Conway. Trump asked Bobby to have a press conference afterward and make an announcement. There he told reporters he'd been asked to chair a new commission on "vaccine safety and scientific integrity." And he had agreed.

Bobby was willing to give Trump a chance to prove himself. He told an interviewer afterward that he believed the president-elect would come into office "less encumbered by ideology or his obligations than anybody that has won the presidency since at least Andrew Jackson." He described Trump as "very thoughtful on the issue. [He] has some doubts about the current vaccine policies. He says his opinion doesn't matter, but the science does matter, and we ought to be reading [and] debating the science. . . . Everybody ought to be able to be assured that the vaccines that we have are as safe as they can possibly be."

Within hours, a Trump spokesperson qualified Bobby's statements, saying the president "is exploring the possibility of forming a commission on autism. However, no decision has been made," and Trump was said to be discussing "all aspects of autism with many groups and individuals."

The behind-the-scenes blowback was apparently swift. "One of the things that happened," Bobby revealed a few years later, "is that Pfizer immediately made a million-dollar donation to the inaugural party. Then President Trump put Alex Azar, a handpicked Pfizer lobbyist, in charge of

Health and Human Services. And Scott Gottlieb, who was also handpicked by Pfizer, to head the FDA. We'd had some really good meetings during the first part of the administration, but then these guys came in and said we're not going to talk to you anymore and shut us down."

At the time, Bobby didn't speak publicly about what he'd learned and experienced. When asked about what happened to the commission by STAT News that following summer, Bobby said he'd had no specific discussions about it since probably February. "You'd have to ask the White House," he added. "It may be that it's evolved. I've been told that the president is still interested in this issue."

If the pharmaceutical industry had moved quickly to prevent such a commission from happening, the reaction was just as extreme among Bobby's longtime allies. "I had the environmental community turn against me," Bobby told me, "which was weird because Trump had just won and hadn't done anything bad yet. We certainly had expectations he would be a terrible president, but I met with every president. I'd even met with Bush, who was the worst thing that ever happened to the environment, and the environmentalists didn't complain. With a president you don't like, I always thought it was much better to meet with them than not."

The organizations with which he was connected did not agree. There was already tension with the Hudson Riverkeepers, whose new president "had other reasons to want to get rid of me." This was the first, and oldest, relationship to sever. "I had enough control on the board that, if I had brought it to a vote as I had done fifteen years earlier with Boyle, I could have won again. But I felt like I'd created an institution, and this would mean shattering that institution and then rebuilding it. It felt like ego on my part and seemed like maybe it was better to just relinquish the fight and leave Riverkeeper intact and move on without knowing what I was going to do."

The Natural Resources Defense Council then cut ties with Bobby completely. He'd been affiliated since his entry into the environmental movement in 1985 and written in a journal entry some years earlier: "The people at NRDC were an extraordinary group. I went to retreats with them and they were articulate and idealistic, pragmatic and extraordinarily effective and knew their issues deep into the weeds, and what was possible in Washington and what was not. They could do strategy and fund raising, public speaking, public relations, and litigation, legislation and agitation." But the organization's senior scientist, Gina Solomon, had long expressed concerns about Bobby's involvement with the vaccine issue.

The last domino to fall was Pace University, a decision clearly connected to Bobby's stance on vaccines. "I moved out to Southern California because Cheryl was living here. I wanted to be with her and support her," Bobby explained. Moving East would have been career-ending for Cheryl. "For me, it meant changing things," he said. He tried to hold onto his old system of commuting once a week to New York. He'd leave every Sunday, teach class Monday and Tuesday, and schedule his other business in the city over those two days. "But the school gradually turned against me," Bobby said, "and wanted to push me out. I felt like it was the universe and God sending me directions.

"Trump's advisors turned their attention to dismantling the science safety net and commoditizing and monetizing every aspect of human discourse—adopting policies that will amplify the wealth of billionaires, even as they sicken our citizens and destroy our planet. The Book of Revelation described the Four Horsemen as War, Conquest, Pestilence, and Death. Donald Trump's choice to invite a group of conscienceless oil men to govern the country has brought such chilling metaphors to the foreground as more than an obscure biblical reference."

As Bobby told STAT News that summer, "President Trump's administration is essentially destroying thirty years of my work on environmental issues, and the work of many other people. And I've written extensively about that." He said this had been part of his discussion with Trump in January, and the president well knew Bobby's position. Then Bobby added: "If President Trump asked me to serve on a commission on fracking or on pipelines or on global warming, I would do it." Or, as J. B. Handley put it, "You know that old saying? You'd rather be a warrior in a garden than a gardener in a war. Bobby's the warrior in the garden, for sure."

CHAPTER THIRTY-ONE

COURTROOM DRAMA: FROM DUPONT TO MONSANTO

For Riverkeeper, Bobby had litigated strictly statutory cases, filing dozens of lawsuits accusing violators of the Clean Water Act, the Clean Air Act, the Endangered Species Act, and other landmark laws passed by the US Congress. As outlined earlier, the successful case against Conoco in 2004 marked his move into class-action litigation, where Bobby worked alongside Pensacola Riverkeeper founder and attorney Mike Papantonio.

Then, in 2007, Bobby joined with Papantonio's firm and two others as plaintiffs' counsel in a major court action seeking punitive damages for residents in and around the little town of Spelter, West Virginia. The legal team took on E.I. DuPont, the nation's third largest chemical company, accused of negligence in creating a 112-acre waste site from a zinc-smelting operation and placing local homeowners at higher-than-normal risk of diseases including cancer, cognitive problems, cardiac disease, and lead poisoning.

West Virginia had a unique role in the Kennedy family legacy. It was here in 1960 that Bobby's parents had disappeared for six weeks to campaign for JFK in a pivotal primary race. JFK had bucked his advisors and confronted the religious issue head-on, appealing to voters in the

Mountain State to overcome their prejudice against a Catholic running for president. He'd gone on to win the primary with an astonishing 65 percent of the vote.

As Bobby recalled in delivering the closing argument at the Spelter trial, his father had come home from West Virginia and spoken to his children at the dinner table. Robert Kennedy told them he'd just been in a place "that should be one of the richest states in the country but has some of the poorest people." Bobby remembered for the jurors a conversation he'd later had with his father. about how strip mining was turning the landscape into a "barren moonscape" in order to break the unions in West Virginia.

Now here was the second son, four decades later, pointing the finger at DuPont for enriching itself at the expense of local West Virginia citizens' health. For two months, Bobby and the other lawyers resided in the only hotel in Spelter—right downstairs from DuPont's attorneys - for what became a four-phase trial.

One of the plaintiffs' lawyers was Kevin Madonna. He'd entered law school at the Pace Clinic in the fall of 1993 and ended up doing research for Bobby on *The Riverkeepers* book. "In the fall of '94, the Republicans took control of the House and the whole world changed," Madonna remembers, "so the book had to be almost completely rewritten." This wasn't completed by the time he graduated, and "we made an arrangement where I moved in with Bobby and his family in the house in Bedford and just worked on the book." He lived with the Kennedys for a year, after which Bobby hired him as executive director of the new Waterkeeper Alliance, and in 2000 they formed the Kennedy & Madonna law firm.

The Spelter case proved to be a breakthrough. Over more than ninety years, DuPont's zinc smelter plant had produced more than 400 million pounds of zinc dust, creating one of the worst environmental disasters in West Virginia's history. DuPont eventually cleaned up the property but refused to take any action to address the contamination of nearby communities from the waste site.

On October 10, 2007, a jury ordered DuPont to provide medical monitoring of some eight thousand residents who would receive free diagnostic testing. In the third phase of the trial, the chemical company was ordered to pay $55.5 million in damages associated with remediation for communities impacted by the toxic waste. Then came the fourth phase, where Bobby delivered what one of the other attorneys described as the best closing argument he'd ever heard.

Bobby took the jury through a list of internal DuPont emails revealing the company's indifference to the West Virginia environment. In one, DuPont's in-house counsel hoped that a couple of large storms would wash the mountain of waste into the West Fork River and thus eliminate DuPont's responsibility for cleanup. Bobby accused DuPont of repeatedly misleading the public with "little lies" intended to save money: "They were being cagey. They were being dodgy. They were being coy. They were being clever. That is who DuPont is. They have lost touch with their moral bearings."

DuPont "had the state wired" through close contacts with officials including the secretary of the Department of Environmental Protection (a former DuPont attorney). "This agency was a sock puppet for the company," Bobby said. DuPont routinely massaged statistics to suit its purposes. "This is part of the corporate DNA, these tiny perjuries that illustrate what the corporate culture here is really all about." He worried about the effect of pollution in Spelter on the food chain, saying that some of the emissions "are going to end up on somebody's plate, and somebody's going to get sick."

Pointing to the packed courtroom, where dozens of residents sat shoulder to shoulder, Bobby urged the jury to send a message to DuPont and other industries raping the state's natural resources. "They looked over the green landscape of West Virginia and they saw a commodity," he said. And the jurors could be the ones to bring justice and democracy back to their region.

The eleven-member panel in Harrison County Circuit Court gathered to weigh whether to award punitive damages to ten Spelter residents. The jury recessed after roughly three hours of deliberation, then resumed the next day. The trial ended with the jury deciding that DuPont's negligence required the corporation to pay $196,200,000 in punitive damages—one of the highest compensatory verdicts in West Virginia's history.

This wouldn't be Bobby's last hurrah with DuPont. In 2016, he became part of a legal team working with Robert Billot, a local attorney who'd been pursuing DuPont for going on two decades for poisoning the water of pretty much the entire community of Parkersburg, West Virginia, with perfluorooctanoic acid (PFOA), or C8, used to make Teflon. These "forever chemicals" did not break down but lodged in dust and people's bodies, allegedly causing various cancers and other diseases.

Bobby recalled a turning point of the trial being the release of emails from an in-house DuPont attorney complaining to his son serving in Iraq "about how corrupt the company was and how bad this chemical was," saying he was "trying to get them to act . . . behave in a way that's honorable, and they absolutely refused." The man's videotaped deposition, calling the product "devil's piss," was key in swaying a jury against DuPont, which after losing in three trials ultimately settled with about 3,500 individuals who divided up $671 million.

Billot's crusade would be portrayed by actor Mark Ruffalo in the 2019 film *Dark Waters*. The Kennedy & Madonna law firm would go on to represent more than a hundred municipalities across the country in drinking water contamination cases resulting from PFOAs. In December 2022, after years of pressure, the 3M corporation agreed to cease making the "forever chemicals" within three years; legal experts had estimated future litigation could cost the company more than $30 billion.

* * *

For Bobby, the next target ended up being Monsanto. This closed a circle in his family's environmental legacy, dating back to JFK's staunch support of Rachel Carson that led to the banning of DDT in 1972. The chemical giant, however, had only transferred its nefarious ways and means to another product.

Bobby's joining a trial team, seeking verdicts for individuals who developed non-Hodgkin's lymphoma from using Monsanto's Roundup weed killer, came about serendipitously. After he moved into the house with Cheryl Hines in Malibu, Bobby's backdoor neighbor turned out to be Michael Baum, a prominent attorney who directed his firm's mass tort litigation against harmful pharmaceutical drugs and consumer products.

"I'd known Bobby was an environmental lawyer and read a couple of his books," Baum remembers. "We ended up at a mutual friend's graduation party for my son and the friend's daughter. Bobby was standing off by himself. . . . So I went over and introduced myself. We had a long talk, a mind meld talking about pharma corruption and environmental cases. I casually mentioned looking for new projects, maybe something involving Monsanto. My law partner Brent Wisner and I were doing research about the effect of its pesticides on the bee population. Then toward the end of the year [2015], a woman in our office had a cousin who'd sprayed Roundup around his orchard, and his dog died of lymphoma. We were asked to take a look. There was plenty to look at, and that was the start of our building a case against Monsanto."

Baum and Wisner knew they couldn't take on such a massive legal effort alone. They attended a national meeting of the Executive Committee for Roundup Litigation and decided to join them. "I got the idea that if Bobby and our firm did a media campaign together, it would be a good blend of our interests, along with using his environmental chops and history to help us find cases."

That was in 2016, and Bobby couldn't have been happier at the fortuitous development marking his entry into California's legal waters. He admittedly "grew up hating Monsanto." As he recalls: "After everything that happened with Rachel Carson, my uncle Teddy held hearings on Agent Orange [the toxic defoliant impacting Vietnam veterans] that I attended. I was peripherally involved with another Monsanto product, Nutrasweet (Aspartame), that caused bad health effects. And when my law career started, the second case I did, Monsanto had made the PCBs that General Electric dumped into the Hudson River and put out of business 2,800 fishing families who were my clients. I worked the next thirty years trying to bring GE and Monsanto to justice."

Before PCBs were banned in 1979, Monsanto still had huge inventories and knew it had to somehow get rid of them quickly. Dumping them in a secure landfill would cost a large amount of money. But an explosion of new schools was being built at the time across America for children of the baby boomer generation. Monsanto invented a special caulking material using PCB oil and went to the school construction groups. According to Bobby, "most of the school buildings constructed in the 1970s contain huge amounts of PCBs. And when the heat is on in those buildings or when the sun cooks the windows, the PCBs volatilize. They go into the classrooms and then into the classroom furniture and any kind of foam rubber or rugs or curtains."

A Harvard study released in 2016 calculated that up to 30 percent of American schoolchildren were still being exposed to these toxic chemicals from elementary grades through high school in as many as twenty-six thousand schools serving up to fourteen million students. Bobby has "represented school districts across the country that are suing Monsanto to remove the PCBs from their schools. These have caused a lot of behavioral problems. PCBs are associated with ADD and ADHD and other diagnoses."

Bobby adds: "So when Michael Baum asked if I would join him, I jumped on it. I told him kind of jokingly that if my life were a Superman comic, Monsanto would be my Lex Luthor—because I feel like I've spent my whole existence in a struggle with that company."

* * *

The origins of what would become a groundbreaking verdict against Monsanto's malice dated back to the year DDT became the first pesticide banned in the US The company needed to find a replacement. In 1974, they had patented a chemical called glyphosate, originally used by Stauffer Chemical as a descaling agent for removing mineral deposits from commercial boilers and pipes. In the late sixties, Monsanto's herbicide screening program was testing potential water softener compounds for herbicidal activity. John Franz was assigned analyzing a couple of those chemicals, including glyphosate, and synthesized a version that could act as a strong plant growth inhibitor. Operating as a disrupter of an enzyme pathway plants and microbes relied upon to make proteins, it was believed to kill plants without harming people or animals. Greenhouse testing in 1970 demonstrated its extraordinary herbicidal efficacy, and Monsanto officials quickly saw its potential to revolutionize the herbicide market.

The company began mass-producing glyphosate as the active ingredient in its Roundup agricultural pesticide. Farm workers would strap a tank on their back, carry a hand sprayer, walk down the corn rows during the early part of the growing season, and directly spray weeds that were competing with the corn. Once the corn got a foot or two high, weeds could never catch it.

It was being sprayed very selectively, only at the very beginning of the growing season and not near the harvest season. Which was a good thing, because in 1985 the EPA classified glyphosate as a Class C carcinogen, panicking Monsanto enough to hire a leading toxicologist to concoct a successful challenge to the government's conclusion. "EPA said it would temporarily withdraw the carcinogen classification," Bobby recalls, "but only on condition that Monsanto redo the mice studies. They never did."

Bobby describes what happened next: "In 1994, Monsanto made a development that changed the forty-thousand-year-old business model for agriculture all over the world." Monsanto employees had discovered bacteria that survived Roundup exposure in the sludge-filled waste ponds surrounding their production plant in Luling, Louisiana. "So Monsanto said, let's take the gene out of that sludge bacteria and put it in soybean; if we do that, we can create soybean that is immune to Monsanto!"

They developed Roundup Ready soybeans, the first generation of genetically modified crops. Which was soon followed by Roundup Ready corn and other GMO crops. Bobby continued, "that meant they could now fire all those farm workers and hire one guy with an airplane to spray the entire field and saturate the landscape—and everything green would die except the corn. It became a very economical way at the outset of growing corn, because you could get rid of most of the labor costs and put them instead into chemicals. That's what they did for the next ten years, and Roundup became the bestselling herbicide in the world.

"The real growth spurt began in 2006," Bobby said, "that's when they discovered you could use Roundup as a desiccant," or drying agent, on non-Roundup Ready crops like wheat and oats. "One of the big costs for farmers comes during harvest season, when, if the crops get wet from rain or dew after they're cut, it will cause mold and they could lose all their produce. Also, if you spray it with a desiccant, everything becomes ready for harvest immediately and you can accelerate the process. This became a very profitable thing for farmers to do. About 95 percent of the Roundup ever used has been used since 2006. Roundup sales increased by fifteen times. But the danger was, by spraying it directly on crops at harvest time, you're spraying it on food, not on plants.

"If you look at when celiac and gluten allergies first appeared on a massive scale, it was 2006. I had eleven brothers and sisters and seventy cousins, and I never knew anybody with gluten allergies. Now suddenly everybody was eating Roundup with their wheat. Every piece of non-organic bread, anything that has wheat in it. Before they used it as a desiccant, there was no Roundup-ready wheat. People had soy allergies, and corn had problems, but the wheat was never sprayed. Now Roundup is in our cereals, our beer and wine. It's in the baby foods. It's ubiquitous.

"When the National Institutes of Health tested the urine of pregnant women in 2015–16, glyphosate was detected in 93 percent of the samples. It disturbs your capacity to digest food in a healthy way."

In 2014, Monsanto sold more than $5 billion worth of Roundup. Its use of glyphosate had soared from 11 million pounds in 1987 to nearly 300 million pounds by 2016. It was being sprayed on 90 percent of the corn, soy, and sugar beets produced in the US, and several studies showed that glyphosate is detectible in around 90 percent of the US population. By this time, glyphosate had been associated with a cascade of chronic diseases. It could result in nonalcoholic fatty liver cancer and was also linked to kidney disease, diabetes, rheumatism, eczema, asthma, peanut allergies, and behavioral issues in children.

But when bringing a lawsuit about chemical exposure in the US, an attorney must pass an evidentiary threshold known as the Daubert standard, based on three Supreme Court decisions since 1993. Each of the states has its own version, all basically saying no one can bring science that is speculative before a jury. There is no pure definition of what that means, but there must be a critical mass of scientific studies that associates a particular exposure with a particular disease.

In March 2015, a working group on glyphosate with the World Health Organization's International Agency for Research on Cancer (IARC)—consisting of 17 renowned scientists from 11 different countries—concluded from numerous animal studies that the chemical is "probably carcinogenic to humans" with evidence pointing specifically to the blood cancer known as non-Hodgkin's lymphoma. As an independent body based in France, the IARC was considered the "gold standard" when it came to chemical analyses. California legislation under Proposition 65 specifically says that the state would follow an IARC ruling deeming something carcinogenic. This meant the Daubert threshold could be crossed if a judge thought it warranted.

Bobby recalls: "A judge narrowly said there was just enough science to get us to a jury. Michael Baum had tremendous courage to do this, because it might cost you millions of dollars. You've got to do all these depositions and testing, interview thousands of plaintiffs, hire doctors and scientists. Most firms don't have the money to do that by themselves, and they'll put together a consortium where each plaintiff firm kicks into the fund. If they win, they split the pot.

"So we had these six firms involved, each of them advertising to find clients who suffered from non-Hodgkin's lymphoma after being exposed to Roundup. Farm workers handled many different pesticides, so it's hard to isolate and say your cancer came from the Roundup. The ideal clients for us were home gardeners. Monsanto had gone to them and said it's as safe as aspirin. Their advertising showed people wearing shorts or bathing suits and barefooted. No indication that you should try to protect yourself.

"Between these six firms, there were about six thousand possible cases, all jostling for who gets to try the first one. The small Miller firm, a great group of guys from rural Virginia, won that battle. All their lawyers are members of their own bluegrass sort of band, and they had a hoedown in a barn next to their law firm every Saturday night."

The legal fight that Mike Miller insisted they take up first was on behalf of an African American father of three from just north of San Francisco, where juries are known to be good on chemical cases. Dewayne "Lee" Johnson was a forty-six-year-old groundskeeper and pest manager for all the facilities in the Benecia school district, spraying Monsanto's Ranger Pro (at higher concentrations than Roundup). After his hose broke and sent concentrated glyphosate formulation pouring down his protective suit, he developed a terrible rash and skin lesions that were eventually diagnosed as non-Hodgkin's lymphoma in 2014.

According to Bobby, Johnson had been "a very handsome guy, beautiful wife who adored him and these wonderful kids, and he loved outdoor activities but was so horribly disfigured that he couldn't go in a public swimming pool for fear of horrifying people. He was going to die for sure. It was terminal. There was no treatment for it, just a matter of time, and Mike Miller said we have to get him to trial before we lose him. He needs a settlement so that maybe he can get some treatment."

Then the Miller legal team suffered two setbacks in short order. Mike Miller, wind surfing on Cape Hatteras, suffered severe injuries when he slammed against a pier and could no longer try the case. He passed the ball to his partner and told Bobby and the others they needed to take a much bigger role. Then the partner, while on a conference call with the other firms, suffered a grand mal seizure.

Brent Wisner, only thirty-four, had been leading the legwork on the case for the Baum Hedlund firm. But was he seasoned enough to be the lead trial lawyer on such a major litigation? Baum was sure to go into serious debt to get the case before a jury. Prior to trial, not surprisingly, Monsanto fought back, spending millions in a PR assault against the IARC's determination of glyphosate causing cancer. In August 2017, the Baum team released "The Monsanto Papers," a massive series of damning internal emails and other documents obtained during the discovery process.

This was followed by a Minority Staff Report prepared for members of the House Committee on Science, Space & Technology in February 2018. Titled "Spinning Science & Silencing Scientists," the report accused Monsanto of having "launched a disinformation campaign to undermine IARC's classification" and cited documentation for the upcoming "multi-district litigation court case against Monsanto" clearly showing the company's "decades-long concerted effort to fend off any evidence suggesting potential adverse human health effects from glyphosate." One of the more damning emails revealed an EPA official told a Monsanto higher-up about how the US Agency for Toxic Substance and Disease Registry expected to replicate the IARC findings: "If I can kill this, I should get a medal."

The stage was set, and rookie Brent Wisner got the nod to present the case as colead counsel. Proceedings began in San Francisco Superior Court on July 9, 2018. In choosing a jury, Bobby pointed out, each side was allowed six peremptory challenges to a pool of sixteen. "So if somebody says a person can't be fair to the defendant, either side can say they're out.

Well, thirty-five San Franciscans said, one after another, that they couldn't be impartial because basically this was the most evil company in the world! This shocked Monsanto's lawyers."

Bobby's son Aidan, then sixteen, accompanied him as part of a high school internship. "Mostly it was assistant things like printing papers and helping with minute logistics," Aidan recalls. "The part where I felt I was contributing the most was on jury selection, trying to keep track of reactions and find jurors that were the most favorable without being dismissed—which took an unprecedented forty hours, because they were so favorable to my dad's side."

Dozens of potential jurors were eliminated after expressing negative opinions about the company. While the jury was still being selected, the Berlin-based Bayer corporation announced it was buying Monsanto for $63 billion. The *Wall Street Journal* would later call this "one of the worst corporate deals in recent memory . . . threatening the 156-year-old [aspirin] company's future" (August 28, 2019). The big reason was what had simultaneously unfolded in Northern California.

Bobby served as attorney-of-counsel and met every night with the team to go through the evidence. He brought in celebrity friends including Oliver Stone, Darryl Hannah, Ed Begley Jr., and Neil Young (who had written a song in 2015 about "Roundup's poison tide"). And Bobby posted a daily blog about the case on Instagram (still viewable on the Organic Consumers Association website).

His first report described Wisner opening with the video deposition of Monsanto's ex-toxicology director, who chronicled "their herculean schemes to discredit a series of key animal studies published through the 1990s." Emails were presented nailing Monsanto for considering hiring renowned genotoxicologist Dr. James Parry to evaluate those studies, but when the company didn't like Parry's conclusions that Roundup could cause cancer, they schemed to suppress his findings.

A week later, toxicologist Dr. Christian Portier explained to the jury how Monsanto colluded with the EPA's Office of Programs Chief Jess Rowland. Bobby wrote that Portier "showed how the EPA, with Monsanto holding its coat, cherry-picked glyphosate-friendly studies. . . . [The federal agency] exonerated Roundup based principally on studies ginned up or ghostwritten by Monsanto and its army of biostitutes." Bobby added that watching Monsanto's attorney Kirby Griffiths try to cross examine Dr. Portier "was like watching a man trying to climb a greased pole." Griffiths, according to Bobby, "never got his feet off the ground."

Then came Wisner's redirect of Dr. Portier, where the witness denounced "the backbone of Monsanto's case, [a 2018 study that] concluded with Trumponian chutzpah that glyphosate actually protects humans against non-Hodgkin's lymphoma." Raw data buried in the Andreotti study "showed a statistically elevated risk of T-cell lymphomas, the exact type of cancer diagnosed in our plaintiff Dwayne Johnson." In fact, Wisner showed that Monsanto had known since the 1980s that glyphosate induced tumors in lab animals.

The trial went into a sixth day with videotaped testimony from Monsanto officials. As the terrible rash began to spread across his body, Johnson had naively reached out to Monsanto's Consumer Complaints and Safety Division to ask about any possible connection to his spraying of Roundup. Dr. Daniel Goldstein, who ran the division, never returned Johnson's calls. Under cross-examination by Wisner, he admitted being aware for fourteen years prior to Johnson's exposure of numerous studies linking glyphosate to cancer. Even after the IARC report appeared as Johnson was fighting the early stages, the groundskeeper continued to expose himself to glyphosate because no one from the company responded to his plea and advised him to stop spraying.

The day that Johnson testified, actor Ed Begley Jr. arrived early at the Superior Court building. Bobby set down in his personal journal (July 22, 2018): "I asked Ed why he had traveled all the way from Los Angeles. He told me, 'I'm here because this is a historic event. For the first time ever, Monsanto will have to explain its crimes before a jury. This company has done more to harm families, farmers and food than any company on earth. If you add together the impacts of Monsanto's principal products— DDT, Agent Orange, Saccharine, Dioxin and PCBs - the company has been waging a bioterrorism campaign against our country for fifty years! It'll be a great message if the American justice system still has the vitality to stand up to corporate power and corruption. I want to tell my grandkids I was here!'"

Right before Wisner's closing argument on August 10, where it's decided how much money to request from the jury, Bobby remembers him saying, "'I want to ask them for $300 million. We were like, you can't do that. They are going to consider it overreaching and punish us. He said, no I think they are with us. So he asked the jury for $300 million, and they came back two days later with $289.2 million in damages. . . . He was amazing throughout with a very hostile judge, and the jury ended up loving him."

The twelve jurors had deliberated for three days, before concluding that the Roundup weedkiller caused Johnson's cancer and that the corporation had failed to alert him of the health hazards from exposure. Johnson had rigorously worn his protective equipment and followed all the rules. But there was no warning label on the product about glyphosate, and the jury found that Monsanto "acted with malice or oppression." The company bore responsibility for "negligent failure," because it should have known that its product was "dangerous."

"It's very difficult to use the courts, because these industries have insulated themselves so effectively," Bobby explained. "Our claim against these guys was for improperly labeling their product. If they had told people they could get cancer from it, we would have had no case." The National Trial Lawyers Association awarded Bobby, the Miller Firm, and Baum Hedlund its 2018 designation of "Trial Team of the Year" for their work on *Dewayne "Lee" Johnson v. Monsanto*.

In March of 2019, the team prevailed again in the first federal case against Monsanto, with another San Francisco jury awarding Edwin Hardeman $80 million in damages. The judge overseeing the case later reduced the verdict to $25.2 million. Two months after that, in Oakland, where hundreds of lawsuits filed in California state courts were consolidated, another trial ran for seven weeks. This time, the discussion of how much compensation to seek resulted in what Bobby called "a lot of tension, a lot of arguments, a lot of screaming, and Brent says, 'I'm going to ask them for a billion dollars.'" Once again, other lawyers told Wisner this was too much and would result in a backlash. But Wisner made the final call, "and he gave an incredible closing argument," incorporating one of the Monsanto Papers where a Monsanto scientist referred to the carcinogenicity issue as "the billion-dollar question." A sixty-nine-year-old couple, the Pilliods, was awarded a stunning $2.055 billion - doubling the young lawyer's request. This was the ninth-largest such jury verdict in American history (although the trial judge later reduced the amount to $87 million).

"At this point, Monsanto really wanted to settle," Bobby said in 2019. "Monsanto made a great strategy move, which was to sell itself to Bayer. After we got the first verdict back, Bayer lost one-third of its value. The second verdict, it dropped another ten points, and now it's 50 percent of the value it had been. Bayer bought Monsanto for $63 billion, and the *entire* value of Bayer is now $63 billion, so the value of Monsanto had been completely erased by these lawsuits. And that is gratifying, because it shows that you can ultimately bring some of these companies to justice."

Bobby and Michael Baum went on to deliver the keynote address to members of the European Parliament. In 2020, Bayer reached a deal to settle the majority of roughly 100,000 by-then-pending Roundup lawsuits for $10 billion but failed to gain court approval for a proposed settlement of future claims for $2 billion. The following year, Bayer announced it would stop selling glyphosate-based weedkillers in the US residential market for nonprofessional gardeners but would continue to peddle Roundup to farmers. In June 2022, the Supreme Court rejected hearing two challenges by the company to the successful court cases Bobby had been part of.

Dewayne "Lee" Johnson recorded a rap song about his ordeal. As of this writing, he is continuing treatment to mitigate his symptoms.

Early in 2023, a study authored by twelve California scientists and health researchers determined that children exposed to glyphosate appear to be at increased risk for liver inflammation and metabolic disorders in young adulthood and more serious diseases later in life. The handwriting, it appears, is now on the schoolyard wall for Monsanto.

"Bobby and I became pals in the course of all this," Michael Baum says. "He has a fearless integrity and idealism that he supports with facts and documents and that he'll stick to, notwithstanding what the orthodoxy says. Standing up to the well-organized pro-pharma and pro-pesticide manufacturing industries takes guts and persistence—as well as dealing with the heartache that comes with friends and family members who buy the narrative without looking at the facts."

CHAPTER THIRTY-TWO

CHANGING THE
ENERGY CLIMATE

Bobby first laid out his blueprint for addressing climate change in the last article he wrote for *Rolling Stone* (June 28, 2007) headlined "What Must Be Done." He began not by denouncing the oil and coal companies, but by reporting on a summit he'd observed that May of the nation's top business leaders: "The attendees, gathered at the invitation of Silicon Valley venture capitalists, included CEOs and other top executives from such Fortune 500 corporations as Wal-Mart, Proctor & Gamble and BP. They had been invited to discuss ways to end America's fossil-fuel addiction and save the world from global warming. But in reality, they had come to make money for their companies—and that may turn out to be the thing that saves us."

One of those was DuPont, which Bobby was simultaneously suing for millions on behalf of small-town West Virginians, but which had "cut its climate-warming pollution by seventy-two percent since 1990, slashing $3 billion from its energy bills while increasing its global production by nearly a third." In an article as lengthy as the piece he'd written two years earlier about mercury in vaccines, Bobby sounded an eloquent plea for true free-market capitalism. Fueled by ending government subsidies for oil and coal, green investment would give innovative solutions a fair chance to compete.

His "platform for the planet" listed boosting fuel-efficiency standards for vehicles; establishing a global carbon market that allowed industry to "trade" and drastically slice emissions; placing an immediate moratorium on new coal plants; replacing incandescent light bulbs with new energy-efficient fluorescents; and decentralizing the power grid to let homeowners who install solar panels or wind turbines sell their excess electricity back to their utility. All ideas, more than fifteen years later, that have either come to pass or remain on the drawing board for implementation.

In 2009, following the financial collapse that marked George W. Bush's final days in office, Barack Obama gave Vice President Joe Biden the task of overseeing the new administration's $787 billion stimulus package. This included a $90 billion investment in clean energy. Bobby wrote an article for the online version of Huffington Post, titled "Big Carbon's Sock Puppets Declare War on America and the Planet" (November 25, 2011). He began:

"It is now become *de rigueur* among the radical right wing rhetoricians to characterize any government support of America's green energy sector as wasteful, fruitless, and scandalous. They greeted with glee the collapse of the government supported solar company, Solyndra, America's first major casualty in our race with China to dominate the 'new energy' economy. With Solyndra dying on the battlefield—its marketplace choking on inexpensive Chinese solar panels—the right wing's response was to hoist the white flag and declare defeat in the war for global cleantech leadership. That brand of' 'Can't Do' cowardice is a boon to the carbon and nuclear power incumbents who fund so much of the right wing's activities—but it's bad for America."

Indeed, as Bobby pointed out, the same Department of Energy program that supported the solar effort gave an astonishing $8.3 billion loan guarantee—many times the size of the solar projects—to the Southern Company to build two nuclear power plants. "Nuclear power is an industry with a product so expensive it cannot compete in any version of free market capitalism," Bobby wrote. He told the Commonwealth Club of California in 2011 that solar-thermal power plants like Ivanpah are not only far cheaper to build, but take less than half the time and, once completed, promise free clean energy forever.

Solar panel manufacturer Solyndra had received the first loan guarantee of the Obama administration's Recovery Act but gone bankrupt in 2011, laying off 1,100 workers. During the presidential race the next year, Republicans spent millions on ads attacking Obama over the loan guarantee. But the Ivanpah solar project, for which Bobby held a limited stake in developing, moved ahead to completion in 2014, with hundreds of thousands of sunlight-reflecting mirrors spread across five square miles of federal land. After a rough start over its cost efficiency, by 2020 experts hailed it as a trailblazer for future large-scale solar power production. "The government made money on that [loan] program while also supporting lots of companies creating new technologies," according to UC Santa Barbara professor Leah Stokes. "And a project like Ivanpah, which is riskier because it's new and innovative, is exactly the kind of thing that the federal government should be supporting."

* * *

In 2008, New York Governor Andrew Cuomo had declared a moratorium on hydraulic fracturing, or fracking for short. This was a technique for extracting gas and oil from otherwise impermeable shale rock, by drilling into the Earth and sending a high-pressure mixture of water, sand, and chemicals down into the rock layer to release the fuel inside.

Cuomo had asked Bobby to be part of an advisory panel to determine whether fracking of the state's gas-rich Marcellus Shale formation could resume. Other states such as Pennsylvania had seen booming local economies as drilling rigs sprouted up. Then in 2010, the documentary *Gasland* premiered at the Sundance Film Festival, focused on how fracking in Pennsylvania had contaminated the air, water wells, and surface water with toxic chemicals as well as the potent greenhouse gas methane. All this had led to a number of chronic health problems—and to tap water that, frighteningly, sometimes came out on fire. Josh Fox's film ended up being nominated for an Academy Award.

The year *Gasland* appeared, Bobby headed for Colorado to attend a Natural Gas Solutions Summit 1.0 sponsored by the Aspen Institute think tank. Well attended by both NGOs and industry representatives, among the invitees was the executive vice president of Southwestern Energy, the fifth largest natural gas provider in the US, Mark Boling, who flew in from Houston at the behest of the Environmental Defense Fund (EDF), a New York-headquartered organization that specialized in seeking solutions with business and industry.

"My company was working with them on trying to bring a little sanity to the discussion around hydraulic fracturing, and what everybody thought they knew was the truth," Boling recalls. Although others in his industry denied any problem, Boling believed they were missing the point: "If the wells are not drilled right and sealed properly with cement, of course methane can migrate up into a water well."

Besides being the primary contributor to ground-level ozone pollution, methane is 80 times a more potent contributor to global warming than carbon dioxide over a 20-year period. According to the United Nations Environment Program, methane has accounted for roughly 30 percent of global warming since pre-industrial times. While industrial agriculture is the primary offender, fracking wells may have methane leakage rates close to 8 percent, which would make natural gas worse for the climate than coal.

At the first afternoon's session, by happenstance energy exec Mark Boling took a seat right next to Bobby. "I didn't know him,'" Boling says, "but he started asking me some questions when the presentation was going on, saying 'does this sound right to you?' At the refreshments break, somebody had put up a diagram of a well. Bobby asked me to show him what was going on. I explained, here's why what [the fracking industry] is saying is half-true and here's why it's not."

When the Aspen gathering ended, Boling remembers Bobby asking him to be on that commission in New York, and Boling said he'd be happy to. They met several times at meetings in Albany. "Bobby set me up doing a PowerPoint presentation at one of the colleges there," said Boling. "At the beginning, there were people with signs saying Go Frack Yourself or Frack Off. I still have one of those. But I couldn't say that their concerns were not legitimate, because they were."

Boling had grown up in a staunchly Republican household but discovered a rapport with Bobby that he wouldn't have expected: "I think one of the reasons he and I got along so well is that, whether he's right or wrong, his heart is always in the right place. He's a smart guy, he's got passion and he's got drive."

On March 7, 2012, Bobby issued a public statement to New Yorkers. "Like many environmentalists, my opinion of natural gas has evolved," he said. "I was an early optimist about natural gas's potential to end the catastrophic practice of mountaintop removal coal mining. Cleaner burning natural gas could also reduce American emissions of carbon, acid rain, mercury, ozone, particulates, and other poisons associated with coal. [But] companies engaged in fracking have not been able to meet the baseline challenge of safely extracting natural gas. Instead, they have relied on their capacity to use political clout to capture the agencies that are supposed to regulate them. In this way, they have derailed the kind of strong, rigorous regulation needed to safely extract and deploy gas-generated power and earn public credibility and trust. My current position is that I oppose shale gas extraction by means of fracking unless and until the industry can prove it CAN and WILL be done safely for both human health and the environment."

A year later, the Associated Press reported (March 2, 2013) that Governor Cuomo "came as close as he ever has to approving fracking last month, laying out a limited drilling plan for as many as 40 gas wells before changing course." Those wells would be allowed "in economically depressed Southern New York towns that want drilling and the jobs it promises." But after a series of phone discussions with Bobby, the AP said, the governor had decided to hold off.

Then, in 2014, Bobby's surprise ally Mark Boling agreed to appear on Showtime's *Years of Living Dangerously* TV series, in an episode titled "Chasing Methane." He disagreed with Cornell University Professor Anthony Ingraffea, coauthor of a study maintaining that shale gas production is dirtier than coal due to methane leaks over its life cycle. But when asked about measurements taken by a group of scientists who found that methane leakage in the Los Angeles basin was as high as 17 percent—much higher than what the industry claimed—Boling replied: "Some of those numbers, they certainly concern me." He didn't understand how the industry could claim that the methane emission rate was only 1.5 percent.

In mid-December of 2014, a month after Cuomo's reelection and as Bobby prepared to pull up stakes for the West Coast, the Cuomo administration made its decision. Acting health commissioner Howard Zucker issued the statement: "I cannot support high-volume hydraulic fracturing in the great state of New York." In an interview a few months later, Bobby expressed being "surprised because the oil and gas industry have been able to exert such power that it's unusual for a politician to have the gumption to stand up to them." But Cuomo had seen a new study that revealed that "the financial cost of fracking and the cost to the health of the people of the state exceeded the financial benefits."

Mark Boling later recounted: "I think what Bobby and I both learned from each other is that the categories we tend to put people in saying 'that's an industry person' or 'that's an environmental person' are unfortunate. The problem I saw was that most environmental folks wanted to do whatever was right, as opposed to many of the industry folks wanting to continue to drill wells because it was gonna make 'em money. I knew there were places in the northern part of Pennsylvania where it was repeatedly done wrong. The whole question really was, do people in the community where they live want to run that risk? I think the resounding answer from the people of New York was no."

Shortly before the New York decision, Boling went to the CEO of Southwestern Energy and received the go-ahead to start a new division that allowed him to work full-time with the Environmental Defense Fund on hydraulic fracturing, on ways to identify and reduce methane emissions. He also worked with Riverkeeper on water use issues, which Boling translated into field operations where employees were proud of reducing waste and doing something good for the environment. Under Boling's leadership, Southwestern Energy made history as the first company in the industry to become "freshwater neutral" in its operations.

"Until we're able to get our global energy system to the point it can be totally de-carbonized," Boling believes, "we need to try to reduce our footprint as much as possible . . . the future is being able to say, as we produce natural gas, we do it better than anybody else."

According to Boling, "everything was going well until there was a change in the company's board makeup." The new regime didn't see any point in what Boling was doing. "We continued to butt heads for the next couple of years, [but] finally in 2017, they said you need to stop." Boling left and formed his own company.

Today, Boling continues to run 2CEnergy, a company dedicated to providing low-carbon energy solutions. Of Bobby, he said in 2022: "I love him like a brother." They continue to keep in touch. When Bobby contacted him asking what he might know about a company wanting to drill some wells off the Bahamas, Boling "looked them up and said whoa, these guys are the most thinly capitalized group in the world, and I'd be afraid even if the geology said they were right, which I don't think it does." Bobby says: "We developed a great friendship. And because he came over to our side, Mark and I killed fracking in New York."

By the time of the 2020 presidential campaign, it had become a trend. Democratic candidates Bernie Sanders, Elizabeth Warren, and Pete Buttigieg all called for a nationwide ban on new hydraulic fracturing. California Governor Gavin Newsom placed a moratorium on issuing new fracking permits, in part over concerns about methane leakage, and on April 23, 2021, directed state agencies to make this permanent.

* * *

In April 2017, at the People's Climate March event in downtown LA, Bobby and actress Jane Fonda were the featured speakers. Bobby seized the opportunity to open with an upbeat analysis of how over the past five years, photovoltaic module prices had dropped 80 percent, and analogous home solar financing operations had spread like wildfire. "Three-quarters of California rooftop solar has been innovatively financed, with no money down, including the system I installed on my own home," he said. "NRG Solar leased me a rooftop solar array with zero cost to myself and a guaranteed 60 percent drop in my energy bills for twenty years. Who wouldn't take that deal? And solar costs continue to fall every day."

Then Bobby launched a no-holds-barred speech into the new Trump administration's assault on the environment, the president's cabinet appointments including ExxonMobil CEO Rex Tillerson as secretary of state, and Trump putting the EPA in the hands of Scott Pruitt, "an unctuous acolyte of Oklahoma's factory meat and Big Oil barons." Marking Trump's

100th day in office, everyone marched across a bridge to demonstrate in front of the Tesoro Oil Refinery, whose consolidation plan would make it the largest petroleum refinery on the West Coast.

Now that he was living on the West Coast, Bobby told *The Influencers* TV program, "We've had to evacuate our house twice in two years, very unusual because it wasn't an area that was part of the traditional fire zone. But the fire seasons in California now are two months longer than they historically have been. The same week we were evacuated on the West Coast, we have a summer home on Cape Cod, and that town was struck by the second storm in two years that destroyed a pier that had, prior to the first storm, been there for a hundred years."

All the modeling for climate change indicated this was only the beginning of "major disruptions not just to humanity but ultimately to civilization." Bobby believes that "this is part of the cost that we're paying for our longtime deadly addiction to coal and oil." With a true free market economy, he wrote, "those industries would be paying the cost of global warming to all of us, they wouldn't be able to externalize those costs." The answer was clear: "Impose market discipline on the energy markets, take away the subsidies for coal and oil." The cost of solar had fallen to about 17 cents per kilowatt hour, compared to nuclear at ten times that, coal about seven or eight times that, and gas at triple that. Once again, Bobby hammered home, "If people really believed in free market capitalism— pollution is a subsidy—we could solve this problem overnight."

Bobby saw all of the 2020 Democratic candidates as likely to have decent records on environment. His kids were working for Mayor Pete Buttigieg, his nephew Joe for Elizabeth Warren, and a number of other family members for Joe Biden. He hadn't offered any endorsement as he kept an eye on the primaries. Trump's environmental record had been "a cataclysm," but not an anomaly—"simply the radical acceleration of a process that's been happening in our country and the Republican Party really since 1980."

Bobby had come of age during those past forty years, warning fervently about the decline of democracy and the rise of authoritarian leadership. Where would the nation, where would he, go from here?

CHAPTER THIRTY-THREE

AT WAR OVER VACCINES

After Bobby met with Donald Trump to discuss chairing a commission on vaccine safety, the announcement was called "very frightening" by Peter Hotez, dean of the National School of Tropical Medicine at Baylor College of Medicine. Hotez told the *Washington Post* that "massive evidence" showed no link between vaccines and autism, which he described as "a genetic condition."

Bobby responded by drafting a letter to his critic (January 19, 2017), which in retrospect is worth quoting at some length: "It's not a conspiracy, Peter, it's an orthodoxy. It's exactly like what happened to the Catholic Church during the pedophile scandal. Only a small number of priests were involved in harming children or in the direct conspiracy. But everyone became complicit in the cover-up—the Bishops, the Vatican, the police and the press. The institution became more important than the children it was meant to protect. Decent and well meaning men and women made subtle calculations that silence was the best course and that the victims were sacrifices for the greater good. A lot of smart, moral men put their heads down, ignored the obvious and kept moving forward. Eventually all orthodoxies are cruel, often lethal. They are misogynistic and anti-science. They require blind faith in an often undeserving authority."

Bobby continued: "The orthodoxy that thimerosal doesn't cause autism requires us to ignore or summarily dismiss a giant library of independent science and the testimony of CDC's senior vaccine safety scientist,

Dr. William Thompson, that he was ordered to destroy and manipulate data to conceal the link that the data disclosed in the very studies you continue to cite. It also requires one to ignore the tragic stories of 10,000 mothers who say they saw their healthy child regress into autism after receiving a thimerosal vaccine. This of course is not science. It's anecdote. But at some point one could argue that a large enough accumulation of anecdotal evidence becomes science. After all, isn't that what case studies are? At World Mercury Project we have collected 5,100 of these stories in just one month. I would feel arrogant if I just dismissed these mothers as hysterical women."

Mark Hyman, President Bill Clinton's personal doctor, wrote: "The answer to this question is simply common sense . . . given the simple fact that mercury is toxic, I can come to no other conclusion than this: we should immediately remove thimerosal from vaccines and all other products used in medicine." But, as he notes, "as Voltaire said, 'common sense is not so common.'"

* * *

Yet for Bobby, expanding his public health campaign came with a heavy price. He wrote in his journal: "The press is 99.9 percent a disaster as expected. I'm drummed out of Riverkeeper. The Waterkeepers are furious. My family is distressed. Kathleen wrote me an anguished note saying I'm being marginalized. Dennis Rivera is begging me to change course. So it's bleak, career wise. But on the plane home, two United Airlines attendants thanked me. Both have affected friends. This is a wonderful consolation." Bobby had to do what was right regardless of the consequences.

When Robert DeNiro agreed to do a press conference with Bobby about vaccine safety at the National Press Club in Washington, DC, in February 2017, Bobby wrote to DeNiro's wife that the couple's "courageous participation in our difficult press conference brought hope to tens of thousands of parents, many of whom have written us to express their profound gratitude. It dramatically elevated the national debate over vaccine safety and it has opened countless new doors to me and our movement that we are now exploiting. I know that these gains for our cause have come at a high cost to you and your family. The issue is radioactive. You and Bob have given more than anyone should ever have had to give - in a battle that no one should ever have had to fight—to protect little children. I promise that I will never ask any more of either of you. You have done your part."

In February of 2017, the *New York Times* ran a series of hit pieces against Bobby. They refused to print any defense by Bobby, rejecting his letters. Throughout all of this, Cheryl was Bobby's rock. "Her faith in me gives me courage, peace, and contentment amidst the sturm and drang, slings and arrows. . . . I feel like I'm in an ugly divorce with Riverkeeper, an amicable separation from Waterkeeper, and a beautiful new relationship with these activists. They have been voiceless, and I give them a voice and my organizational savvy."

Bobby explained his position clearly: "I've met with Adam Schiff, Donald Trump, Roger Ailes, Lindsey Graham, Paul Offit, Peter Hotez, Bernie Sanders, Marianne Williamson, Gavin Newsom, Francis Collins, Steve Bannon, Reince Priebus, Kellyanne Conway, Hope Hicks, and a hundred other Democratic and GOP leaders on vaccines. As an environmental leader I met with oil shills from Ronald Reagan and both Bushes to communist leaders including Fidel Castro. I would meet with Lucifer if I thought it might save one child from vaccine injury. My uncle Ted Kennedy was the most prolific and effective senator in US history because he maintained friendships, even those with whom he disagreed. It was Reagan and the pro-corporate Republican Supreme Court justices who created the vaccine mess. Today it is the Democrats who are making it worse. Nobody can predict which party will finally emerge as the redeemer. . . . Nobody knows who will stand up for our children when we finally turn the world. It's always unwise to unnecessarily burn bridges that you might someday need to cross."

* * *

Bobby had literally been working for years on the memoir *American Values: Lessons I Learned from My Family*, and at long last the book was ready for publication in the spring of 2018. In a C-Span televised talk Bobby gave to a packed house at the Free Library in Philadelphia in mid-May 2018, he began by answering how the book became part autobiography and part history.

"I started out writing this book ten years ago," Bobby said. "I wrote probably 70 percent of it in the first year, then put it down for ten years. I wrote three other books, and then my publisher got angry at me." The audience laughed. "Originally, Tim Duggan at HarperCollins heard me talk about wilderness expeditions that I'd gone on with my father on all the big western whitewater rivers and asked me to do a book on that, and the project evolved where I was writing something of a memoir. But I ended up writing a book targeted for my children and their cousins. I have six kids and they have 105 Kennedy cousins"—more laughter— "who are the progeny of the twenty-nine grandchildren of Joseph and Rose Kennedy, with whom I was raised essentially communally on the Cape."

Early in 2018, Bobby's organization had begun working to rebrand. That September, the World Mercury Project became Children's Health Defense (CHD). Denied access to major media because of his views, Bobby continued to write regularly for the publication *Eco-Watch*, "until their board started taking too much heat and said, no more talk about mercury in vaccines." So the fledgling CHD organization started publishing "Kennedy News & Views."

At the time, the latest brouhaha centered around the MMR (measles, mumps, rubella) vaccine manufactured by Merck, a second shot of which had been recommended for all kids in 1989. Bobby didn't think that the vaccine by itself could cause autism, because kids commonly received up to nine other shots at the same time, and the MMR was a live virus vaccine that didn't contain mercury. But Bobby saw other problems with the push for nationwide school mandates amid a measles "outbreak" that the CDC said in May 2015 had 880 confirmed cases across twenty-four states, the largest number reported in twenty-five years and since the disease was declared eliminated in 2000. But as Redwood put it, "when you look at the molecular genomics of the virus, it was vaccine strain measles they were getting. In other words, the vaccine was bringing about the measles."

And the mumps component appeared to be badly flawed. As Bobby chronicled on the CHD website (April 4, 2019), "Rather than protecting a generation of American children from mumps infection in childhood, the vaccine has merely postponed the onset of the virus to older age groups, putting them at much greater risk." Some 150 outbreaks with

9,200 cases of mumps had been reported by the CDC between January 2016 and June 2017. Mumps after puberty was known to reduce fertility in men, with reduced testosterone and sperm counts. The number of college campuses reporting such outbreaks had surged, and a mumps contagion devastated the crew of a naval ship despite all the personnel having received the MMR shot as well as an immediate booster. Two whistleblowers had asserted in a lawsuit that Merck falsified its test data and engaged in other fraudulent activities, representing "a falsely inflated efficacy rate for its mumps vaccine."

So Bobby took to the stump in various locales where MMR mandates were under consideration for schoolkids and adolescents, including Washington State, California, and New York. In May 2019, he became one of the listed attorneys for a group of anonymous parents in Williamsburg, Brooklyn, filing a religious objection lawsuit in New York State Supreme Court over an emergency order from Mayor Bill de Blasio mandating all unvaccinated individuals to come forward for their measles shots. "We are confident that no American court will allow government bureaucrats to force American citizens to take risky pharmaceutical products against their will," Bobby said in a statement on behalf of CHD.

When members of Bobby's family sent a letter to the *New York Times* to say that they disagreed with him on vaccines, Bobby explained: "We were taught at the dinner table. One person would take one side and someone else would take another side and we would have it out. It doesn't mean we don't love each other." Bobby reflected back on how his parents encouraged arguments among his brothers and sisters every night at the dinner table. They would raise topics. "I remember my father saying, should we get out of Vietnam? Is it good to take LSD? Does God exist?" It was a Kennedy family tradition that had begun with his grandfather: "Should we join the war in Spain? Is fascism good for Italy?" Whatever it was, the children were encouraged to take extreme points of view and defend them. "I understand we need debate," Bobby added. "That's what we ought to be doing."

Despite the media, the government, and his family, Bobby had no intention of abandoning his mission to defend children. According to Redwood, "we were getting inundated with calls from all these states imposing vaccine mandates for schools and taking away exemptions. There was no way to handle this with our small staff. So what do we do? We've got to have chapters in each state that we support, like with Waterkeeper. I remember saying to Bobby and the CHD board, "Five years from now, I want CHD to be a household name, like the March of Dimes when I was a kid.'"

In September 2019, Mary Holland came aboard CHD full-time, as general counsel working with Bobby. By the beginning of the next year, the organization had fifteen people working full- or part-time.

* * *

Late in 2017, a small group arrived to talk with Bobby about aluminum in vaccines. Just before this, on December 1, a new study confirmed that aluminum might be a cause of Autism Spectrum Disorder (ASD), given its own medical abbreviation as the numbers of impacted children continued to increase. A peer-reviewed study published in 2012 had already indicated that aluminum toxicity targets the body's mitochondria, with evidence that aluminum's use as a vaccine adjuvant can lead to permanent detrimental alterations of the brain and immune system. According to Martha R. Herbert, a physician and retired assistant professor of neurology at Harvard Medical School, the "toxicity [of mercury] can be even worse in the presence of aluminum, which . . . has toxicity issues of its own," and now scientists hypothesized that some children with autism might suffer from a genetic change that causes them to accumulate aluminum in their bodies. "This all being the case," Herbert wrote, "why are we still putting mercury in vaccines—or in any medical product (roughly 169 consumer products including eyedrops and nose drops still contain thimerosal)—and how can we bring ourselves to stop doing this?"

Many of the people gathered in Bobby's conference room had kids with autism. The main presenter was an entrepreneur who'd started an online organic food and now cosmetic company called Thrive Market. He and others discussed how mercury remained in the flu vaccines and aluminum levels in the vaccine had increased.

"So there's a synergistic relationship between mercury and aluminum that nobody has ever studied," Bobby said. "Boyd Haley [a retired chemistry professor and expert on mercury toxicity] put mercury in a petri dish of neuronal cells of human brain tissue, and the mercury killed something like 70 percent of them. Then he put aluminum in, and it killed 20 percent. Then he combined mercury and aluminum, and it killed a 100 percent. So the child may be getting a flu shot with mercury the same day that they're getting a shot with aluminum. We're playing with forces that nobody understands and nobody wants to study, and nobody wants to look at the collateral damage from vaccines. All they want to see is, does it produce the antibody response?"

"Many people are aware of only two black-and-white options: you are either pro-vaccine, or anti-vaccine," Herbert noted, but "I ask you to consider that, at minimum, there is a third alternative: you can be pro-vaccine and at the same time seek to improve the vaccine program." This shouldn't be a radical stance. "People who advocate for safer vaccines should not be marginalized or denounced as anti-vaccine," Bobby wrote, and, with "a broad consensus among research scientists that thimerosal is a dangerous neurotoxin," there needs to be an effort "to dissuade the press from accepting the tired claim that anyone who questions thimerosal safety is 'anti-science.'"

* * *

Bobby's forebears had a long history of taking unpopular stands on issues they believed in. Bobby once wrote in his private journal: "Grandpa Joseph's staunch opposition to WWI damaged his relations with many close friends and biz partners caught up in war fever. Grandpa believed the war would only benefit industrialists and tyrants and destroy democracy and European civilization. His prescience about the predicted outcome informed his initial opposition to WWII prior to Pearl Harbor. Long before anyone knew of the death camps, his frantic efforts as US Ambassador to Britain, to find safe haven for German and Austrian Jews earned him ridicule of American newspapers who derided him as the 'Zionist Charles McCarthy' and helped destroy his relationship with FDR and Secretary of State Cordell Hull."

So what should we be debating? "Okay, I'm going to dive down the worm hole," Bobby replied. "I quickly discovered that vaccines, unlike other medicines, are never really safety tested against a true placebo. People find that astonishing. The reason for it is as an artifact of the Center for Disease Control's legacy as the Public Health Service, its predecessor health agency. That was a quasi-military agency, which is why people at CDC still have military ranks like surgeon general. The vaccine program was initially launched as a national security defense against a biological attack on our country. The regulators and the Pentagon wanted to make sure that if the Russians attacked us with anthrax or some other bio-weapon, they could quickly fabricate and deploy a vaccine to 200 million people without regulatory impediments. They realized that if we call it a drug, we're going to have to safety test it and that usually takes two to five years. So let's call it something else. We'll call it a biologic and then exempt biologics from having to do safety testing."

He took a deep breath, then went on: "Later the four companies that make all seventy-two vaccines currently mandated for our children received another gift that has contributed to the lack of safety assessments and lack of safety concerns with vaccines." He went on to explain that, in the 1980s, the diphtheria, pertussis, and tetanus (DPT) vaccine was found to be killing some children and causing brain damage in a large number of others. And the industry was getting sued.

"So the pharmaceutical company now called Pfizer went to Congress and said they were getting out of the vaccine business because they were paying twenty dollars in damages for every dollar they made on the DPT vaccine. They knew that it's impossible to make a completely safe vaccine. A certain amount of people are going to suffer grievous enough injuries that nobody will ever make more money on vaccines than the companies are going to have to pay in damages. So they were planning to stop making vaccines unless they received complete blanket immunity from liability.

"The Democratic Congress, Republican President Reagan, all of them take lots of money from these pharmaceutical companies—they were the number one contributor. And they passed a law in 1986 called the Vaccine Act, which offers blanket immunity. So no matter how grievous your injury or your child's injury, no matter how toxic the ingredient, no matter how sloppy the protocols, no matter how negligent that company, you cannot sue them for redress. There is no discovery, no depositions, no medical malpractice, no class actions, zero consequences if they kill you or injure you."

Bobby continued: "I'm still a Democrat, but I'm really angry about this . . . [when] Reagan signed the 1986 law into effect, I think he was probably trying to do something good, and the Democrats who voted for it believed that this law would somehow make it easier for people to recover injuries quickly, and that they would keep people from getting sick. But when we had the ruling that extended complete blanket immunity to the companies for all product design defect claims, it was the Republicans on the Supreme Court that voted to do that.

"The pharmaceutical companies took a look at the landscape and said, Holy cow, now we have a product where there's no liability—whereas for every medical product the biggest cost is paying liability at the back end. Now that's scrubbed and there's no reason to safety test - a huge cost for every other medication that they have now avoided. And there is no marketing cost and zero advertising cost, because the government is going to mandate this product to 78 million children. No market or legal forces, no consequence of giving them a dangerous vaccine. And that's a gold mine. If you can get a vaccine on the CDC schedule, it's worth a billion dollars a year typically to your company.

"So the gold rush happened. All these companies rushed on CDC and captured the agency. That started in 1989, three years after the Vaccine Act became law. If you look at what happened that year, we suddenly had an epidemic of chronic disease . . . Asthma exploded. Another allergic disease, anaphylactic eczema, which I never heard of, is now ubiquitous." Bobby continued, "Peanut allergies in my generation, or the one just before this, were about one in 1,200. Today it's one in twelve kids. When the vaccine schedule changed in the late 1980s, everything changed. We're subjecting children now to a toxic assault that has never happened in history. As of 2007, 54 percent had a chronic disease, up from 6 percent in the forties, fifties, and early sixties."

As we have seen, Bobby's son Conor suffered from a peanut allergy and severe asthma attacks. In 2011, Skyhorse had published a book by historian Heather Fraser titled *The Peanut Allergy Epidemic*, written after her son had an anaphylactic reaction to peanut butter. Fraser puzzled over the fact that "more than four million people in the United States alone are affected by peanut allergies, yet there are no reported cases in India, a country where peanuts are the primary ingredient in many baby food products."

Bobby had never made any connection between vaccines and allergy until he came across Frazer's book around 2015. "It then became clear," he told me later, "the aluminum is put in the vaccine to prompt an allergic response to the antigen, but it also prompts an allergic response to anything else that is even in the ambient air. There are studies that show that children who are vaccinated have thirty times the allergic rhinitis. The indicator is, if a kid gets an aluminum [containing] shot the same day that there's an outbreak of Timothy weed, that could theoretically prompt a permanent allergy to Timothy weed. If there is a peanut oil excipient in the vaccine, or if the mother is breast-feeding and eats peanut butter, you could get an allergy to peanut proteins. There are a lot of plausible explanations that have been validated by numerous scientific studies, and Heather Fraser did a good job of laying out that science."

Bobby and his wife Mary had helped start a nonprofit called the Food Allergy Initiative, later to become Food Allergy Network. One of the partners was David Koch of Koch Industries, a global conglomerate that Bobby had mounted a number of lawsuits against, but whose coowner had a child the same age as Conor with severe food allergies. They had approached a group of scientists to conduct studies into how to potentially treat and cure these. In experiments on rats, the scientists took the aluminum adjuvant from the hepatitis-B vaccine and combined it with peanut, dairy, or latex proteins to induce food allergies. "We've giving that adjuvant five times to every kid, beginning the day they are born," Bobby realized.

He also came to learn: "There are more pharmaceutical lobbyists on Capitol Hill than there are congressmen, senators, and Supreme Court justices combined, and they put out twice the money of the next biggest lobbyists, which are oil and gas. They are notoriously corrupt. Every one of the four companies that make all the vaccines now mandated for our children have felony convictions. In the last ten years, Sanofi, Merck, Pfizer, and Glaxo paid $35 billion in penalties and damages, civil and criminal penalties for falsifying science, for defrauding regulators, for lying to doctors and manipulating and bribing and blackmailing—and for killing thousands of people."

Merck manipulated its clinical trial data for Vioxx, a pain reliever that the company knew caused heart attacks. Thousands of avoidable and premature deaths occurred before Merck pulled the drug off the market in 2004. Bobby recalled: "When we sued Merck, we got documents that were flow charts from their bean counters that said we're going to kill a certain number of people, but we will still make more money even if we have to pay damages to the families of all those people we kill. So we should go ahead and do this. . . . But Merck didn't warn them, so they could weigh the risks against possible benefits. It lied to them, like so many other pharma companies who have lied and lied and lied to the American people.

"Doesn't it require a kind of cognitive dissonance to believe that the company that would cause 100,000 heart attacks to make a buck on Vioxx suddenly has found Jesus when it comes to vaccines, where by law they have no liability?" asked Bobby. "All those drugs they got caught for, it was through private attorneys, plaintiffs' attorneys, representing injured clients who sued the company. They came across documents in the course of discovery that showed criminal behavior, brought those documents to the US Attorney's office, and said you need to prosecute this company."

That can't happen with vaccines because the 1986 act prevents lawsuits. And by injuring people, the explosion in chronic disease has made those companies the richest in the world. The same companies selling that measles vaccine are now selling DTP and Hep-B vaccines, giving you seizures or epilepsy or autism or food allergies or asthma, and then selling your child who is sick for life Albuterol inhalers, Coban inhalers, Adderall, Ritalin, Concerta, anti-seizure medication, diabetes drugs, rheumatoid arthritis drugs. These companies are making $50 billion a year selling vaccines and another $500 billion a year selling the treatments for the chronic diseases being caused by those vaccines. They have monopoly control. They make money on both ends and then give a big chunk of it to hired guns in Washington to make sure the gravy train never stops.

Then along came COVID.

CHAPTER THIRTY-FOUR

"ICH BIN EIN BERLINER"

Early in 2019, Bobby met with US Congressman Adam Schiff. "I went through a PowerPoint, where I showed him all the science on vaccines. I thought it was insurmountable. He listened, nodded his head, smiled endearingly, and then a week later sent a letter to [Facebook's Mark] Zuckerberg demanding he censor any criticism of vaccines." After Bill Gates made donations to a PAC controlled by Schiff, in February 2019 the powerful Democratic Party congressman wrote not only to Facebook, but to Google, Amazon, and Pinterest, "Vaccines are both effective and safe," Schiff assured them in urging vigilance against dissenters. "There is no evidence to suggest that vaccines cause life-threatening or disabling disease." Schiff claimed no connection even though, as US Congressman Bill Posey states, "the CDC's own studies have demonstrated a link between increased thimerosal exposures and the development of vocal and motor tics, which are generally recognized as autism-like features." But, as Posey continued, the "alarming association has been downplayed to the public."

March 3, 2019, from Bobby's journal: "On Wednesday, Adam Schiff asked Amazon to start censoring any criticism of vaccine safety or of the government's increasingly authoritarian health policies. Amazon dutifully responded by removing *Vaxxed* - a film about government corruption—from its streaming services. Meanwhile, Facebook, at Schiff's request, blocked sharing of photos of vaccine-injured children and Google deplatformed videos about vaccine science. Is there anyone in the Democratic

Party who thinks that it is wise, or consistent with American values, for elected Democrats to be encouraging these all-powerful corporations to censor criticism of pharmaceutical products or government policies? Schiff complained to Bill Maher on Tuesday that Trump's presidency is costing us democracy, 'drip by drip.' Can anyone credibly argue that Schiff's own actions this week didn't push us across a new threshold into tyranny?"

Two weeks later in Connecticut, Bobby was invited to the Statehouse in Hartford, where the Democrats were "proposing to abolish religious exemptions for vaccines." There he had a scheduled debate with five Yale Medical School physicians and virology professors, all of whom "came with heavy vaccine industry pedigrees." Upon arrival, he received word that all five had cancelled at 3 a.m. "Right now I'm thinking," he wrote on March 19, "how blatant, how anti-democratic and fundamentally anti-American the censorship has become. I'll be thinking today of all the moms who are silenced when they try to tell their stories. They can only win by silencing us."

Bobby tried explaining to his sister Kerry in May 2019 that all he wanted was vaccine protocols "applying the same rigor required for every other drug. If Pharma would agree to test vaccines against a placebo group, I'd shut down CHD and go back to Waterkeeper full time." In the Morgensen et al. 2017 retrospective study financed by the Scandinavian governments, "the first to examine health outcomes in large populations vaccinated with DTP comparing them to unvaccinated populations," Bobby pointed out that the scientists "were astonished to discover that DTP is in all probability killing more [African] children than Diphtheria, Tetanus, and Pertussis combined. Nobody noticed all the deaths because the children were dying of a variety of diseases that no one associated with the vaccine—malaria, anemia, bilhartsia (Schistosomiasis), dysentery, etc. The vaccine had destroyed their immune systems."

The declaration by the World Health Organization of a global coronavirus pandemic on March 11, 2020, marked the beginning of an unprecedented situation. Dr. Anthony Fauci, head of the National Institute of Allergy and Infectious Diseases (NIAID), quickly emerged as the avuncular adviser to American families instructing lockdowns of schools and businesses, social distancing, and eventually masking in public places.

Hydroxychloroquine, an older drug and the first proposed treatment for COVID-19, found its emergency use authorization (EUA) rescinded two months later as President Trump's declared Operation Warp Speed established the race for a vaccine through fast clinical trials. By law, vaccines could not receive EUAs if there was an existing, effective remedy. So all other potential remedies were subjected to unprecedented propaganda campaigns aimed at discrediting them. Radio, TV, statements from public officials all came to the same conclusion. Only a vaccine could be the solution.

To the great relief of the majority of Americans, the Pfizer and Moderna biotech companies forged ahead with massive funding from the government and the Bill and Melinda Gates Foundation. No vaccine using their experimental messenger-RNA technology had ever been licensed, but unlike traditional vaccines that could take several years to develop, the pharma companies and the government claimed that the new variety was particularly suited to quick development and mass production. And, as with the childhood vaccine schedule, Congress made it so that the companies would be exempt from liability.

Bobby expressed skepticism about all of this and, amid a growing pervasive culture of fear, was soon excoriated like never before as anti-vaxxer public enemy number one. At the same time, vaccine proponent Bill Gates was telling Chris Wallace on *Fox News Sunday* (April 5, 2020): "It is fair to say things won't go back to truly normal until we have a vaccine that we've gotten out to basically the entire world."

So the battle lines were drawn, between proponents of the vaccination mantra and those considered fringy and dangerous conspiracy theorists. Besides the "Truth" interviews he began in 2019 for the Children's Health Defense website, Bobby had more than 700,000 followers on Instagram, where he wrote in April 2020: "The *Daily Mail* today reports that it has uncovered documents showing that Dr. Anthony Fauci's NIAID gave $3.7 million to scientists at the Wuhan Lab at the center of Coronavirus leak scrutiny. According to the British paper, 'the federal grant funded experiments on bats from the caves where the virus is believed to have originated.'" The story was soon picked up by *Newsweek*, describing so-called "gain of function" research undertaken in a collaboration between laboratories at the University of North Carolina, Fort Detrick, and the virus's point of origin in Wuhan, China.

On May 3, CHD posted a video of Bobby's closing remarks at a vaccine conference focusing on a prospectively grim future where technocrats like Gates supported unprecedented tracking, tracing, and surveillance on acquiescent human beings. A few days after that, Bobby published a lengthy piece on the Children's Health Defense website headlined "The Brave New World of Bill Gates and Big Telecom." He began by recounting how police in Malibu had begun ticketing Point Dume surfers a thousand dollars apiece for using the ocean during the pandemic quarantine. And he threw out some questions for readers to consider: "Was this merely an appalling police judgment at which we will laugh post-quarantine? Or does anyone else feel that this is the first wave of compliance and obedience training for something more permanent?"

On May 20, Bobby set down: "At the Quarantine's outset, I warned that despots and billionaires would leverage the crisis to ratchet up surveillance and authoritarian control. They would transform Amerika into a National Security State, engineer the final liquidation of America's middle class and transfer the wealth to a plutocracy of digital and Pharma billionaires. Sure enough, at breakneck speed we have devolved from the world's exemplary democracy to a militarized tyrannical police state squirming under the heel of the Big Data, Big Telecom, the Medical Cartel and the Military/Industrial/Intelligence Apparatus. Quarantine has permanently shuttered 100,000 small businesses, cost 38 million jobs and 16 million Americans their health care. The Super Rich are fattening on the bones of the obliterated middle class. According to a report by Americans for Tax Fairness, billionaires are watching their wealth compound beyond imagination."

Less than ten days later, Facebook removed an interview Bobby conducted with Polly Tommey ten minutes after it was posted. Bobby knew that "every statement [could] be sourced to peer reviewed scientists." But according to Facebook, it was "vaccine misinformation." Bobby wondered in his journal: "Where are all my friends, the liberal advocates for free speech?"

His longtime friend and business partner Chris Bartle put it like this when I asked him what he believed. "I don't fully understand much of what he's saying," Bartle said, but "I watched his journey through his environmental law practice. . . . He really knows what he's talking about in terms of his advocacy on science. When people question what he's doing, I tell them: 'I saw him go after Anaconda Copper when they lied about everything they were doing and had done, and the same with General Electric, and Entergy and nuclear plants on the Hudson. . . . So if Bobby thinks there's a large entity out there lying for venal reasons, he's probably right—because he's been right every other time I've known about it.'"

On July 16, 2020, RFK Jr wrote in his journal: "In his 1942 classic, *The Screwtape Letters*, Christian writer and theologian C. S. Lewis (*The Chronicles of Narnia*) recounts a series of letters and conversations in which the Devil questions his nephew about the variety of strategies for corrupting and recruiting souls. In the passage they settle on fear of illness as the most potent formula for motivating negative human behaviors. Lewis believed that both Hell and Heaven exist in our minds, that our moral courage dictates the relative size of the spaces that each of those places occupy within our thoughts and ultimately, the character of the world in which we dwell."

* * *

Almost sixty years earlier, in the midst of the Cold War and some five months before his death, John F. Kennedy had traveled abroad to give a short inspired speech at the recently erected Berlin Wall on June 26, 1963. He had gone there to reaffirm American support in the struggle against the Communist threat. He told an estimated 450,000 citizens of West Berlin that the wall was "an offense not only against history but an offense against humanity, separating families, dividing husbands and wives and brothers and sisters, and dividing a people who wish to be joined together." The president went on to speak of how "freedom is indivisible, and when one man is enslaved, all are not free." And he concluded with the resounding words: "All free men, wherever they may live, are citizens of Berlin, and, therefore, as a free man, I take pride in the words '*Ich bin ein Berliner*.'" The crowd went wild.

A year later, ten-year-old Bobby had journeyed with his parents to Berlin "to dedicate a memorial to Uncle Jack." They visited the Berlin Zoo and saw the American bald eagle that JFK had presented to the mayor and West German chancellor. "I couldn't help but be struck," Bobby would recall in *American Values*, "by this symbol of American freedom being kept in a very small cage." Mayor Willy Brandt took them to the Berlin Wall, "where I looked up at rolls of barbed wire spanning menacing guard towers, from which Communist sharpshooters studied us with shark eyes." Bobby went on: "Once again, giant crowds gathered and engulfed us whenever our family appeared. Seventy thousand Germans jammed the plaza beneath the Rathaus Schoeneberg during the dedication of Jack's memorial, showing their affection for America and their hope for US leadership."

No one, least of all Bobby, could have anticipated that on August 29, 2020, he would stand before an even larger crowd of thousands of Germans sprawled across the Unter den Linden Plaza and beyond, addressing the largest demonstration in Europe protesting measures taken in response to COVID-19. Bobby had come to Berlin with Mary Holland to launch Children's Health Defense Europe, photographed with German organizers in front of the Brandenburg Gate, where his uncle had spoken in 1963.

When asked to be the last-minute keynote speaker, he accepted. Berlin authorities tried to ban the demonstration on several occasions, including on the day before it was slated to begin. According to one account, "the media attempted to dissuade the public from reaching the capital. Many buses planned from abroad cancelled scheduled trips." But because Germany had been scarcely affected by COVID, they weren't able to impose a general health emergency order. Citizens responded by making individual requests, while lawyers appealed to a federal court.

The scene was reported as follows: "Excitement was high when an influencer dropped the scoop of Kennedy's arrival on social media. The next morning, it was announced that the federal court had ruled in favor of the protesters and that the event would move forward. Protesters poured in early in the day and more than three thousand police were deployed in the city. They carried out numerous roadblocks and pushed back several dozen buses. . . . In several streets, protesters were surrounded and 'confined' for hours without being allowed to move. Fortunately, things remained peaceful, and many attendees began to sing, '*Liebe polizei macht die strasse frei*' ('Dear policemen, free the street')."

A huge crowd was still gathering as word spread of Bobby's imminent appearance. With police preventing some of the processions from joining the others at the victory column, they couldn't be counted as a single compact mass. They swarmed around forty podiums and rebroadcast screens, as well as in dozens of alleys and nearby streets. As one observer put it: "Could anyone find a stronger symbol than the presence of Robert F. Kennedy to defend fundamental freedoms in the face of a new totalitarian agenda?"

Told he would be giving the keynote address in the afternoon, Bobby drove over in the morning, "just to be near the speaker stand," he remembers. "I like to listen to what other people say and get some ideas of what to talk about. I never think about what I'm going to say until the day, and it's better for me to be kind of spontaneous. But then the police threatened to shut it down, so they rushed me up onto the stage."

Bobby stood holding a microphone, wearing a blue shirt, checked tie, and gray slacks, with a long-haired young male translator in a T-shirt and jeans alongside him. He began by predicting (not wrongly) that the media would report he'd been speaking in front of five thousand Nazis. "I look at this crowd and I see the opposite of Nazism," he went on. "I see people who love democracy, who want open government; people who want leaders that are not gonna lie to them, leaders who will not make up arbitrary rules and regulations to force obedience of the population. We want health officials who do not have financial entanglements with the pharmaceutical industry, who are working with us and not Big Pharma. We want officials who care about our children's health and not about pharmaceutical profits or government control."

Bobby continued: "I look at this crowd and I see all the flags of Europe. I see people of every color. I see people from every nation, every religion, all caring about human dignity, about children's health, about political freedom. This is the *opposite* of Nazism!"

Watching one of the home-style videos that appeared afterward on YouTube, I couldn't help but be mesmerized by the crowd's mounting enthusiasm. "Governments love pandemics," Bobby continued, "for the same reason they love war—because it gives them the ability to impose controls on the population that the population would otherwise never accept, to create institutions and mechanisms for orchestrating and imposing obedience."

Bobby electrified the crowd as he spoke: "It's a mystery to me that all of these important people like Bill Gates and Dr. Fauci have been planning and thinking about this pandemic for decades, so that we would all be safe when it finally came. Yet now that it's here, they don't seem to know what they're talking about. They seem to be making it up as they go along. They're inventing numbers. They cannot tell you what the case fatality rate of COVID is. They cannot give us a PCR test that actually works. They have to change the definition of COVID on the death certificates constantly to make it look more and more dangerous.

"The one thing that they're good at is pumping up fear," Bobby told the crowd. "Seventy-five years ago, Hermann Göring testified at the Nuremberg trials. He was asked how he had made the German people go along with all this. And he said, it was easy. It's got nothing to do with Nazism, it has to do with human nature. You can do this in a Nazi regime [or] in a socialist regime [or] in a communist regime, in a monarchy or a democracy. The only thing a government needs to make people slaves is *fear*. If you can figure out something to make them scared, you can make them do anything you want."

He paused a moment before remembering: "When my uncle came here, he proudly said to the people of Germany, '*Ich bin ein Berliner.*' The crowd erupted in cheers, as their forebears had two generations ago. "All of us who are here today can proudly say once again, *Ich bin ein Berliner.* Today, once again, Berlin is the front line against totalitarianism."

Bobby went from periodically waving his index finger at the audience to using his whole hand for emphasis. There was another pause as the crowd gradually quieted. "I want to say one more thing. They haven't done a very good job of protecting people's health. But they've done a very good job at using the quarantine to bring 5G into all our communities. And to begin the process of shifting us all to a digital currency—which is the beginning of slavery. Because if they control your bank account, they control your behavior.

"We all see these advertisements on television saying, 5G is coming to your community, it's gonna be a great thing for all of you [and] change your lives. And it's very convincing, I have to say. I look at those ads and I think, that's great, I can hardly wait till it gets here—because I'm gonna be able to download a video game in six seconds instead of sixteen seconds!"

The facetiousness wasn't lost on many who laughed. "And is that why they're spending five trillion dollars on 5G? No. The reason is for surveillance and data harvesting. It's not for you and me, it's for Bill Gates and Mark Zuckerberg and Jeff Bezos and all the other billionaires. Bill Gates says that his satellite fleet will be able to look at every square inch of the planet twenty-four hours a day. That's only the beginning. He also will be able to follow you on all your Smart devices through biometric facial recognition, through your GPS. You think that Alexa is working for you. She isn't working for you, she's working for Bill Gates spying on you!

"The pandemic is a crisis of convenience for the elites who are dictating these policies," Bobby continued to the cheering crowd. "It gives them the ability to obliterate the middle class, to destroy the institutions of democracy, to shift all our wealth to a handful of billionaires to make them richer while impoverishing the rest of us. And the only thing between them and our children is this crowd that has come to Berlin, telling them today: you are not gonna take away our freedoms, you are not gonna poison our children. We are going to demand our democracy back. Thank you all very much for fighting!"

Bobby waved as the whistles and shouts and applause cascaded, and he walked off the back of the stage with his translator.

In April 2021, Bobby would learn from a story broken by the news agency AFP that after his appearance, Germany's domestic intelligence agency announced that it would begin monitoring the top leaders of the group that invited him. The spy agency accused COVID protesters of trying to "permanently undermine trust in state institutions and their representatives."

* * *

After the event, it was reported in the international press that the demonstration had ended in violence, with German security officers having to intervene to disperse neo-Nazi extremists who had stormed the Reichstag. This had in fact happened, but *not* at the rally where Bobby spoke. The violence occurred at a separate and much smaller event. Other reports also merged the two gatherings, stating that participants "wore T-shirts promoting QAnon . . . white nationalist slogans and neo-Nazi insignia."

In the US, one of the only outlets even to mention the Berlin rally, and Bobby's presence there, was *Daily Kos*, a widely read liberal progressive news and activist website. Under a headline reading "Anti-Vaxxer RFK Jr. Joins Neo-Nazis in Massive Berlin 'Anti-Corona' Protest," a blogger using the pseudonym "Downeast Dem" went on to describe the event as being "organized by right-wing extremist organizations, including the AFD party and various anti-Semitic conspiracy groups as well as the neo-Nazi NPD party." Witnesses told a starkly different story—a story of hundreds of thousands of peaceful protestors listening to speeches by thoughtful, inspiring, and impassioned leaders.

This was too much for Bobby. On August 30, the Correll Law Group representing him sent a letter to *Daily Kos*'s managing editor demanding a retraction. His event had been sponsored by Querdenken 711 [Lateral Thinkers], "a broad-based, peaceful citizens' movement for freedom, peace and human rights. Among the speakers were doctors, lawyers, clergy of every persuasion, soldiers, police officers, parents, children, world-famous athletes, a regional politician from the Green Party, human rights activists and leaders of nonprofit associations and associations for the protection of children. . . . Mr. Kennedy's speech was a screed *against* Nazism. There was no evidence of Nazi iconography anywhere near the protest. The principal stage ornament was a giant mural of Mahatma Gandhi."

The lawyer's letter described the mischaracterization as "malicious intent," but *Daily Kos* refused to remove or correct the defamatory statements. They took the position that Bobby's beef was really with the blogger but wouldn't reveal his or her identity. Bobby filed a petition with the Westchester County Supreme Court seeking to force the disclosure, and in April 2022 a justice ruled in his favor. As of this writing, the matter remains pending.

On August 31, 2020, Bobby wrote a letter to the Waterkeeper Board: "That bittersweet moment has arrived for me to resign as your president.

While the COVID quarantine has idled most of the planet, it has some-how had the opposite effect on me. My work at Children's Health Defense has devoured all my bandwidth and I am neglecting my executive duties at the very moment that Waterkeeper needs an energetic, outspoken leader."

Bobby said he was immensely proud of what they had created together. "I relish my memories of all the trials we've endured in pursuit of those victories, the relentless anxieties, the laughter, the adventures, the charac-ters and the tattoos. I think of the legion of malevolent scoundrels we've sued. My dreams overflow with the thousands of miles of magnificent waterways that I've been privileged to paddle or travel with each of you over forty years; the mangroves, the muskies, the Spanish moss, the school-ing salmon, the shrimp, crayfish and bluecrab and yellow perch, the calv-ing glaciers and all that flowing water from the Himalayas to the Tetons, from the Andes to the Arctic, from Bimini to Homer, from Bhutan to the Jordan and from Lake Ontario to the Futaleufú. Your friendships have enriched my life and I'm especially grateful for the time that each of you has taken to share with me those magical stretches of water to which you have devoted your lives. There is so much wisdom and courage in this cohort and so many genuine heroes. In a life filled with incomprehensible wonders, it had been my greatest gift to serve this cause with each of you."

That October, Bobby recorded a speech, which was broadcast to fifteen countries, in which he announced that Children's Health Defense was launching a daily journal called *The Defender*. "We are going to weaponize information for you," he said, "tell you what the newest science is, take all the information censored somewhere else and reprint it in our publica-tion. We are going to be the enemies of censorship. We welcome the caul-dron of debate, because the free flow of information is the only thing that allows governments to develop rational policies in which self-governance will actually work and triumph."

He concluded; "You are on the frontlines of the most important battle in history, the battle to save democracy and freedom and human liberty and human dignity from this totalitarian cartel that is trying to rob us simultaneously in every nation in the world of the rights that every human being is born with. Thank you for your courage, your commitment, and your brotherhood. . . . I will see all of you on the barricades."

CHAPTER THIRTY-FIVE

BEHIND THE CENSORSHIP
OF RFK JR.

It is instructive, in considering what happened to Bobby during the pandemic, to reflect back upon the experience of Upton Sinclair—muckraking journalist, media critic, and eventual politician. Sinclair's *The Jungle*, an exposé of the meatpacking industry published in 1906, resulted in the country's first Pure Food and Drug Act and the creation of the FDA. But few people know just how many powerful forces Sinclair took on.

The Jungle had been rejected by six publishers (Macmillan called it "gloom and horror unrelieved"), so Sinclair self-published what became a global bestseller. In what must be some kind of record, he went on to write ninety books in his ninety years, both fiction and nonfiction. His mother's family was rich, and the contrast between wealth and poverty troubled him and became his major theme.

Like Bobby a century later, Sinclair fought against Big Oil, Big Coal, Big Food, and Big News Media. He was on the front lines of calling for regulation and safety regarding drugs as well as food. He also took aim at organized religion as a "source of income to parasites and the natural ally of every form of oppression and exploitation."

Sinclair also excoriated journalism (specifically the Associated Press) as a "wet nurse for all other monopolies." He offered his investigative reporting into matters like union busting to a number of newspapers but was almost completely ignored. Sinclair told that story in the 1919 book *The Brass Check: A Study of American Journalism*, calling for a law that any newspaper printing a false statement ought to be required to give the same prominence to a correction or pay a large fine. He challenged those charging him with inaccuracy to review his published facts and sue him if they found he'd been wrong. Nobody did, but Sinclair was denied access to mainstream media to refute the charges. Most newspapers refused to review *The Brass Check*, and the *New York Times* wouldn't even run paid advertisements for it. Four years after its publication, Sinclair's book gave rise to the first code of ethics in journalism. It's incredible how little has changed in 100 years.

"In the course of my twenty years career as an assailant of special privilege," Sinclair said, "I have attacked pretty nearly every important interest in America." He became tremendously popular internationally, while being maligned for decades at the highest levels of mainstream American society. Eventually, MGM Studios created three newsreels depicting his supporters as seditious Socialists; William Randolph Hearst did the same in print.

In 1934, when Sinclair ran for governor of California as a Democrat,the first smear campaign to target a major candidate was orchestrated against him, using quotes from his novels as if they'd issued from his own mouth. He lost the election but played an important role in changing the state's rigid conservatism.

"Returning to writing, Sinclair reinvented himself as a historical novelist," the *New York Times* said in his obituary. "*World's End*, in 1940, would be the first of eleven 'Lanny Budd' novels, in which Sinclair's young protagonist roams the world, meeting leaders both good (Roosevelt, Churchill) and evil (Hitler, Mussolini)." The one about fascism earned him a Pulitzer Prize.

When Sinclair died in November of 1968, his obituary concluded by calling him "that rarity in the literary world, a man of action as well as of ideas."

* * *

Let us, then, compare and contrast. There are a couple of obvious dichotomies. Almost a century after enactment of the Pure Food and Drug Act came the Public Readiness and Emergency Preparedness Act (PREP Act) providing liability immunity "to certain individuals and entities against any claim of loss caused by, arising out of, relating to, or resulting from the manufacture, distribution, administration, or use of medical countermeasures." First passed in 2005 after 9/11 and the anthrax scare, it included vaccine manufacturers in March 2020 after the COVID-19 emergency was declared. By this time, the Food and Drug Administration had basically become a captive agency of the pharmaceutical companies it was designed to regulate. The fact-checking code for journalists was also hijacked during the pandemic, when Facebook and Google hired "FactChecker" (Politifact) to censor vaccine misinformation.

Sinclair may have had many of the same enemies as Bobby, but his muckraking predated the advent of television and seems light years distant from our current world of computers and social media. Still, John D. Rockefeller—son of a patent medicine salesman—spread as many tentacles to the medical/pharmaceutical realm as he had to oil. America's first billionaire sloganized "a pill for an ill," a kind of premonitory of Bill Gates.

Stifling Bobby's right to exercise free speech has emerged from three parallel paths—the TV advertising blitzkrieg undertaken by the pharmaceutical industry, Bill Gates's slow-but-sure financial takeover of traditional media outlets, and the merger of big tech giants Facebook and Google with the government, making it a First Amendment violation.

TV pharma ads are so ubiquitous now that it's hard to believe that commercials for prescription drugs remain illegal everywhere but the US and New Zealand. In the 1980s, drug companies here began talking about marketing not just to doctors and pharmacists, but directly to consumers. FDA rules about including long slow scrolls of side effects on TV ads crimped that desire until 1997, when a "clarification" implied that the companies only had to list the major risks and places where people could get more information (even just the recommendation to talk to your doctor).

Thus, the floodgates opened. The $300 million spent by pharma on TV ads in 1997 doubled the next year. By 2000, it was $1.57 billion. By 2016, it was $6.6 billion. In 2020, the industry accounted for 75 percent of the total TV ad spending. A Yale University study found that some

three-quarters of the ads were for chronic conditions like arthritis and diabetes, illnesses that guarantee return business. Those marketing expenses
are all tax-deductible and, of course, have represented a bonanza especially in revenues to network news divisions, which were otherwise losing
ground to cable and satellite channels.

Bobby's entry almost two decades ago into raising questions about
this lucrative realm represented a threat to the pharma/media paradigm.
Initially, after his exposé on autism rates and mercury in vaccines appeared
in 2005, he was lionized like a prizewinning muckraker. Jon Stewart
praised him on *The Daily Show*. "Let's get you running for public office,"
Joe Scarborough enthused on MSNBC.

But things went sour fast. The most egregious happened at ABC. Bobby
told the story in *The Defender* in January 2022, shortly after his book *The
Real Anthony Fauci* was published, and CNN anchor Jake Tapper used
his Twitter feed to label Bobby as "dangerous," a "menace," a "liar" and
"grifter," "unhinged," and more.

In 1999, Tapper had written a long article for Salon about the new
generation of Kennedys, in which he touted Bobby as "almost a mini-me
of his father" and "the favorite to assume the family mantle" in political office. By 2005, Tapper had become a senior producer with ABC's
news division. Bobby wrote that Tapper "saw an early draft of [his] *Rolling
Stone* story [on autism and vaccines] and proposed that, in exchange for
exclusivity, he would do a companion piece for ABC timed to air on the
magazine's publication day."

Tapper spent several weeks working on the story with Bobby, "on fire
with indignation."

The day before the piece was to air, "an exasperated Tapper called
[him] to say that ABC's corporate officials ordered him to pull the story.
The network's pharmaceutical advertisers were threatening to cancel
their advertising."

But ABC had already been promoting the planned exposé, and the
sudden cancellation "disappointed an army of vaccine safety advocates and
parents of injured children who deluged the network with a maelstrom of
angry emails. In response, ABC changed track and publicly promised to air
the piece. Instead, following a one-week delay, the network duplicitously

aired a hastily assembled puff piece promoting vaccines and assuring lis-
teners that mercury-laden vaccines were safe."

Two pharmaceutical advertisements bracketed the news segment.
When Bobby called Tapper to complain, "he neither answered nor
returned my calls."

Bobby reflected in 2022: "During the sixteen intervening years, Pharma
has returned Mr. Tapper's favor by aggressively promoting his career." Each
episode of his news show on CNN opens with "Brought to you by Pfizer."
During the pandemic, the show became "the go-to pulpit" for Dr. Fauci,
"a safe place . . . to hit all Jake's reliable softballs out of the park."

Borrowing a page from Upton Sinclair, Bobby concluded: "What Jake
Tapper does is the opposite of journalism. Tapper, instead, aligns himself
with power and makes himself a propagandist for official narratives and a
servile publicist for powerful elites and government technocrats. No wonder his fury at those who challenge their narratives."

Ironically, it wasn't one of TV's progressive brethren, but former Fox
News chief Roger Ailes who leveled with Bobby about what was going
on. "I didn't agree with him politically, but we were friends," Bobby says.
"We'd spent a couple months together in a tent in Africa when I was eighteen. He had a personal experience with vaccines when a kid who was
close to him was injured, so he knew it was true. A couple of years before
Roger died, when he was still at the height of his powers, I'd helped make
a documentary film [*Trace Amounts*]. I didn't expect to play it on Fox, but
I asked if I could come on and talk about it.

"And Roger said, 'I can't let you do that.' In fact, he said, 'if any of my
hosts allowed you on their show, I would get a call from Rupert [Murdoch,
who owned Fox] within ten minutes, and I would have to fire them.' He
went on to tell me that during certain parts of nonelection years, 70 per
cent of the advertising revenue on their evening news comes from pharma,
typically about seventeen out of the twenty-two ads."

And there is more than one way to skin a cat, as billionaire Bill Gates
knows all too well.

A *Columbia Journalism Review* (*CJR*) exposé revealed (August 21,
2020) that Gates steered over $250 million to the BBC, NPR, NBC,
Al Jazeera, ProPublica, *National Journal, The Guardian*, the *New York*

Times, Univision, Medium, the *Financial Times, The Atlantic*, the *Texas Tribune*, Gannett, *Washington Monthly, Le Monde*, Center for Investigative Reporting, Pulitzer Center, National Press Foundation, International Center for Journalists, and a host of other places. To conceal his influence, Gates also funneled unknown sums via subgrants to other press outlets.

Those media gifts, wrote CJR reporter Tim Schwab, mean that "critical reporting about the Gates Foundation is rare." In April 2020, as COVID-19 began to severely impact the US, corporate media networks rolled out the red carpet for Gates to advise everyone on how to handle it. He was hosted by CNN, Fox, PBS, CNBC, BBC, MSNBC, CBS, *The Daily Show* and *The Ellen Show*. The *Washington Post* called Gates a "champion of science-backed solutions." The *New York Times* hailed him as "the most interesting man in the world." The fashion magazine *Vogue* wondered, "Why Isn't Bill Gates Running the Coronavirus Task Force?" Gates even starred in a Netflix docuseries released a few weeks before COVID struck the US. Titled *Pandemic: How to Prevent an Outbreak*, the series was produced by Sheri Fink, a *New York Times* correspondent who formerly worked at three Gates-funded organizations: Pro Publica, the New America Foundation, and the International Medical Corps.

Gates had particularly focused on gifting to press outlets that market themselves as trustworthy, reliable, and free from corporate influence. His contributions targeted public TV and National Public Radio in the US and *The Guardian* newspaper in the UK. Contributing about $2.5 million every other year, Gates funds NPR's "Global Health Beat" and *The Guardian*'s Global Development page. The BBC, another recipient of Gates's largesse, obligingly runs regular stories that conceal vaccine risks and "debunk" stories that disclose risks or expose corruption as "conspiracy theories."

In addition to the 250 million previously mentioned, since 2012, Gates's Foundation has granted some $25 million to media outlets, specifically requiring journalist recipients to submit "success stories" about the international vaccine program that he funds. His foundation holds an annual seminar on "Strategic Media Partnerships" for compliant journalists and also invests in journalism training programs. Experts coached in Gates-funded programs write columns that appear in media outlets from the *New York Times* to the Huffington Post. And millions more have gone, ironically, to the Bureau of Investigative Journalism.

The Gates Foundation is the largest owner of the "FactChecker" used by Facebook and Google to purportedly sniff out vaccine misinformation. Gates also funds an army of "independent" fact-checkers including those at the Poynter Institute and Gannett —which use their platforms to "silence detractors" and to debunk "false conspiracy theories" and "misinformation," such as charges that Gates has championed and invested in biometric chips, vaccine IDs, satellite surveillance, and COVID vaccines.

* * *

It is striking to consider the lockstep strides that Gates and his longtime colleague Anthony Fauci had made together. On February 5, 2020, as COVID-19 closed in on America, the Gates Foundation said it would commit up to $100 million for coronavirus research and treatment efforts. "The new approach I'm most excited about is known as an RNA vaccine," wrote Gates in the *Washington Post*. That vaccine had been developed by Dr. Fauci's NIAID in conjunction with a little-known company called Moderna, which had never brought a product to market but soon began receiving millions from the federal government and the Gates Foundation. The company's short history included having been the beneficiary of lucrative secret contracts with the Pentagon's biological warfare research arm— the Defense Advance Research Project Agency (DARPA)—contracts which Moderna failed to disclose, even though disclosure was required by law.

In 2013, DARPA awarded Moderna up to $25 million to establish a messenger-RNA platform, and Britain's pharmaceutical giant AstraZeneca surprised the world by entering into a $240 million partnership with Moderna, toward licensing the new technology for treating a number of ailments. That was the most money any company had ever been willing to spend on drugs not yet tested on humans.

Two years later, Moderna signed a collaborative development agreement with Dr. Fauci's Vaccine Research Center, and the Gates Foundation proceeded to pour millions into Moderna's pipeline. The company boomed in value while concealing or obfuscating both its research and its funding sources. In February 2016, the prestigious scientific journal *Nature* published an op-ed highly critical of Moderna for not publishing any peer-reviewed papers about its technology. Later that year, Moderna received another award of up to $125 million to accelerate a vaccine effort against the Zika virus, this time from a different arm within Health and Human Services (HHS), the Biomedical Advanced Research and Development Authority, or BARDA, Office of Preparedness and Response. Moderna produced the vaccine in a mere ten months, but it never came to market because it simply wasn't needed. Had Dr. Fauci ginned up the Zika scare to obtain $1.9 billion in emergency funding toward combating its spread onto American shores?

In July 2017, Moderna added Moncef Slaoui to its Board of Directors. He simultaneously served as an interim board member of the Coalition for Epidemic Preparedness Innovation, a public-private partnership backed by the Gates Foundation, Britain's Wellcome Trust, and the World Economic Forum. In May of 2020, Slaoui would be selected to head President Trump's Operation Warp Speed vaccine delivery program.

In 2018, the founder and CEO of Germany's BioNTech, Dr. Ugur Sahin, delivered a keynote speech at a conference in Berlin, where, as the *New York Times* reported, he "made a bold prediction. Speaking to a roomful of infectious disease experts, he said his company might be able to use its so-called messenger RNA technology to rapidly develop a vaccine in the event of a global pandemic." BioNTech would soon partner with Pfizer, which, in 2020, won the COVID vaccine sweepstakes along with Moderna.

A few seemed to have a sixth sense that something worth censoring was just around the corner. As early as 2016, World Economic Forum founder Klaus Schwab called for a global health pass to be enforced by implantable microchip technology. In 2017, Seth Berkley, the longtime director of Gates's GAVI nonprofit, proposed in *The Spectator* that "anti-vaxxers" should be excluded from social media. On April 26, 2018, buried deep in a document titled "Strengthened Cooperation against Vaccine Preventable Diseases," the European Commission first called for vaccine passports. Early in 2019 came a road map document outlining specific plans to implement the proposal.

Six months *before* COVID-19 erupted, the US intelligence-linked think tank InfraGard raised the specter in a research paper that "the US anti-vaxxer movement would pose a threat to national security in the event of a 'pandemic with a novel organism.'" The paper claimed that the movement's prominent figures were "aligned with other conspiracy movements including the far right" as well as campaigns run by Russia's Internet Research Agency.

The *New York Times* reported (November 22, 2020) that in November 2019 Dr. Fauci's Vaccine Research Center at NIH "agreed to stage a war game of sorts the following spring, a mock pandemic with a virus unknown to Moderna to see how quickly the company could come up with a vaccine." Moderna ended 2019 with a $514 million net loss. According to a year-end filing with the Securities and Exchange Commission, its accumulated deficit at the time was $1.5 billion. That wasn't to be the case for much longer. The neophyte biotech company boasted four billionaires by the end of 2020.

* * *

Bobby reflected in 2022: "My father once told me, people in power lie, and the job of a reporter is constant skepticism, constant scrutiny, and to ask the difficult questions, and that simply is not happening. When I was a boy, it was unthinkable that an American liberal would acquiesce to censorship. It was axiomatic that all the Nazi atrocities had begun with censorship and silencing of critics of government policies. Our civics teachers taught that the free flow of information—even inconvenient truths—was the lifeblood of democracy."

In describing his concern that history would record that men calling themselves "liberal" had led the clamor for censorship, Bobby cited a quote of Harry Truman warning: "Once a government is committed to the principle of silencing the voice of opposition, it has only one way to go, and that is down the path of increasingly repressive measures, until it becomes a source of terror to all its citizens and creates a country where everyone lives in fear."

According to Bobby, until the pandemic, "this had not happened before in American history. There had been some kinds of voluntary censorship, for example, during the McCarthy era, and small outbreaks of censorship, but right now essentially the government uses these social media platforms as surrogates to censor speech that challenges government policies."

During the first months of the pandemic, he and his colleagues at Children's Health Defense raised questions about everything from the wearing of masks and closing of schools to the emphasis on mRNA vaccine development as opposed to the far less expensive (and shown to be effective before being shunned) treatments Ivermectin and Hydrochloroquine.

CHD's President Laura Bono remembers how the pattern perpetuated itself in social media: "After Bobby wrote his first article questioning the epidemic and the motives behind it, people were hungry for it and flocked to the website, so many that it went down. Obviously he'd hit a nerve. His second article was about Gates and Fauci, a preamble to *The Real Anthony Fauci* book. That's when Mailchimp, the marketing platform which we'd been using for years to send out our newsletters, cancelled our account out of nowhere. We'd been backing things up for months prior to that, so we didn't lose any sign-ups, and within a few days we were back up. Then Facebook shut down our donations link, long before they took away our complete platform. And PayPal wouldn't let us collect any money through them, so we had to go through Stripe, which has been fine."

I well remember my shock when Bobby first mentioned to me that Google (which owns YouTube) was also a pharmaceutical company. Two subsidiaries of Google's parent company, Alphabet, marketed and manufactured vaccines. One was Calico, an anti-aging drug company; the other, Verily, teamed with pharma to conduct drug and vaccine clinical trials.

In 2016, Alphabet signed a $715 million deal with GlaxoSmithKline (GSK) to create Galvani, designed to develop bioelectronic medicines and vaccines as well as mine medical information from Google customers. (The Customer Services President of Google sits on the board of Merck.) According to Bobby's research, "GSK's contract to mine data about people's health looks at, for example, your heart rate when you're holding your cell phone and what Alexa or Siri can find out about you living in your home. Then to use that data to develop drugs and vaccines, and directly market them to you."

In 2018, Google invested $27 million in Vaccitech to make vaccines for flu, prostate cancer, and the MERS virus. Vaccitech billed itself as "the future of mass vaccine production" and, two years later, started work on its own COVID vaccine.

Today, Bobby asserts, while Google claims to provide politically and commercially neutral searches, it systematically manipulates search results. Its algorithms censor negative information about COVID vaccines and positive information about therapies that might compete against vaccines. "It's called shadow banning," Bobby says. "They tell their users that search results are neutral and that they are based upon previous traffic, but it's not true. They actually hand-manipulate the search results so that people cannot find information about, for example, vitamins, chiropractors, functional medicine, integrative medicine. And any criticism of the vaccine programs is buried so deep that it will not appear on your screen even when you are directly looking for it."

Just before the pandemic, the *Washington Post* reported on Facebook's soaring pharma advertising revenues (March 3, 2020): "Ads promoting prescription drugs are popping up on Facebook for depression, HIV, and cancer. Spending on Facebook mobile ads alone by pharmaceutical and health-care brands reached nearly a billion dollars in 2019, nearly tripling over two years. . . . Facebook offers tools to help drug companies stay compliant with rules about disclosing safety information or reporting side effects."

Also in March 2020, Zuckerberg emailed Dr. Fauci to offer his help against COVID-19. "Dr. Fauci has since appeared on multiple livestreamed interviews on Facebook with Mr. Zuckerberg," according to the *New York Times*. Over the course of the first pandemic year, Facebook would use its nonprofit arm, the Chan Zuckerberg Initiative, to fund a research center tool for tracking the spread of COVID-19 while processing 165,000 free tests for the virus.

That summer, Children's Health Defense filed a lawsuit (August 17, 2020) in San Francisco Federal Court charging Facebook, Mark Zuckerberg, and three fact-checking outfits with censoring truthful public health posts and for fraudulently misrepresenting and defaming CHD. According to the CHD press release, "Facebook acknowledges that it coordinates its censorship campaign with the WHO and the CDC. While earlier court decisions have upheld Facebook's right to censor its pages, CHD argues that Facebook's pervasive government collaborations make its censorship of CHD a First Amendment violation."

The complaint continued: "Facebook has insidious conflicts with the pharmaceutical industry and its captive health agencies and has economic stakes in telecom and 5G. . . . The WHO issued a press release commending Facebook for coordinating its ongoing censorship campaign with public health officials. That same day, Facebook published a 'warning label' on CHD's page, which implies that CHD's content is inaccurate, and directs CHD followers to turn to the CDC for 'reliable, up to date information.'

"The lawsuit also challenges Facebook's use of so-called 'independent fact-checkers'—which, in truth, are neither independent nor fact-based— to create oppositional content on CHD's page, literally superimposed over CHD's original content, about open matters of scientific controversy. . . . CHD asks the Court to declare Facebook's actions unconstitutional and fraudulent, and award injunctive relief and damages."

As Bobby would later explain on Megyn Kelly's podcast, "They've created a government-corporate cartel that controls all our communications, and that's really dangerous. For Democrats advocating censorship, the underlying concern they have is legitimate. Because of the power of social media, inflammatory and dishonest characterizations have a way of amplifying on the Internet in a way they wouldn't with conventional newspapers or news sources. The algorithms used to keep us on the site also have the side effect or fallout of polarizing opinion and making it more extreme. You can censor pedophilia and incitements to violence, but when you get outside of those and a couple of other narrow categories, as a society we have to figure out some other way than censorship."

Early in 2023, the Twitter files and other information being released showed without a doubt that the White House and other government agencies were colluding with social media sites and other news organizations to censors criticism of government policies. As award-winning journalist Matt Taibbi testified before Congress (March 9, 2023): "We learned Twitter, Facebook, Google, and other companies developed a formal system for taking in moderation 'requests' from every corner of government:

the FBI, DHS, HHS, DOD, the Global Engagement Center at State, even the CIA. For every government agency scanning Twitter, there were perhaps 20 quasi-private entities doing the same, including Stanford's Election Integrity Project, Newsguard, the Global Disinformation Index, and others, many taxpayer-funded."

Bobby and CHD had already filed a nearly 100-page complaint in a US District Court in Texas, suing several large legacy media companies who had joined the so-called Trusted News Initiative (TNI). Named defendants are the *Washington Post*, the BBC, the Associated Press, and Reuters, partnered up with tech companies Facebook, Google, Microsoft, Twitter, YouTube, and LinkedIn. (James C. Smith, founder of the Reuters Fact Check Foundation before he retired in 2010, has also served as a Pfizer Director since 2014.)

While the TNI "publicly purports to be a self-appointed 'truth police' extirpating online 'misinformation,' in fact it has suppressed wholly accurate and legitimate reporting in furtherance of the economic self-interest of its members," the suit alleges. As Bobby explained it to former Fox News host Tucker Carlson, "the more motivating purpose of the cartel was revealed in one of the memos we have obtained from BBC. . . . It said although we ostensibly are all rivals with each other, the existential threat to all our business models comes from thousands of independent news sites who are not only providing all this content but also diminishing trust in our organizations, and the way we can destroy them, 'stamp them out and choke them,' is to deny them platforms on the social media sites." Misinformation is acknowledged by the TNI as a euphemism for any statement that departs from official government orthodoxy.

Not long after Bobby appeared on his show to discuss his run for president, Carlson was suddenly fired by Fox News—"five days after he crosses the red line by acknowledging that the TV networks pushed a deadly and ineffective vaccine to please their pharma advertisers," Bobby pointed out on Twitter. The Tweet received over twelve million viewers.

Upton Sinclair, in his previously discussed 1919 exposé on journalism, described the formation of the Associated Press. Yes, it was about a group of newspapers working together, but only those that would pay to join would receive the content, which of course meant that the AP controlled all the content. In its coverage of Bobby's announced presidential run, the AP's unnamed source characterized his work as "misleading and dangerous." The more things change, the more they stay the same.

> *I intend to do what little one man can do to awaken the public conscience, and in the meantime I am not frightened by your menaces . . . I know that our liberties were not won without suffering, and may be lost again through our cowardice. I intend to do my duty to my country.*
> —Upton Sinclair, Letter to Louis D. Oaks,
> Los Angeles Chief of Police, May 17, 1923

CHAPTER THIRTY-SIX

UNDRESSING THE EMPEROR

"What on earth prompted you to take a hand in this?"
"I don't know. My . . . my code of morals, perhaps."
"Your code of morals. What code, if I may ask?"
"Comprehension."

—Albert Camus, *The Plague*, 1947

It's not inconceivable that another part of Bobby's growing comprehension revolved around a personal issue—the problems with his voice that first appeared in the early 2000s. In 2020, he'd been preparing a lawsuit against flu vaccine manufacturers and began looking into the ones he'd received annually up until 2005, when he first became concerned about vaccines. The Pace University infirmary was immediately adjacent to his office, so in the 1990s when it came time for the annual flu vaccine, he'd simply walk over and roll up his sleeve.

"The flu shots at that time were loaded with mercury," he told me, "and when I was looking through the manufacturer's inserts and writing down all of their listed side effects, spasmodic dysphonia was prominent on the list. There is no way of knowing, but it makes sense because doctors would always ask me, was there a trauma that occurred at the time my voice changed? I couldn't remember any, but the trauma may have been the mercury I got from the flu shots."

Back in March of 2017, in the wake of President Trump asking him to run a vaccine safety commission, Bobby had traveled again to Washington with Lyn Redwood and his fellow attorney Aaron Siri. At Trump's behest, The National Institutes of Health had agreed to see Bobby's PowerPoint presentation. Dr. Anthony Fauci and his boss Francis Collins were among the officials present. So was Reed Cordish, the president's assistant for Intragovernmental and Technology Initiatives.

The *New York Times* later reported (February 26, 2022) that Dr. Fauci "at the instruction of the Trump White House, spent an hour listening to Mr. Kennedy give a briefing on childhood vaccines." Dr. Fauci was quoted reminiscing: "As soon as the first slide went up, I raised my hand—I said, 'Bobby, there's no data.' He said, 'I never get a chance to offer the facts, so I want to make a presentation, but I don't want to be interrupted until I'm finished.'"

Bobby remembers the "congenial but confrontational meeting about vaccine safety" somewhat differently. "At some point [Fauci] said, 'let us present our side.' I said, 'with all due respect I know what your side is, so let us finish this PowerPoint and then we'll hear it.' We lambasted him at the meeting. I was going through all the peer-reviewed studies, the main point being that none of the 72 doses of the 16 different vaccines effectively mandated for American children had ever been tested in a pre-clinical, placebo controlled trial. The definition of placebo is an inert substance, but what these are tested against is another toxic vaccine. For example, with the DPT vaccine trial . . . as long as they were in the so-called placebo group, then Dr. Fauci's agency declared it safe." (DPT immunization "might induce a fatal disorganization of respiratory control in susceptible infants," according to a 1987 study published in the *American Journal of Public Health*.)

"Afterward I said, 'Dr. Fauci, for many years you've said that I'm not telling the truth when I say this. So I'm asking now for you to show us one study for any of these vaccines that have been tested against a placebo.' He made a display of looking through his file box there and he said, 'I'll send them to you.' And that's the last I heard."

Fauci told the *Times* that he'd walked Bobby out of the conference room. "I said, 'Bobby, I'm sorry we didn't come to any agreement here. Although I disagree factually with everything you are saying, I do understand and I respect that deep down you are really concerned about the safety of children.' I said that in a very sincere way."

Bobby again has a different recollection: "Afterwards Dr. Fauci buttonholed me in the hallway. He said, 'Keep doing what you're doing, it keeps us all honest.'"

Bobby and Siri took that advice to heart and sent a legal demand to Health & Human Services asking for the clinical trial data and also for the biannual report it was supposed to submit to Congress detailing improvements in vaccine safety. "HHS didn't come up with anything and we had to file a federal lawsuit. After extended litigation, they came back around and said, 'Yeah, there aren't any.'" The case was eventually dismissed.

* * *

Writing his book *The Real Anthony Fauci* had its seeds in Bobby's dialogue with a former law partner, John Morgan, who subscribed to government orthodoxies. "We had a discussion during the pandemic, through email and text, where we would send each other scientific studies with some commentary. It struck me that that's what we should be doing as a nation, have continual debate about government policies, which had been absent."

Bobby had been aware of Dr. Fauci and crossed paths with him for much of his adult life, "because my family has these deep entanglements with the public health agencies going back generations." President Kennedy had been instrumental in establishing many of the enabling statutes at the NIH. Other relatives were closely associated with Fauci. The [Eunice] Kennedy Shriver Institute [of Child Health and Human Development] was funded in part by Fauci's NIAID agency and Johns Hopkins. Ted Kennedy had been chair of the Senate Health Committee for several years "and essentially defended Fauci's budget year to year." Bobby's sister Kathleen Kennedy Townsend had been lieutenant governor of Maryland, where NIH is located.

So it was not with relish that, sometime in the fall of 2020, Bobby spoke to Tony Lyons of Skyhorse Publishing about doing a book that would expose the truth about the man who'd just become a household name, so popular that dolls were being fashioned in his image ("America's doctor" as the *New Yorker* fondly described him). Bobby had known for some time, he told me, that "the image Tony Fauci was selling people—this kind of avuncular, trustworthy, devoted public servant—was not real. He was a shill for the pharmaceutical industry." And now, during the COVID-19 pandemic, the policies promoted by Dr. Fauci were in Bobby's eyes costing thousands of lives and sending democracy into a death rattle.

This kind of chronicle was sure to win Bobby even greater enmity among a public that purchased "In Fauci We Trust" coffee mugs and "I heart Fauci" throw pillows. As Bobby would point out, to most of his fellow Democrats Dr. Fauci "seemed to offer a rational, straight-selling, science-based counterweight to President Trump's desultory, narcissistic bombast. Navigating the hazardous waters between an erratic president and a deadly contagion, Dr. Fauci initially cut a heroic figure, like Homer's Ulysses steering his ship between Scylla and Charybdis."

Caught between that proverbial rock and hard place, the therapeutic drugs Hydroxychloroquine (HCQ) and Ivermectin didn't have much of a prayer up against the Fauci-Gates multi-billion-dollar vaccine project. HCQ had been around for over sixty years and been approved as safe and effective (and cheap) against a variety of illnesses. Bobby remembered: "During my many childhood trips to Africa, I took HCQ daily as a prophylactic against malaria, a ritual that millions of other African visitors and residents embrace." In 2015, the Nobel Prize Committee for Medicine gave its annual award to the men who discovered Ivermectin, a multifaceted drug deployed against some of the world's most devastating tropical diseases.

Numerous studies showed benefits of HCQ used for early outpatient treatment of COVID. The rub for most progressives was that President Trump had endorsed its use before Operation Warp Speed pushed vaccine development to the forefront. A little-known law in the US prohibits emergency use of vaccines (meaning to short-circuit the customary safety precautions) when pre-existing drugs like HCQ and Ivermectin are available and known to work against a particular disease. Bobby learned in 2020 that Dr. Fauci had funded studies of HCQ designed to fail by overdosing and administering the drugs too late to effect recovery. Meanwhile, an expensive and often toxic therapeutic called Remdesivir—in which Dr. Fauci and Gates had a financial stake—became the "drug of choice" for people admitted to a hospital with COVID. Scientists showed Bobby how Remdesivir tests had been skewed to show positive results, when in fact it was known to cause liver failure and respiratory failure and did not meet efficacy endpoints in a trial conducted during the Ebola outbreak.

Skyhorse wanted to publish Bobby's exposé during the first part of 2021. But the more Bobby dug into Dr. Fauci's history, the more he uncovered, the more outraged he became, and the more the book kept expanding. "I didn't know what he did with AIDS until I started researching the book," he says. He would devote four chapters and more than a hundred pages to ferreting out what happened during the early years of the AIDS epidemic in the 1980s and 1990s.

"Dr. Fauci was suppressing treatments like he's doing now," Bobby told one interviewer (*Spin Magazine*, January 17, 2022). With "repurposed medications that are virtually all patent expired, the profits to pharma are zero because they cost pennies, so all these generic companies make them. Doctors were using these drugs extremely effectively against AIDS. But Dr. Fauci said there's only one solution, AZT. It had been developed and then discarded by the National Cancer Institute as too toxic for treating cancer. They were testing it as a chemotherapy drug that you use for two weeks. But when Dr. Fauci took over the AIDS crisis, he inherited AZT and he chose to hide the results of those drug tolerance studies from the public. The only way to make AZT appear to be somewhat effective was to give blood transfusions to the people getting it. After he anointed it with legitimacy and got FDA approval, AZT started being administered to people who were HIV-positive but not sick at all. Fauci told gay people to use it for the rest of their lives, which of course were dramatically shortened by it. Anybody who takes a chemotherapy drug for life is going to die very quickly."

Even more shocking to Bobby was learning about experiments on hundreds of HIV-positive black and Hispanic children in New York's foster care system between 1998 and 2002. "These experiments were the core of Dr. Fauci's career-defining effort to develop a second generation of profitable AIDS drugs," Bobby would write. The kids basically became "lab rats, subjecting them to torture and abuse in a grim parade of unsupervised drug and vaccine studies." Investigative reporter Celia Farber, who conducted field research for the 2004 BBC documentary *Guinea Pig Kids*, recounted to Bobby having found a mass grave at Gates of Heaven cemetery in Hawthorne, New York, with over a hundred wooden coffins that she learned each contained more than one child's body. "That was the most heartbreaking thing I came across," Bobby told me.

As Bobby's investigation unfolded, his next surprise was how deep the connection ran between Fauci and the biodefense industry that burgeoned after 9/11 and the subsequent anthrax scare. NIAID's biodefense research funding soared from less than $100 million in fiscal 2001 to $1.75 billion from the Bush administration two years later. Fauci announced a "brain trust" of civilian and military scientists to develop new vaccines and drugs. "We're trying to do research to protect the population, not to make weapons," he said. But Eileen Choffner of the National Academy of Sciences penned an article for *The Bulletin of Atomic Scientists* warning that "these laboratories might become a pathogen-modification training academy or biowarfare agent 'superstore.'"

Perhaps inadvertently, that seems to be what transpired. Fauci's agency in 2011 began funding what is called "gain of function" research, the effort to "weaponize" a virus by breeding souped-up pathogens that are more deadly, more virulent, and more transmissible in humans than their wild cousins. Dr. Fauci justified this as an effort to anticipate and improve crisis preparedness in the event of an accidental or deliberate release of such a manufactured virus from a laboratory. In 2014, President Obama declared a moratorium on these controversial studies, which stayed in place through his tenure in office, until being lifted in the early days of the Trump administration.

Bobby recounted: "Fauci had to keep doing it during the moratorium, otherwise he would've lost his military funding. So he moved the coronavirus research offshore to Wuhan and worked with the Chinese military scientists. And he continued an existing grant funding Ralph Baric's study at the University of North Carolina. They developed what's called 'seamless ligation,' which Baric dubbed the 'No-See-Um process.' It's a way of hiding the human tampering on the coronavirus. Baric taught it to the Chinese."

If all this sounds too outrageous to possibly be true, it's been well documented in scientific publications and by a few intrepid reporters. (See, for example, Nicholson Baker's "The Lab Leak Hypothesis," *New York Magazine*, January 4, 2021.) Early in 2023, a leaked classified study by the Department of Energy concluded that the virus likely escaped from the Wuhan Institute of Virology. This doesn't mean there was a conspiracy between the US and China to deliberately release COVID-19. It does derail what Dr. Fauci and his compatriots fought hard to claim, that the virus originated naturally in a Wuhan animal market.

Some critics have speculated that Bobby's persistent skepticism traces back to the assassinations of his uncle and father. David Nasaw, a retired history professor from City University of New York, recalled interviewing Bobby in 2011 for a book about his grandfather, after which Bobby handed him a copy of *JFK and the Unspeakable* with its conclusions that the president had been murdered by rogue members of the intelligence community. "This type of thinking penetrated Kennedy early on," Nasaw speculated to the *Boston Globe*.

There may be an element of psychological truth to Nasaw's hypothesis, at least in the sense that it sowed a reasonable amount of mistrust in the government. That, coupled with his forty years of experience with corrupt agencies and greedy corporations, led him to question authority and power. But Bobby is nothing if not an indefatigable researcher and an independent thinker, with a focus on the details. He's been reading the science for decades, and he understands it. *The Real Anthony Fauci*, with 2,194 footnotes, backs every factual assertion with reference QR codes placed throughout the book that take readers directly to the documents, so they can check them for accuracy.

"I had no idea that the CIA was involved in pandemic response. That shocked me," Bobby told me. It is not inconsequential that the last section of the book bears subheads on how "The CIA Dips in Its Toe," "How War Games Became Instruments for Imposing Obedience," "Laying Pipe for Totalitarianism," "The Triumph of the Military/Intelligence Complex: Intelligence Agencies and COVID-19," and "COVID-19: A Military

Project." Nor is it surprising that, as publication approached, the mainstream and social media powers stepped up their campaign against its author.

* * *

Bobby's uncle JFK had said in 1962 that American libraries "should be open to all—except the censor. We must know all the facts and hear all the alternatives and listen to all the criticisms. Let us welcome controversial books and controversial authors. For the Bill of Rights is the guardian of our security as well as our liberty."

As JFK's nephew began assembling one of the most controversial books of our time, the vaccine rollout had just gotten underway in December of 2020. President-elect Biden, his running mate Kamala Harris, and Trump's Vice President Michael Pence all received their shots publicly.

At the time, Bobby continued to be a powerful presence on social media. He had more than 300,000 followers on his Facebook page, some 215,000 on Twitter, and a whopping 800,000 on Instagram. But Twitter then announced it would remove claims that vaccines intentionally cause harm or are unnecessary, as well as "conspiracy theories" about their adverse effects. "In the context of a global pandemic, vaccine misinformation presents a significant and growing public health challenge," Twitter said in a blog post. "Starting next week, we will prioritize the removal of the most harmful misleading information, and during the coming weeks, begin to label Tweets that contain potentially misleading information about the vaccines."

At the end of that first year of the pandemic, Bobby's niece, Kerry Kennedy Meltzer, MD, published an op-ed in the *New York Times*. She wrote: "I love my uncle Bobby. I admire him for many reasons, chief among them his decades-long fight for a cleaner environment. But when it comes to vaccines, he is wrong. . . . His concern that the COVID vaccine is potentially unsafe, and hasn't been properly tested, is widespread, and dangerously wrong."

Meltzer, an internal medicine resident physician in New York City, said that she and her colleagues welcomed the "historic" vaccine rollout aimed at saving lives. "It's hard to express how momentous it is to receive the COVID vaccine," she added. She received immediate backup from her aunt Kerry, Bobby's sister, who tweeted that the op-ed was excellent and urged people to "get your vaccine. For yourself, your family, your friends, and your country." Mariah Kennedy Cuomo added her voice, calling the op-ed itself "life-saving."

In January 2021, investigative journalist Glenn Greenwald wrote a column on censorship citing a CNN interview with former Facebook security official Alex Stamos, describing "the need for social media companies to use the same tactics against US citizens that they used to remove ISIS from the Internet—'in collaboration with law enforcement'—in order to 'get us all back in the same consensual reality.'"

That same month, baseball great Hank Aaron died in his sleep a few weeks after getting his shot to show other Black Americans that the vaccine was safe. Bobby described this in posts on Facebook and Twitter as "part of a wave of suspicious deaths among elderly closely following administration of COVID vaccines." His reasoning was that the Fulton County coroner, who initially issued a statement that Aaron had *not* died from the vaccine, responded to Bobby's further query that "we declined jurisdiction and never saw a body." The new Biden White House's COVID-19 Digital Director Clarke E. Humphrey then wrote an email (released publicly in January 2023) requesting Twitter staff "get moving on the process for having [Bobby's post about Aaron] removed ASAP."

In February, Facebook announced that "we are expanding our efforts to remove false claims on Facebook and Instagram about COVID-19, COVID-19 vaccines, and vaccines in general during the pandemic," updating a "list of false claims" after consulting with the WHO and other "leading health organizations." With Bobby's post, Facebook added a warning that it offered no context and "could mislead people."

Instagram had been acquired by Facebook for a billion dollars in 2012. On February 11, the Instagram account where Bobby had been posting daily 300-word articles for several years was de-platformed with no advance notice—for "repeatedly sharing debunked claims about the coronavirus or vaccines," Facebook said in a statement. Bobby wrote in response: "Every statement I put on Instagram was sourced from a government database, from peer-reviewed publications and from carefully confirmed news stories. None of my posts were false. The mainstream media and social media giants are imposing a totalitarian censorship to prevent public health advocates, like myself, from voicing concerns and from engaging in civil informed debate in the public square. This is a formula for catastrophe and a coup d'état against the First Amendment, the foundation stone of American democracy."

Bobby reflected in 2022: "I kind of knew they'd probably throw me off Instagram if I kept poking the bear, but what else was I gonna do. Now I realized, I've lost close to a million followers and that was a very valuable list. I have no way of ever finding those people, and it would have been better for me just to dial back and not post anything but pussycats and unicorn pictures until the censorship died down. So with Twitter, I started posting innocuous stuff so they wouldn't get rid of me."

The day before his Instagram account was taken down, Bobby gave an interview to NewsGuard, a website that tracks online misinformation. Its board members included former CIA Director Michael Hayden and the nation's first secretary of Homeland Security, Tom Ridge. Under the bylines of cofounder Steven Brill and several other NewsGuard staffers, a transcript appeared in *Newsweek* (March 1, 2021) accompanied by the website's rebuttals to each of Bobby's statements.

Asked about Dr. Fauci, Bobby had told the interviewer how, "since he took over in '84, chronic disease has increased from 12.8 percent of the US children population to 54 percent. . . . Why did autism go from one in ten thousand in my generation to one in every thirty-four people today? Why did food allergies go from one in 1,200 to one in twelve? Why did asthma and autoimmune diseases and arthritis all explode in our genera-tion, or in our kids' generation?" For Bobby, Dr. Fauci "has not been a suc-cess. He has been an abject failure." A man who "failed his way to the top."

In March, Children's Health Defense released an hour-long documentary called *Medical Racism: The New Apartheid.* The film chronicled the history of minorities being targeted for unethical experiments such as the forty-year-long Tuskegee Study by the CDC conducted on nearly four hundred African American males with untreated syphilis, resulting in the unnecessary deaths of one-quarter of the men before the scandal was exposed by Senator Edward Kennedy's health subcommittee in 1972.

The film also explored the pharmaceutical industry's "drug dumping" in modern-day Africa, which Bobby would write about in his book. *Medical Racism* ended with Bobby asking the viewers to beware of the COVID-19 vaccine narrative. The documentary was decried by critics as a blatant attempt to keep an already skeptical African American community from getting the shots.

On March 21, Bobby was listed in the number two spot among the "Disinformation Dozen," who were labeled responsible for almost two-thirds of anti-vaccine content circulating on social media platforms. The forty-page report was released by a new nonprofit with offices in London and Washington calling itself the Center for Countering Digital Hate. (The Center's Communications Director, Lindsay Moran, had published a memoir in 2005 titled *Blowing My Cover: My Life as a CIA Spy*.) A *Newsweek* analysis determined that the Disinformation Dozen boasted a combined six million followers on Facebook, Instagram, and Twitter, platforms on which they remained active.

Three days later, the attorneys general from a dozen states called on the social media giants to enforce their stated policies and ban repeat offenders. "Anti-vaccine misinformation continues to spread on your platforms, in violation of community standards," wrote Connecticut Attorney General William Tong, who spearheaded the initiative. The day after that, Facebook's Zuckerberg, Jack Dorsey of Twitter, and Sundar Pichai of Google appeared before a congressional hearing titled "Disinformation Nation: Social Media's Role in Promoting Extremism and Misinformation."

In *Vanity Fair* (May 13, 2021), a long profile appeared written by Keziah Weir and headlined "How Robert F. Kennedy Jr. Became the Anti-vaxxer Icon of America's Nightmares." Weir's most recent article was a puff piece titled "Bill Gates Loves Burgers and Bridge" (February 12, 2021). Before that, she had lambasted Tony Lyons in an obvious hit piece that was strewn with factual errors. Weir chose not to reveal that she'd applied for the assistant to the publisher position at Skyhorse back on May 29, 2013, where she wrote: "Skyhorse in particular seems to value the parts of the business I most admire. I would relish the opportunity to be part of the dynamic publishing house."

A personal note is appropriate here. Skyhorse has been my primary publisher since 2008, two years after its founding, when an editor with whom I'd worked before expressed interest in a memoir I'd coauthored for former Minnesota Governor Jesse Ventura. My literary agent had previously sent queries to some twenty mainstream publishers, all of whom declined with the general consensus being that Ventura was passé. A lot of readers apparently didn't think so, because three of the five books I eventually wrote with Ventura became *New York Times* bestsellers.

In the years since, Skyhorse went on to publish a number of controversial books that other publishers had cancelled or avoided for not being politically correct, including memoirs by Michael Cohen, Woody Allen, Allen Dershowitz, and Garrison Keillor. Lyons gained an increasing reputation as a maverick, even bringing out two books simultaneously during the pandemic that argued for and against wearing masks. He told *The Guardian* that maybe the role of publishers is to encourage people "to read things they disagree with, that make them angry, but ultimately to learn things that help them bridge the gap to what they thought they hated but may find some nuance in."

In July 2021, President Biden had weighed in concerning the "Disinformation Dozen," flat-out saying that "they're killing people." US Surgeon General Vivek Murthy described the Dozen as "an imminent and insidious threat to our nation's health." The White House began pressuring Facebook about posts allegedly spreading COVID-19 misinformation, and, after imposing a 30-day ban, Facebook and Instagram removed the accounts of Children's Health Defense on August 16. Bobby released a statement saying: "Facebook is acting here as a surrogate for the federal government's crusade to silence all criticism of draconian government policies." His personal Facebook page, with its almost 247,000 followers (the *New York Times* reported), was still up. The paper did a survey and pointed out that "other Facebook pages dedicated to Children's Health Defense, including those of its California, Florida and Arizona chapters, also remain online and have thousands of followers." Was the *Times* lobbying for further censorship?

In September, the *Washington Post* reported: "YouTube is taking down several video channels associated with high-profile anti-vaccine activists including Joseph Mercola and Robert F. Kennedy Jr., who experts say are partially responsible for helping seed the skepticism that's contributed to slowing vaccination rates across the country. As part of a new set of policies aimed at cutting down on anti-vaccine content on the Google-owned site, YouTube will ban any videos that claim that commonly used vaccines approved by health authorities are ineffective or dangerous."

* * *

By early October of 2021, Bobby was still adding material to *The Real Anthony Fauci*. CHD's executive director, Mary Holland, sent him a ten-point list of reasons why it should come out right away. There were concerns that Amazon, which had received tens of thousands of pre-orders, kept having to push back the scheduled date of availability—and either as a business decision or under pressure from Bobby's enemies might even decide they were justified in refunding the pre-orders.

Tony Lyons remembers: "That entire year was such an incredible process. As I learned more and more about the book, I began to feel it was the most important we'd ever published and wanted to be sure we did the very best job we could. I was desperate to have it come out quickly, because I felt its arguments were so important to the national discourse. But Bobby doesn't cut corners. He was completely invested in making sure it was impossible to impeach what was in the book. He believed that the media would attack him and the arguments made in the book and try to prove that it was littered with misinformation."

CHD had a small army of fact-checkers making sure all the citations were included and correct. And in the middle of October, after nearly twelve months of sixteen-hour days, Bobby turned in the book. Skyhorse staff worked around the clock to produce, print, and publish the initial 200,000 copies. Chapter One, the first hundred pages, was titled "Mismanaging a Pandemic," with subsections on Hydroxychloroquine, Ivermectin, Remdesivir, and vaccines.

The same week that the book became available to readers, the surgeon general's office released a twenty-two-page pamphlet titled "Confronting Health Misinformation." As ABC News put it, "The guide provides a road map for vaccinated people to talk to unvaccinated people who have bought into conspiracy theories or lies that spread on the Internet about the COVID-19 vaccine."

Lyons later commented: "This tool-kit would frankly be the envy of any dictator in history. It includes an arsenal of censorship tools to make it unlikely that anyone could find the book in a search, that no one believes it, and that most venues refuse to carry it, let alone promote it. The pamphlet describes how they would partner with city, state, and federal agencies; big tech companies; try to induce private companies to join them; engage nonprofits and hire PR firms—all the things they were going to do to stifle dissent, avoid debate, and destroy democracy."

Soon after this, on November 15, Bobby appeared for a full hour on Tucker Carlson's *Fox Nation* - the only big news show that would let him on. Carlson described Bobby to his viewers as "one of the smartest and most articulate chroniclers of the erosion of our civil liberties in this country. This is an exhaustively reported book that I can't recommend strongly enough." Bobby told *Spin Magazine*: "Tucker and I don't agree on 90 percent of the things he talks about. But he loves the First Amendment, and he's willing to stand up for it. There's nobody else in the media who's doing that." The night he appeared, 45,000 of Carlson's watchers bought copies of *The Real Anthony Fauci* on Amazon. "Then Amazon created a special designation for it," according to Lyons, "which carried a kind of 'warning label' saying that for accurate and up-to-date information, people should go to the CDC website."

But the book kept selling like hotcakes—110,000 copies by the end of its first week. "It was a slugfest," Lyons adds. "While the book was shut down in every conceivable way, we were really fighting hard to get the word out, using a whole team of people. Bobby was doing dozens of alternative media interviews each day, as were a whole group of people from CHD." Lyons, who had never done interviews for a book, teamed up with activist Sofia Karstens and gave dozens of interviews about the censorship of the book. "Notwithstanding the unprecedented censorship, de-platforming, and vilification of Bobby, he was quickly becoming a national hero."

Did the success of the book surprise Bobby? "Yeah, it was interesting because we weren't allowed to advertise for it in the *New York Times* or other major papers, and there were no reviews anywhere," he said a year later.

"The articles attacking me wouldn't even mention the name of the book in the *Washington Post,* etc. People would be de-platformed for mentioning it on Instagram. The bookstores boycotted it, Barnes and Noble and all the independents. So virtually all the sales were on Amazon."

At the time Bobby went on Tucker Carlson, the book was already near the top of Amazon's bestseller list. Lyons expected it could sell between 100,000 and 200,000 copies and was confident it would reach number one on the *New York Times* list. That never happened. Even though sales considerably outpaced the nation's other bestsellers going into December (Will Smith's memoir and *The 1619 Project*), the *Times* first placed *The Real Anthony Fauci* at number seven and never higher than number five. "They claimed there were a lot of bulk sales, which was absolutely untrue," according to Lyons. "The *New York Times* was simply curating its list in yet another form of censorship." At the *Wall Street Journal,* the book was initially rated number one on Kindle, but not even listed among the leading hardcovers. By the end, the media simply ignored the book. Lyons said it was the most censored book that he'd ever been involved with in thirty years of publishing. It was likely the most censored book of the pandemic.

By the beginning of February 2022, according to BookScan, nearly 390,000 hardback copies had been sold, plus 185,000 e-books and 142,000 audiobooks. The book was in its twelfth printing, and Skyhorse ordered another 150,000 copies made. Well into 2022, *The Real Anthony Fauci* remained an Amazon bestseller and had received thousands of customer reviews, 90 percent of them five stars. Adding up the various sales figures a few months later, the book sold over a million copies in the United States alone and was translated into fourteen languages. Bobby donated all his royalties to Children's Health Defense.

The *Associated Press* reported (December 15, 2021): "According to data from Similarweb, a digital intelligence company that analyzes web traffic and search, Children's Health Defense has become one of the most popular 'alternative and natural medicine sites in the world,' reaching a peak of nearly 4.7 million visits per month. That's up from less than 150,000 monthly visits before the pandemic." The group had launched an Internet TV channel and started a movie studio, while opening branches in over twenty states, in Canada, Europe, and Australia. The book had an incredible impact and began to sway public opinion.

In January 2022, a judge in Texas blocked the Biden administration from enforcing vaccines on federal workers, and the Supreme Court had voted down the government's attempt to mandate vaccines for large businesses.

"Bobby is kind of the Salman Rushdie of America," as bestselling author and executive protection specialist Gavin De Becker puts it, "in that you have a book people really wanted to read despite the controversy." Rushdie's 1988 book *The Satanic Verses* inspired a fatwa against him. In Bobby's case, silence didn't exactly prove golden for his critics. As De Becker points out, "The last hit piece on Bobby by the *New York Times* doesn't even list the name of the book. This was certainly the first major bestseller in my lifetime that wasn't reviewed by any corporate media company."

Tony Lyons shakes his head in wonderment: "It's a beautiful story. A book that some of the most powerful people on Earth tried to shut down sold more than a million copies. But the other side of the story is that the best exposé of a corrupt official in the history of the United States wasn't mentioned or reviewed in any mainstream newspaper." They did everything they could to pretend *The Real Anthony Fauci* never existed and turned it into the most censored book that Lyons had ever been involved with in thirty years of publishing.

What's true of all the evils in the world is true of the plague as well. It helps men to rise above themselves. All the same, when you see the misery it brings, you'd need to be a madman, or a coward, or stone blind, to give in tamely to the plague.

—Albert Camus, *The Plague*, 1947

CHAPTER THIRTY-SEVEN

GATHERING OF
THE TROOPS

In 2022, Bobby wrote *A Letter to Liberals,* a hundred-page book with thirty additional pages of source notes. He opened with a quote from Dr. Fauci on *Meet the Press*, where Fauci brashly stated that attacks on him "quite frankly, are attacks on science."

Bobby began the book by responding to Fauci's claim: "It is troubling enough that our country's leading public health technocrat . . . would make such a narcissistic and scientifically absurd statement. The more serious concern is that the majority of my political party—the Democrats—and the mainstream media generally accept Dr. Fauci's assertion as gospel. . . . It is my hope that this short book will remind all Americans that blind faith in authority is a feature of religion and autocracy, but not of science and democracy."

Acknowledging that his open questioning of government policies for managing the pandemic "has made me a pariah, primarily in liberal circles," Bobby proceeded to deconstruct "the key canons of the reigning liberal mythology and throw down this gauntlet to the liberal intelligentsia to defend their assumptions on the battlefield of scientific debate." The ensuing pithy chapters covered all aspects of the narrative on COVID vaccines along with analyses of other measures (masks, social distancing, school closures, and lockdowns).

At the same time, Bobby was at work on another book titled *The Wuhan Cover-Up*, scheduled for publication in late 2023. As Bobby described it: "It links the deep involvement of the intelligence community historically with bioweapons development, beginning with [the post-World War Two] Operation Paperclip and the Japanese bioweapons program, and showing a straight line to how the intelligence agencies and the Pentagon continue to be the principal funder to Anthony Fauci of bioweapons technology disguised as vaccine development." Also coming out in 2023 is a book coauthored by Bobby and Dr. Brian Hooker, *Vax-Unvax: Let the Science Speak*, describing some sixty peer-reviewed studies comparing vaccinated versus unvaccinated populations.

Children's Health Defense remained a growing force, with twenty chapters globally and *The Defender*'s daily reports translated now into six languages. In June 2022, CHD flooded the FDA and CDC with more than 1.3 million emails advocating against the COVID shots. Its legal team has more than fifty active lawsuits on censorship, challenges to vaccine mandates; CHD won a case in 2022 against the Federal Communications Commission on safety guidelines for 5G and wireless.

And while media criticism of Bobby and CHD continued to appear, the *New York Times* also published a front-page article (August 1, 2022) detailing a major shift in parents' attitudes about the impact of the pandemic on their children. Opening with a description of a protest in Orinda, California, the *Times* observed that most of the activists "had never been to a political rally before. But after seeing their children isolated and despondent early in the coronavirus pandemic, they despaired. . . . On Facebook, they found other worried parents who sympathized with them. They shared notes and online articles . . . about the reopening of schools and the efficacy of vaccines and masks. . . . Nearly half of Americans [now] oppose masking and a similar [percentage] is against vaccine mandates for schoolchildren, polls showed. What is obscured in those numbers is the intensity with which some parents have embraced these views."

In late October 2022, CHD held its first annual two-day conference in Tennessee. It was billed as "The Path Forward: Uniting to Create a Better World." Bobby gave several talks, alongside a number of MDs, legal experts, and others who had gained notoriety (and scorn) since

COVID-19 emerged almost three years earlier. The gathering at the Knoxville Convention Center was sold out, with seven hundred ticket holders and a list of standing-room-only hopefuls. It was an opportunity to participate in what seemed to be a growing national movement.

A "Great Resist Session" was led by Bobby and Catherine Austin Fitts, a highly unlikely ally. She was the former managing director of the investment firm Dillon, Read, Inc. and had served under George H. W. Bush as assistant secretary of Housing and Urban Development. As an insider who had witnessed government fraud firsthand, Fitts ended up leaving all that behind and becoming a major voice warning against a coming tyranny of centralized control through the forced use of digital currency.

"We've been in denial for many years as a country," Fitts told those assembled. "The first thing we need is transparency. We're at a crossroads where we either let go of the Constitution or fight to preserve it. Whatever the darkness is, face it." She went on to speak of the injection of vast sums into the economy fueling inflation ("my fear is this will make it that much harder for small business and farmers to function"), and also of the alliance forming between CHD and the Weston A. Price Foundation, which focuses on nutrition and health.

Bobby offered a preview of his next book, describing how "they have created an engine called PPR, Pandemic Preparedness Response, a new industry that they've been patching together since 2000 and COVID brought to its maturity." Through programs like the Global Virome Project, Predict, and the USAID, the plan is to "go out and inventory and break down the genome of every virus in the world. This is a hoax, there's no reason in the world to do this. But there are a lot of illegitimate reasons, because every one of those viruses is a potential bioweapon." Outlining what *The Real Anthony Fauci* had revealed about pandemic scares around Zika and swine flu, he seconded Fitts's expressed concern that "ultimately it's about authoritarian governance. The whole purpose of fear is to dismantle the capacity for critical thinking, manipulate people to have us invite our own slavery—because we're so scared that we believe only the government can help us."

The beginning of their relationship dates to when Fitts read Bobby's memoir, *American Values,* and realized the parallels between their upbringings. "It was actually a remarkable experience," Fitts told me. "His family was much more wealthy and successful than mine, but my parents had both started out life as serious Democrats who deeply believed in progressive values. My father was a surgeon who, when he came back from the war, led the redesign of all the emergency rooms around the country. If you came to our dinner table, just like Bobby describes his, you were challenged to discuss ideas. As a four-year-old, I had to think about, what's your contribution to civilization? Reading *American Values* transported me back to where the family is engaged in the wider society and community and really cares, and they compete on their contribution. The great thing about the training he got at the dinner table is that it taught him how to get things done in the real world."

Fitts spent many years working on building an investment model that aligns the financial system with the environment. She has also had extensive litigation experience. "It's a sea of corruption," she says, "but every once in a while I'd meet someone who really believed in and understood the power of the law, and had incredible integrity and was willing to do the unbelievably hard work that it takes. I kept listening to Bobby's interviews and realized, this may be one of those people."

Fitts sat down with *The Real Anthony Fauci* on a weekend day off, figuring she'd just read the first chapter and then go sailing. "I couldn't put it down, read for the whole day. I didn't know the half of it, had no idea how bad it was. Then I said I'm going to do everything I can to promote this book." She herself bought ten copies at a time to give away. Her Solar network, which consists partly of heartland types "who are very conservative and very Republican," devoured it.

Fitts also related to what both she and Bobby had dealt with in the power structure. "He has grown up and navigated where there are snakes in the grass everywhere. He has more patience than I do, but I admire that. I watched the realm of real men totally engrossed in making money on money, which destroys all real things and they think that's cool. Then Kennedy is focused totally on everything that's real—the animals, the plants, the water. He just keeps laser-beam attention on what others have forgotten, as if they're in a trance. I watched him describe how he very reluctantly got into the vaccine issue, because he's trying to protect what's real. Look at the integrity in his choices, because he knows how dangerous and painful they are."

Fitts added: "I love money and finance. But all real equity value comes from life. If you destroy life, you've got nothing. Why is this guy the only real man on the planet?" She raised the question with a laugh, but I could tell that she was dead serious.

* * *

As the conference neared its end, Bobby gave a talk invoking American history and the "deep, deep suspicion among our most visionary leaders, Republican and Democrat, about the rise of misplaced corporate power and its capacity to undermine not only democracy, but our humanity and all of our other values." He used quotes from Jefferson and Lincoln, Teddy Roosevelt and Eisenhower, as well as FDR and Mussolini. He traced his becoming conscious of the dichotomy through his environmental battles against the oil and chemical companies and agricultural conglomerates. He told the stories of the Riverkeepers and how he came to the autism-mercury campaign through the courageous mothers attending his talks. And then he arrived at a striking parable—his family's "weird relationship" with Cuban Premier Fidel Castro.

Bobby began by recalling the "central rivalry between my uncle and Fidel Castro," a rivalry that came close to resulting in nuclear Armageddon. Thirty years later, in 1993 amid ongoing energy shortages, "the Cubans still had to rely on the Russians for oil, and they didn't want to do that anymore. So they built a nuclear power plant, and it was the same model as the Russians built in Chernobyl. And they were about to turn it on."

At the time, the Cubans harvested sugar cane and then burned it, causing ubiquitous fires and smoke for large parts of the year. But the US had thermal generator technology, originally designed in Germany, able to turn that heat into energy. Bobby traveled to Cuba with a couple of economists and agronomists. "We begged Castro not to turn on that nuclear generator and instead told him that we'd help him purchase these generators for every sugar cane plant in Cuba. He agreed. So that thing is sitting empty. I've been in it."

Bobby visited Castro a number of times after that. He took his family down in 2014, right before Castro died, his wife Cheryl and his kids, and they spent a day with him in his house. Bobby described the meeting: "The day before, we went to the Cuban National Museum. There they have a boat called the *Granma*, which Castro brought into Cuba from Mexico, where he'd been exiled.

They landed in the Sierra Maestra mountains. There was an ambush because they'd been betrayed, and only twelve men survived. But throughout the region they built up guerrilla columns. In collaboration with other groups in the foothills and on the plains and the urban resistance, eventually they marched into Havana and overthrew Batista in January 1959.

"My youngest son Aidan was I think thirteen at the time and had studied a lot of Cuban history. He asked Castro, 'How did you choose who would go on that boat, on the *Granma*? Did you do it by their communist ideology?' And Castro said, 'No, we just chose the smallest people, because we wanted to fit as many as possible.'

"So it was little people who won that revolution. And we're all little people here, but we have big hearts and we have a lot of intensity. And American politics is driven by two forces. One is money and the other is intensity. We had a revolution in this country that nobody believed we could win. And we have another one now, and we're going to win it."

Bobby continued Castro's theme of what it takes to succeed in a revolution on the final morning of the conference. Many attendees had departed, but those who remained received a Kennedy lesson in grassroots organizing, which Bobby called "the most difficult thing you could do in advocacy." With the exception of Greenpeace and the Sierra Club, he explained, most of the environmental groups are top-down organizations and with reason: "The ones that work at the grassroots have a hard, hard time keeping people together."

Bobby told a story of having once spent two weeks in a tent with Peter Bahouth, who had been the executive director of Greenpeace, during the campaign to stop Mitsubishi from building the world's largest salt mine in the gray whale nursery halfway down Mexico's Baja peninsula. Ultimately, grassroots organizing had "generated a million letters and petitions and demonstrations" and killed the project. But when he'd asked Bahouth what it was like running Greenpeace, Bobby never forgot the reply: "It's like being in charge of 1.2 million people, and the only thing they have in common is they all hate authority."

Which meant they were difficult to unite—skeptical, turf-conscious, independent people who'd been lied to before. You can't just tell activists what to do. "Their minds are tuned onto critical thinking, and you have to explain everything to them." It requires patience, social intelligence, and a deep understanding of the subject matter to keep these gifted, talented individuals marching in the same direction. But in building solidarity and membership in an insurgent movement, you had to beware of "deliberate manipulation and infiltration" from the forces out to destroy you. The key was to "discourage drama and backbiting" while building ways to work together and "treating other groups who have a different approach and constituency with respect and support."

Bobby gave the example of the NRDC's founder John Adams, his first mentor in the environmental movement, whose chief competitor was the Environmental Defense Fund. Although Adams "wanted to beat them in the race" to be number one, he made a point of never speaking badly about his rival. At the same time Adams was willing "to take a backseat, push other people forward . . . let them take the front line if they're good." Which made NRDC the most effective environmental group in the world.

"A lot of people will say, 'Well, why don't all these groups consolidate and make one movement?' But that's not necessary. And it actually isn't desirable. The competition and the ferment and the existence of other groups makes the pie bigger and it becomes a more dynamic movement. Diversity is important because each of these groups attracts different constituencies. You could not have had a successful civil rights movement if it was just Martin Luther King. You needed Malcolm X and the Black Panther Party and CORE and SNCC that were competitive and oftentimes highly critical of each other, but all marching in the same direction."

Bobby went on: "The role that we want to fulfill at CHD is to be above the fray, the ones people come to when they need equanimity. We're reliable with the information we give and try to be a resource. We make our lawyers available to theirs, to school them and share briefs and complaints and experience. We try to break down the science to make it readable, comprehensible and accessible to the public. In other words, to weaponize it for our movement and make sure that people aren't saying things that are untrue."

Bobby explained that "one of the disinformation techniques that the other side uses is to seed our movement with apparent issues that aren't really issues, like 'it's not really a vaccine.' PCR tests are another issue. There's literally nothing we can do about that from a litigation standpoint. So let's focus on things where we can actually move the ball up the hill."

Bobby noted that there are more than three hundred physicians and PhD scientists on CHD's advisory board to run things by. "If we make a mistake, which we're going to, we correct it immediately and are grateful for people saying, 'You got that wrong.' Science is dynamic, always changing, and the information is constantly being updated." And it has to be.

"We have a filtration system from which the activism emerges, which is usually from the terrible tragedies that change the trajectory of people's lives," Bobby said. "A child or a friend gets injured and suddenly they start studying the issue. Their eyes open up and they find all of us at the bottom of the rabbit hole, and their horizons begin to expand. Many of them are immensely talented doctors, lawyers, scientists, pharmacists, business people. You need to be patient with people, and you need to cut them slack. If somebody is ideologically impure, that's okay, as long as they're on this side of the line . . . "

Bobby concluded with the following inspiring remarks: "Without your actions we won't win, but God is in charge of the outcome. The only thing we're in charge of is our own conduct. We need to wake up every day and say, 'Reporting for duty, Sir.' We need to be strategic." As an analogy, Bobby recalled that, before he switched to rugby, he'd been on the rowing crew in college for a year. "Your back is to the horizon. You have no idea where you're going. You have one job, which is to pull that oar, but there's a guy in front of you who's making all the decisions. You just have to have faith that you're heading in the right direction. That's what we have to do. We're in charge of our own little space, and we try to make it as compatible and as kind and encouraging of humanity and dignity and compassion and justice and all those things that we devote our lives to.

"People often say to me, 'Isn't it terrible what's happened to you, that you've lost your family and your friends and your jobs and your income.' And I say, 'No, it's a privilege to be part of this movement. It's a gift that I get to be part of something that's larger than myself.' It's a battle and we get to be the leaders in it. They can take stuff away from us, but they cannot take our character. Part of what you are doing here is not building success on the outside, but character on the inside. And when you have that, you become resilient and indomitable. Because no matter how many times they knock you down, you're going to get up and have the spiritual strength to go at them again.

"And that is what ultimately will defeat them."

CHAPTER THIRTY-EIGHT

THE INSTILLING
OF VALUES

Bobby has six children—four boys and two girls—two with his first wife, Emily, and four with his second wife, Mary. While they were growing up in New York, he turned away from a number of opportunities to run for state attorney general, governor, and US senator. Instead, he took them along on trips ranging from fishing expeditions in Florida to Native American reservations to whitewater rivers in Latin America.

He has always made family a top priority in his life. Now the children are all adults, each pursuing their unique paths. One is a filmmaker, one a fashion model, another is finishing law school. Bobby has said of raising them: "My role is not to control, but to encourage them. I think children need to develop their own sense of self and confidence. Most of my kids went through periods of revolt against me, which I welcome. I like when they argue with me, and I have one in particular who does this all the time. It's really important that we develop a generation that understands the importance of critical thinking, and that fear can diminish our capacity for that. It's important to raise brave children and instill courage and risk taking, if we are to continue to have a democracy."

The oldest, his namesake Bobby III born in 1984, recalls of his youth: "He tried to instill in me to have courage under pressure, to sacrifice personal comfort for the greater good, to face adversity merrily, and to always carry more than my own weight—in Tennyson's words, 'to strive, to seek, to find, and not to yield.' We always had highly interesting and intelligent people coming over. We'd never sit and just have dinner with somebody, it was 'we're going salamander hunting at midnight.' But at the same time, my dad was working eighteen hours a day, while he taught himself to be a beautiful water color artist and memorized a poem a week."

The youngest son, Aidan, born in 2001, recalls: "My dad follows a pretty stoic value set—to be present, not allow yourself to be too internally affected by external things you can't control; to place more emphasis on personal growth and development than superficial markers of progress; and to try and make lighter and easier the lives of people around you. We grew up going to church every Sunday, but the emphasis wasn't so much about church doctrine as to give us a sense of some higher mission beyond ourselves. I think my dad has a very clear moral compass and is also analytical and intentional about his decision making. He tries in each moment to do 'the next right thing.'"

His daughter Kathleen (Kick), born in 1988, says: "I've learned a lot from my dad lately about perseverance. I think it's amazing that he doesn't quit. And he's also really funny. That's something a lot of people don't know about my dad. They think he's a really serious guy, kind of scary and intense, but his sense of humor is incredible. When there were bumps, he always picked me right up."

Finn, born in 1997, says: "The primary values I've learned from my father are curiosity and critical thinking, to challenge convention—the fights where you meet the most resistance are the most worthwhile. The purpose of life is a search for existential truths. My father once told me that he would feel much more comfortable with himself if he led a life that had the most positive impact on the greatest number of people. The stories we heard around the dinner table every night were about people who had lived their lives along those lines."

Bobby's daughter Kyra, born in 1995, set down about a family excursion: "The wild excitement of the rapids, the thrill of jumping off high cliffs into cold water, the long hikes in the woods, the sweet smell of the campfire smoke and the way it sticks to your skin and your clothes.

Looking up into the stars one night, I realized that my very life was the result of those adventures."

Conor Kennedy, born in 1994, was entering his third year of law school in 2022 when he took his father aside. He told Bobby that he was planning to go away for a while but asked that his dad not inquire why or where. Conor was gone for three months. He recounted upon his return:

"Like many people I was deeply moved by what I saw happening in Ukraine over the past year. I wanted to help. When I heard about Ukraine's International Legion, I knew I was going. I told one person where I was, and I told one person there my real name. I didn't want my family or friends to worry, and I didn't want to be treated differently there. Going in, I had no prior military experience and wasn't a great shot, but I could carry heavy things and I learned fast. I liked being a soldier, more than I had expected. It was scary, but the rewards for finding courage and doing good are substantial. The people I fought alongside were the bravest I've ever known."

At the dinner table prior to his departure, Conor and his father agreed that Putin was a malevolent man but debated the need for American involvement. Bobby reflected later: "The bottom line for me is that we have built a military-industrial complex in this country that feeds on a continual state of war. Conor felt he shouldn't be arguing unless he was willing to go there and risk his own skin. I'm glad he didn't tell me beforehand because I would have been worried sick. When he called and told me where he'd been, Conor said, 'Dad, you taught me to stand up for what I believe in.' And I just said, 'Thank God you're home. And I'm proud of you.'"

EPILOGUE: TOWARD
HEALING THE DIVIDE

As Bobby's presidential announcement speech moved forward, it was as though time were running backward. Fifty-five years ago, at another moment of "unprecedented polarization," his father had entered the race as a longshot challenger to the Democratic party establishment then in power. Those who had been Robert Kennedy's friends in the "New Frontier" were now working for the Johnson White House and opposed him. Even the universities and most of the Kennedy family's Hollywood allies were against his running against Eugene McCarthy.

The late senator's son recently reflected: "The only people he had were poor white people in rural areas like Appalachia, poor blacks in the Delta and Harlem, Hispanics in East LA, and the Native Americans. But that hopelessness in his campaign gave him the freedom to tell the truth to the American people." And by June 1968, Kennedy was on his way to the Democratic nomination, and likely the presidency, when he was assassinated.

This recollection, bringing tears to the son's eyes as well as many in the audience, proved a poignant segment of a speech that highlighted pivotal and revolutionary moments in American history—from the midnight ride of Paul Revere to his uncle JFK's decision to "scatter the CIA into a thousand pieces" after the disastrous Bay of Pigs invasion. Bobby's eloquent rendering caused one pundit to realize that "this generation has been robbed of the historical understanding [that he] is determined to bring back," an understanding that would make them "far less easy to manipulate."

So did he frame himself in emulation of his father, a political leader who sought to appeal not to "the darker angels of our character," but to "transcend their fear and their bigotry and their anger and see themselves as part of a noble experiment [that] helps them to find the hero that we all have in each of us." Bobby himself had *been there*—faced his own darkness and found inspiration from the fishermen on the Hudson River and among the farmers of North Carolina and beside the Native Americans at Standing Rock. He had felt their frustration and their pain as the merger of state and corporate power accelerated, threatening "to poison our children and our people with chemicals and pharmaceutical drugs, to strip-mine our assets, to hollow the middle class and keep us in a constant state of war."

Bobby was victimized by a media monopoly more concerned with pleasing its advertisers and protecting government narratives than examining his claims of insidious corruption and greed. So why should we trust Bobby Kennedy Jr. to make a difference? Through all that he has suffered, through all his personal tragedies, he has developed spiritual strength, and moral courage. As his father incorporated compassion into the fiber of his being through the anguish of losing his brother, Bobby learned the hard way and emerged as a man with a heart—a man of conscience and conviction, loyalty and values and, above all, a man with an abiding concern for the generations that will follow him.

He is a rebel with a cause: salvaging democracy as the "last best hope" for mankind, as Abraham Lincoln once said of America. Bobby has paid a mean price for refusing to compromise his commitment to what he believes is right. Friends and family have abandoned him. His livelihood as a public speaker has been stripped away. Against the grain of acquiescent society, he has fought with a religious fervor the slow but steady deterioration of our children's bodies and minds at the hands of Big Chemical, Big Pharma, and Big Data. And in this quest, he is definitely not alone. His true allies are not beholden to party affiliation, but to those who might realize that while Democrats battle Republicans, elites are consolidating their power at the expense of the American people. His objective, as he expressed it, "will be to make as many Americans as possible forget that they are Republicans or Democrats and remember that they are Americans."

Bobby has continually reached across party lines and cultural divides to find practical solutions and convince people to do the right thing. Often by forging strong friendships, he has fought and won against some of the most powerful corporations on the planet and the most captured agencies. The thing about Bobby is that he's always open to ideas, to dialogue and debate about complex issues. He works to find solutions that could benefit all parties and, once again, he has never seen capitalism as at odds with responsible stewardship of the environment.

Author David Samuels writes (*Tablet*, April 24, 2023): "The collision he's about to cause between the world of official group-think and the world of normal-speak—where most Americans weigh what might be best for themselves and their children—can only be good for American democracy, and for the American language."

With a memory like a steel trap, Bobby is a student of science and the statistics behind it all. This was clearly evidenced during his announcement speech in Boston. But he remains, at his core, a "nature boy"—the kid who thrills to the majestic swoop of a falcon landing on his arm or the tug of a striped bass at the end of his fishing line. Such are the mysteries that help sustain him. As the poet Langston Hughes once wrote:

I've known rivers ancient as the world and older than the flow of human blood in human veins.

My soul has grown deep like the rivers . . .

Is it too much to hope that the American soul might yet find healing across the divide and become again the exemplary nation once looked to by the world?

Late in April 2023, Bobby announced that effective immediately he would go on leave from all his responsibilities with Children's Health Defense. He said that leaving CHD was "absolutely crushing, because the people working at CHD are amazing freedom fighters." Bobby Kennedy has dedicated his life to the fight against government corruption and corporate greed and he has never tired. After repeatedly deciding against running for public office, Bobby is running for president to take back our freedom from the corporate kleptocracy. He believes that the issues the nation faces today are so grave and the future of mankind so imperiled that he has no alternative but to run. He wants to serve, he must serve, in any way he can. "It would be fatal," Martin Luther King Jr. once said, "for the nation to overlook the urgency of the moment." And whatever it takes, Bobby will face this moment.

ACKNOWLEDGMENTS

I must begin by expressing my gratitude to Bobby himself, with whom I spent many hours talking for this biography. The idea to undertake writing the book was neither his nor mine. The need had been raised to our mutual publisher, Tony Lyons, by colleagues of Bobby's in the public health community—for a book that might reveal the remarkable lifetime of accomplishments of a man often vilified by mainstream media today.

I accepted the assignment long before Bobby's decision to run for the presidency, and over the course of the past year he trusted me enough to share intimate details of his life that he'd never spoken about to other journalists. I hope I have done justice to the remarkable saga, so far, of a man for whom I have the utmost respect.

My deepest thanks to Tony Lyons and his editorial staff at Skyhorse, in particular Hector Carosso and Stephan Zguta. Also especially to Randolph Severson, a psychologist/therapist friend in Dallas who read and provided incisive and inspiring feedback for drafts of each chapter, as he has done for several of my books (*The Life and Ideas of James Hillman* and *My Mysterious Son*). I deeply appreciate the enthusiastic responses of another early reader, Sofia Karstens, as well. And for his ongoing wisdom and guidance, my friend Orland Bishop.

Bobby's son Robert Kennedy III offered much-valued help in reaching out to a number of his father's longtime friends and associates whom I might otherwise not have been able to interview. I was also so glad to sit down for conversations about their father with all six of Bobby's children—Bobby, Kick, Conor, Kyra, Finn, and Aidan. And on Bobby's home front, my thanks to his devoted assistants Sue Paradise and David Whiteside for their help in facilitating the biography, and to Wilbur and Maria. Among other things, David was indefatigable in searching out the best possible photos from Bobby's archive for the insert section that appears in this book.

For fact-checking of accuracy on the public health-related chapters, I am indebted to Lyn Redwood, and Mary Holland. My thanks to Ken McCarthy, who pointed out the censorship parallels with Upton Sinclair and was instrumental in suggesting that a biography needed to be written. Also, thanks to Helen Buyniski for her help with that research.

Finally, in memory of my lifelong friend Jessie Benton, who was always there with words of encouragement, forthrightness, and support. Her death in 2023 leaves a void that can never be replaced.